AF358473

Climate Changes and the Impacts on Power and Energy Systems

Climate Changes and the Impacts on Power and Energy Systems

Editors

Younes Mohammadi
Aleksey Paltsev
Boštjan Polajžer
Davood Khodadad

Basel • Beijing • Wuhan • Barcelona • Belgrade • Novi Sad • Cluj • Manchester

Editors
Younes Mohammadi
Umeå University
Umeå
Sweden

Aleksey Paltsev
Umeå University
Umeå
Sweden

Boštjan Polajžer
University of Maribor
Maribor
Slovenia

Davood Khodadad
Umeå University
Umeå
Sweden

Editorial Office
MDPI AG
Grosspeteranlage 5
4052 Basel, Switzerland

This is a reprint of articles from the Special Issue published online in the open access journal *Energies* (ISSN 1996-1073) (available at: https://www.mdpi.com/journal/energies/special_issues/798OOMRX71).

For citation purposes, cite each article independently as indicated on the article page online and as indicated below:

Lastname, A.A.; Lastname, B.B. Article Title. *Journal Name* **Year**, *Volume Number*, Page Range.

ISBN 978-3-7258-2635-3 (Hbk)
ISBN 978-3-7258-2636-0 (PDF)
doi.org/10.3390/books978-3-7258-2636-0

Cover image courtesy of Younes Mohammadi

Contents

Editorial

Climate Change and the Impacts on Power and Energy Systems

Younes Mohammadi [1,*], Boštjan Polajžer [2], Aleksey Palstev [3] and Davood Khodadad [1]

[1] Department of Applied Physics and Electronics, Umeå University, 90187 Umeå, Sweden; davood.khodadad@umu.se
[2] Faculty of Electrical Engineering and Computer Science, University of Maribor, 2000 Maribor, Slovenia; bostjan.polajzer@um.si
[3] Department of Ecology and Environmental Science, Umeå University, 90187 Umeå, Sweden; aleksey.paltsev@umu.se
* Correspondence: younes.mohammadi@umu.se

Citation: Mohammadi, Y.; Polajžer, B.; Palstev, A.; Khodadad, D. Climate Change and the Impacts on Power and Energy Systems. *Energies* **2024**, *17*, 5403. https://doi.org/10.3390/en17215403

Received: 28 October 2024
Revised: 29 October 2024
Accepted: 29 October 2024
Published: 30 October 2024

This collection, extracted from the Special Issue *"Climate Change and the Impacts on Power and Energy Systems"*, features ten papers that address key topics related to the resilience and adaptation of electrical power and energy systems in response to climate change. Among the ten published papers, three are review papers [1–3] and the remaining seven are research articles [4–10], with the final paper authored by the Guest Editorial team.

The following is a brief summary of these papers, intended as a reference for readers of this collection.

With rising greenhouse gas emissions, the thermal demands of buildings are increasingly tied to climate conditions. Raimundo and Oliveira [4] examine Heating, Ventilation, and Air Conditioning (HVAC) energy requirements for various building types in Mediterranean climates, finding that energy efficiency can be enhanced through the strategic use of insulation and shading. Buonanno et al. [5] investigate photovoltaic (PV) energy forecasting using machine learning and weather models, demonstrating that linear models can effectively refine predictions when data are limited. Gómez-Beas et al. [6] utilize stochastic flow analysis to optimize mountain-based run-of-river hydroelectric plants, revealing that the impacts of rainfall can be mitigated with storage solutions to significantly enhance operational efficiency.

Osawa [7] evaluates residential energy configurations incorporating a bidirectional Electric Vehicle (EV) power supply, suggesting that battery storage and Vehicle-to-Home (V2H) systems reduce emissions and improve cost-effectiveness, particularly for households with variable parking durations. Tratnik and Beković [8] analyze the impact of Slovenia's abolition of net metering, finding that aggregators and battery storage enhance cost savings and energy self-sufficiency in the absence of net metering. Wang et al. [9] conduct a land eligibility study related to Greece's decarbonization efforts, concluding that solar installations have a greater potential impact than wind installations, despite spatial limitations.

In Scandinavia, particularly Norway and Sweden, Mohammadi et al. [10] explore the potential impacts of climatic indices, such as the North Atlantic Oscillation (NAO) and the Atlantic Meridional Overturning Circulation (AMOC), on winter temperatures. They assess how these impacts influence electrical power systems, specifically in terms of renewable energy generation and increased power demands during colder winters. Cooper et al. [1] review anomaly detection in power system state estimation, discussing the potential of Artificial Intelligence (AI) and data-driven approaches to meet the evolving needs of future smart grids. Ekonomou and Menegaki [2] focus on building energy usage, emphasizing the need to transform energy systems by employing sustainable practices to withstand climate impacts. Finally, Moriarty and Honnery [3] critically evaluate both conventional and geoengineering solutions to prevent catastrophic climate change, stressing the urgency of reducing global consumption through accelerated policy measures.

In summary, this Special Issue collectively underscores the critical need for adaptable, resilient power and energy systems, along with innovative climate strategies, to ensure energy sustainability amid escalating climate challenges. We extend our gratitude to the Academic and Managing Editors, as well as the reviewers, for their valuable contributions. Editing these papers has been a rewarding experience. Given the breadth and importance of this topic, we are pleased to present a second edition, titled *"Climate Change in Power and Energy Systems: Challenges, Innovations, and Solutions"*, and look forward to continued contributions from researchers worldwide.

Author Contributions: Y.M., B.P., A.P. and D.K.; writing—review and editing. All authors have read and agreed to the published version of the manuscript.

Funding: This research was funded by the Kempe Foundation (*Kempestiftelserna*) (https://www.kempe.com/, accessed on 1 November 2022), grant (funding) number JCK22-0025.

Conflicts of Interest: The authors declare no conflict of interest.

References

1. Cooper, A.; Bretas, A.; Meyn, S. Anomaly Detection in Power System State Estimation: Review and New Directions. *Energies* **2023**, *16*, 6678. [CrossRef]
2. Ekonomou, G.; Menegaki, A.N. The Role of the Energy Use in Buildings in Front of Climate Change: Reviewing a System's Challenging Future. *Energies* **2023**, *16*, 6308. [CrossRef]
3. Moriarty, P.; Honnery, D. Review: The Energy Implications of Averting Climate Change Catastrophe. *Energies* **2023**, *16*, 6178. [CrossRef]
4. Raimundo, A.M.; Oliveira, A.V.M. Assessing the Impact of Climate Changes, Building Characteristics, and HVAC Control on Energy Requirements under a Mediterranean Climate. *Energies* **2024**, *17*, 2362. [CrossRef]
5. Buonanno, A.; Caputo, G.; Balog, I.; Fabozzi, S.; Adinolfi, G.; Pascarella, F.; Leanza, G.; Graditi, G.; Valenti, M. Machine Learning and Weather Model Combination for PV Production Forecasting. *Energies* **2024**, *17*, 2203. [CrossRef]
6. Gómez-Beas, R.; Contreras, E.; Polo, M.J.; Aguilar, C. Stochastic Flow Analysis for Optimization of the Operationality in Run-of-River Hydroelectric Plants in Mountain Areas. *Energies* **2024**, *17*, 1705. [CrossRef]
7. Osawa, J. Evaluation of Technological Configurations of Residential Energy Systems Considering Bidirectional Power Supply by Vehicles in Japan. *Energies* **2024**, *17*, 1574. [CrossRef]
8. Tratnik, E.; Beković, M. Empowering Active Users: A Case Study with Economic Analysis of the Electric Energy Cost Calculation Post-Net-Metering Abolition in Slovenia. *Energies* **2024**, *17*, 1501. [CrossRef]
9. Wang, Q.; Gontikaki, E.; Stenzel, P.; Louca, V.; Küpper, F.C.; Spiller, M. How to Decarbonize Greece by Comparing Wind and PV Energy: A Land Eligibility Analysis. *Energies* **2024**, *17*, 567. [CrossRef]
10. Mohammadi, Y.; Palstev, A.; Polajžer, B.; Miraftabzadeh, S.M.; Khodadad, D. Investigating Winter Temperatures in Sweden and Norway: Potential Relationships with Climatic Indices and Effects on Electrical Power and Energy Systems. *Energies* **2023**, *16*, 5575. [CrossRef]

Article

Investigating Winter Temperatures in Sweden and Norway: Potential Relationships with Climatic Indices and Effects on Electrical Power and Energy Systems

Younes Mohammadi [1,*], Aleksey Palstev [2], Boštjan Polajžer [3], Seyed Mahdi Miraftabzadeh [4] and Davood Khodadad [1]

[1] Department of Applied Physics and Electronics, Umeå University, 90187 Umeå, Sweden; davood.khodadad@umu.se

[2] Department of Ecology and Environmental Science, Umeå University, 90187 Umeå, Sweden; aleksey.paltsev@umu.se

[3] Faculty of Electrical Engineering and Computer Science, University of Maribor, 2000 Maribor, Slovenia; bostjan.polajzer@um.si

[4] Department of Energy, Politecnico di Milano, Via Lambruschini 4, 20156 Milano, Italy; seyedmahdi.miraftabzadeh@polimi.it

[*] Correspondence: younes.mohammadi@umu.se or mohammadi.yunes@gmail.com; Tel.: +46-0738209544

Citation: Mohammadi, Y.; Palstev, A.; Polajžer, B.; Miraftabzadeh, S.M.; Khodadad, D. Investigating Winter Temperatures in Sweden and Norway: Potential Relationships with Climatic Indices and Effects on Electrical Power and Energy Systems. *Energies* **2023**, *16*, 5575. https://doi.org/10.3390/en16145575

Academic Editors: John Boland and Abu-Siada Ahmed

Received: 22 March 2023
Revised: 10 July 2023
Accepted: 22 July 2023
Published: 24 July 2023

Abstract: This paper presents a comprehensive study of winter temperatures in Norway and northern Sweden, covering a period of 50 to 70 years. The analysis utilizes Singular Spectrum Analysis (SSA) to investigate temperature trends at six selected locations. The results demonstrate an overall long-term rise in temperatures, which can be attributed to global warming. However, when investigating variations in highest, lowest, and average temperatures for December, January, and February, 50% of the cases exhibit a significant decrease in recent years, indicating colder winters, especially in December. The study also explores the variations in Atlantic Meridional Overturning Circulation (AMOC) variations as a crucial climate factor over the last 15 years, estimating a possible 20% decrease/slowdown within the first half of the 21st century. Subsequently, the study investigates potential similarities between winter AMOC and winter temperatures in the mid to high latitudes over the chosen locations. Additionally, the study examines another important climatic index, the North Atlantic Oscillation (NAO), and explores possible similarities between the winter NAO index and winter temperatures. The findings reveal a moderate observed lagged correlation for AMOC-smoothed temperatures, particularly in December, along the coastal areas of Norway. Conversely, a stronger lagged correlation is observed between the winter NAO index and temperatures in northwest Sweden and coastal areas of Norway. Thus, NAO may influence both AMOC and winter temperatures (NAO drives both AMOC and temperatures). Furthermore, the paper investigates the impact of colder winters, whether caused by AMOC, NAO, or other factors like winds or sea ice changes, on electrical power and energy systems, highlighting potential challenges such as reduced electricity generation, increased electricity consumption, and the vulnerability of power grids to winter storms. The study concludes by emphasizing the importance of enhancing the knowledge of electrical engineering researchers regarding important climate indices, AMOC and NAO, the possible associations between them and winter temperatures, and addressing the challenges posed by the likelihood of colder winters in power systems.

Keywords: winter temperatures; Atlantic Meridional Overturning Circulation (AMOC); weakening; North Atlantic Oscillation (NAO); Singular Spectrum Analysis (SSA); electrical power and energy systems

1. Introduction

1.1. Problem Description

Climate change has heightened global concerns, imposing a comprehensive understanding of its regional effects to develop effective adaptation strategies. In Scandinavia,

Sweden and Norway face severe winters that rely heavily on stable electrical power and energy systems to meet specifically heightened heating demands. However, studies such as [1,2] indicate that climate change may induce substantial alterations in the Atlantic Meridional Overturning Circulation (AMOC), as a crucial component of global oceanic circulation. The AMOC includes a northward surface warm water flow (upper 1000 m) of North Atlantic drift, which is balanced by the southward cold deep flow (1000–5000 m) [2–4]. It plays an essential role in climate by transporting heat, freshwater, and carbon [5–7]. AMOC-associated poleward heat transport substantially contributes to the North American and continental European climates [8,9]. The Gulf Stream (GS), in contrast to other western boundary currents, is expected to slow down because of the AMOC weakening. North Atlantic Oscillation (NAO) is also another key climatic index. According to a traditional definition, it is "the difference of normalized sea level pressure anomaly between Iceland and the subtropical eastern North Atlantic" [10]. The changes in AMOC/GS, in terms of weakening, have the potential to impact winter temperatures in the European climate [11] and possibly the regions of Scandinavia. The impact of NAO on the AMOC and/or climate change is also probable [12,13]. Consequently, investigating the possible influence of AMOC/NAO on Sweden and Norway's winter temperatures becomes imperative to assess the vulnerability of their power and energy systems. Hence, there is a knowledge gap in analyzing historical temperature data, examining AMOC/NAO variations, and evaluating their potential effects on power generation and energy systems.

1.2. Literature Review

Several studies have been conducted on the AMOC and GS patterns and trends. The variability in the AMOC is credited to wind forcing (interannual time) and to geostrophic forces (interannual to decadal scales) [14]. Increased freshwater fluxes from melting Arctic Sea and land ice can make "open-ocean convection" and "deep-water formation" weaker in the Labrador and Irminger Seas, leading to AMOC weakening [11,15]. While one study [15] has suggested that the AMOC has weakened over the past 13,000 years, and another study [16] suggested slowing on faster timescales, there is insufficient data-based evidence to support a conclusion of AMOC-weakening strength over the 20th century in a long-term view [17] or the last 50 years [14]. Some studies have shown long-term trends [18,19]; however, combining sparse data and large cyclic variability may also cause an improper understanding [20]. Later, several high-resolution modeling studies, constrained with limited data, suggested that the detected AMOC weakening at 26° N from 2004 is mainly due to natural variability and that anthropogenic forcing has not yet produced a substantial AMOC weakening. In addition, direct observations of the AMOC in the South Atlantic fail to demonstrate an anthropogenic trend unambiguously. Moreover, under a higher scenario (RCP (Representative Concentration Pathway) 8.5) in CMIP5 (Coupled Model Intercomparison Project Phase 5) simulations, the AMOC will likely weaken over the 21st century [21], with a decline ranging from 12 to 54% (with uncertainty in the AMOC behavior projections). Another study [22] predicts a possible AMOC decline between 34 and 45% over the 21st century. According to this study, in a lower scenario, such as RCP4.5, CMIP5 models forecast a 20% AMOC weakening within the first half of the 21st century, followed by a subsequent stabilization (minor recovery). The projected AMOC weakening will be counteracted by deep ocean warming (below 700 m), which will be disposed to make the AMOC strong. The saltiness transport versus observations in the models, as a criterion of AMOC stability, showed complicated situations.

However, some argue that coupled climate models require correction for the known bias and that AMOC variations could be even larger than the gradual decreases predicted by most models, explaining if the AMOC were to entirely shut down and "flip states". Any AMOC slowdown could result in less heat and CO_2 absorbed by the ocean from the atmosphere, which is positive feedback to climate change [21].

Zhang et al. [23] analyzed data obtained from temporally homogenous two-satellite merged altimeter observations from 1993 to 2016 and inferred that the transport, maximum surface speed, and meridional location of the GS exhibit negative linear trends east of 61° W at the 95% level, although they are small and not significant between 72 and 61° W. Additionally, the weakening trend of GS in the 1993–2016 range is combined with a southward-shifting path, which is associated with the NAO decline in 2010 and a 30% reduction in the AMOC, indicating the link between NAO, AMOC, and sea level.

Andres et al. [24] verified that the mean GS transport at 68.5° W within 2010–2014 is almost 10% weaker than that observed by a moored array in the late 1980s. The sixth assessment report of the intergovernmental panel on climate change (IPCC) [25] has stated that GS collapse is unlikely, and although GS decreases with a weakening in the AMOC, it will not shut down in a warming climate. Climate models confirm that GS weakening in the 21st century is due to global warming [26]. It is suggested that the changes in the GS strength are related to the variations in the AMOC, and the GS will likely weaken due to the weakening of AMOC in a warmer climate [26–29]. Chen et al. performed ocean general circulation model (OGCM) experiments and concluded that AMOC weakening was caused by a global warming-induced surface freshening of the high-latitude North Atlantic, leading to the GS weakening [27]. While IPCC is uncertain about the GS behavior as studied in [27], the GS weakening is highly likely during the latter part of the 21st century [1]. Another study [30] proposes that AMOC contributes 25% to maintaining a temperature climate in North-Western Europe. On the other hand, the AMOC has experienced an unprecedented decline over the past century as well as around 2009–2010. Regarding some models stated in [30], this weakening by 2100 is 5 to 40% of the historical average state of a separate model; while others predict 15 to 60% for the same period [30]. It is also suggested that the GS is one of the reasons for the AMOC weakening [31]. However, having a proper model to observe AMOC is important. For example, the study in [32] shows that the eddy-rich ocean model VIKING20X is capable of representing realistic forcing-related and ocean-intrinsic trends. A potential slowing of the AMOC, of which the GS is one key component, because of increasing ocean heat content and freshwater-driven buoyancy changes, could have dramatic climate feedback as the ocean absorbs less heat and CO_2 from the atmosphere. This slowing would also impact the climates of North American and European climates, as stated in [21].

The major effects of a slowing AMOC are expected to be colder winters and summers around the North Atlantic Ocean to the Norwegian Sea and small regional increases in sea levels on the North American coast [33]. Refs. [34,35] estimated, on a global scale, that the weakened AMOC will cause a 0.2 °C cooling in the global mean sea surface temperature (SST) by 2061–2080. An increase in the frequency of winter extremes due to AMOC weakening is investigated in [36]. The possible link between AMOC anomalies and colder winters around 2009–2011 in Europe was studied in [37–39]. Later, in [40], the link is understood with more evidence. In [41], the impact of AMOC weakening on the Europe winter climate concluded a large temperature decrease; however, the analysis is general.

The impact of NAO on the AMOC and/or climate change has been studied in many works. In [12], the relation between NAO, AMOC, and large-scale climate is mentioned. A positive phase of NAO (NAO$^+$) strengthens AMOC for timescales bigger than 20–30 years. The study in [13] showed that one European blocking event (which is less movable) and three NAO$^+$ events contributed to the two heatwaves of July and August 2018.

Although some works have been conducted on the AMOC variations [42–44] and NAO concepts [45,46], there is still a need to go further and firstly analyze historic temperature data in particular countries (Norway and Sweden, in our case), and, secondly, investigate any protentional relationships with climate indices such as AMOC and NAO. The impact of colder potential winters on humanity, especially regarding electrical power and energy systems, is another needed topic to be considered in terms of energy consumption and generation, peak electrical loads, electrical grid planning (including renewable energy sources), security of electricity supply, power grid resilience, and buildings' energy planning. One evident illustration is the rise in electricity consumption during colder winters due to the substantial usage of electricity for heating. Additionally, extreme weather conditions like powerful winds and storms can lead to operational disturbances, potentially resulting in power outages. Section 5 of this paper conducts an extensive literature review on these matters to emphasize the significance of this topic for professionals in electrical power and energy system engineering.

1.3. Contribution and Paper Organization

The contributions of this paper are made in a way to answer the following questions: What are the latest observations of AMOC and its components? Any evidence for AMOC/GS weakening as local or long-term? What are the long-term trends in winter temperatures for Sweden and Norway? Is there any upward/downward trend for the temperatures as highest, lowest, or average in the different winter months, and if so, how is it for selected locations in Sweden and Norway? Can any evidence be found to show some protential similarity between AMOC variations and the winter temperatures in the mid to high latitudes? Does another climate index, NAO, have a possible impact on winter temperatures in the studied locations? Can possible colder winters affect the electrical power and energy systems, and are the multidisciplinary researchers prepared to address the colder winters, whether caused by the investigated climatic indices in this study or other factors like winds or sea ice changes, and associated changes in the electrical power and energy systems? In order to address these questions, we analyzed the latest measurements of AMOC and its components. Then, we selected six locations to investigate winter temperatures in terms of highest, lowest, and average values over a span of approximately 50 to 70 years. Two locations in northern Sweden were chosen due to their historically very cold winters in the past, making it important to predict any potential colder winters in those areas. Additionally, four locations in Norway, ranging from northern to almost southern regions, were selected based on data availability. This selection allows us to assess the potential impact of climatic indices on the entire coastline of Norway, which is expected to be more susceptible to the effects of the indices compared to other regions in the country. Our results, obtained from analyzing long-term trends of temperatures, yearly averages of the climatic indices and temperatures, and lagged correlations between winter AMOC and temperatures as well as between winter NAO and temperatures show a stronger possible link of NAO-winter temperatures (particularly December) for northwest Sweden and coastal areas in Norway, with more confidence for most of the Norwegian sites. The results also confirm that plans in the face of colder winters in those countries must commence for the different aspects and parts of the electrical power and energy systems.

The remainder of the paper is organized as follows: Section 2 presents the latest datasets on AMOC variations from the Rapid Climate Change (RAPID) monitoring program [42], the winter temperatures' dataset extracted from the Norwegian Climate Service Centre and the Swedish National Knowledge Centre for Climate Change Adaptation for the selected sites [47,48], and daily variations in the NAO index, based on 1000 hPa pressure height, obtained from [49]. Section 3 describes the signal processing methods used in this study. The results of the variability of AMOC (and its components), winter temperature variations over the selected locations, the possible similarity between winter AMOC and winter temperatures, and the potential impact of other variables, particularly NAO, on AMOC and/or temperatures are presented in Section 4. Section 5 states some findings on the potential impact of colder winters on the operation of electrical and energy power systems, and, finally, Section 6 concludes the paper.

2. Dataset and Selected Locations

To examine the trend of the AMOC, the most recent daily (a daily aggregation on the half-day measurements is performed) time-series data of AMOC, and its components (at 26.5° N line) are utilized from the RAPID monitoring program [42]. The dataset is the daily measurements in Sverdrup (Sv (1 Sv = 10^6 m^3 s^{-1})) from 7 April 2004 to 10 December 2020. However, it is important to note that while the RAPID dataset has facilitated a better understanding of the AMOC complexities, some literature, such as [50,51], has identified certain limitations and biases associated with it. For instance, Sinha et al. [50] suggest that the estimated variability at 26.5° N is robust on seasonal–interannual timescales, but the presence of geostrophic transport results in a significant mean bias with minimal variability. McCarthy et al. [51] also mention that AMOC mooring arrays of RAPID (and SAMBA (South Atlantic Moored Buoy Array at 34.5° S)) have limited coverage on continental shelves and face challenges in observing deep ocean flows. Nevertheless, this study relies on the advantages presented by utilizing the RAPID dataset for AMOC investigations, as reported in [43,44], among the other relevant studies. These data estimate the strength of the overturning circulation in the North Atlantic at 26.5° N. As shown in Figure 1, the northward red arrow is a schematic of warm surface flow (top 1000 m) of North Atlantic drift, balanced by the southward blue arrow regarding deep cold flow (1000 to 5000 m). Both red and blue arrows together make AMOC. However, GS as a key component of AMOC, by default, flows northward. The dataset of daily NAO index variations since 1950 is obtained from [49], with values derived from 1000 hPa pressure height. Additionally, to examine potential correlations between climatic indices and temperatures in mid to high latitudes, six locations are selected, as depicted in Figure 1 and described in Table 1.

December, January, and February are selected as the winter months for the countries mentioned in Table 1, and their temperatures are extracted from the Norwegian and Swedish centers for climate adaptations [47,48]. The initial time resolutions for temperature measurements are 1 h, 2 h, 3 h, 6 h, and half-day.

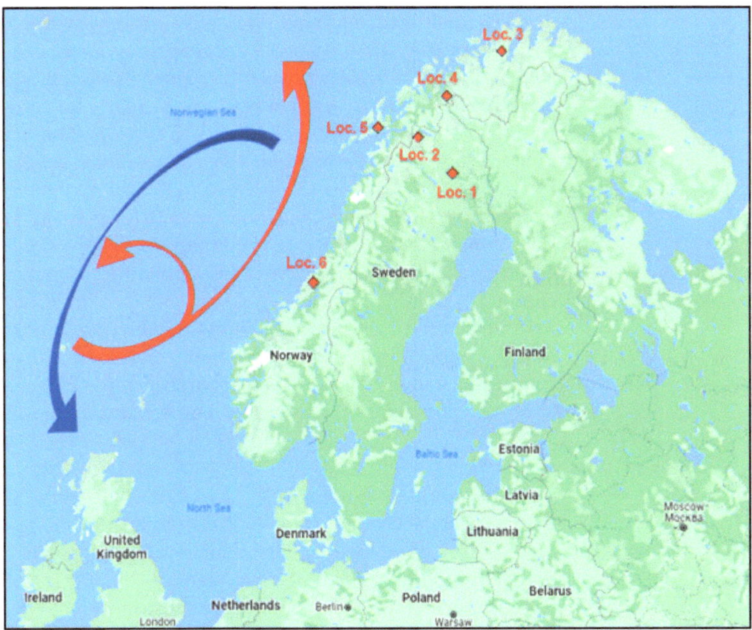

Figure 1. The geographical representation of the six selected locations in Northern Sweden and Norway (Loc. 1 to 6). The part of the northward red arrow (warm surface flow of AMOC) and the southward blue arrow (deep cold flow of AMOC) are marked. However, the real circulation of the arrows is from the Antarctic Ocean to the Greenland Sea and back.

Table 1. Candidate temperature measurement locations.

| Location | City/Country | Measurement Period | | Observed | Measurement |
		From	To	Years No.	Station
1	Kiruna/Sweden	1958	2020	63	Kiruna Flygplats
2	Katterjakk/Sweden	1970	2019	50	Katterjåkk
3	Fruholmen/Norway	1955	2022	68	SN94500
4	Torsvag/Norway	1956	2021	66	SN90800
5	Tromsø-Langnes/Norway	1965	2022	58	SN90490
6	Nordøyan/Norway	1951	2021	71	SN75410

In order to create an integrated dataset for winter temperatures, the temperature values are aggregated daily (as average/maximum/minimum). It is important to note that the authors of this study first considered the daily temperature measurements and found that they had a limited impact on the correlation analysis. Hence, to ensure a smoother representation of temperature variability and better alignment with the daily AMOC and/or NAO time series, an averaging with a running 10-day window is then applied to temperature variations. This approach is effective in reducing the discontinuities in the data resulting from connecting different months over the years. The choice of a "10-day time window" was based on the consideration that each month typically consists of approximately 30 days. While alternative window sizes such as 5, 15, or 20 days were also considered, they did not impact the analysis of long-term trends or yearly averages. The selected 10-day window allows for a more meaningful comparison of winter temperature variations with AMOC and/or NAO, particularly for correlation analysis.

Examining the criteria of average, lowest, and highest temperatures is ascertaining potential disparities in their respective temperature trends. It is important to consider that the lowest temperatures recorded at midnight may have originated from colder initial conditions, whereas the highest temperatures may not have, or to a lesser extent. The three winter months were examined individually to conduct a comprehensive analysis and ascertain the most distinct month and potentially the most impacted month in terms of climatic indices. This analysis holds particular significance for regions such as northeast Sweden and the coastal areas of Norway, where the characteristics of winter months vary in terms of cooling intensity, wind patterns, and other contributing factors. However, to provide a comprehensive assessment and encompass the entirety of the winter season, the analysis also includes the collective temperatures across all winter months.

3. Methods

The results presented in [52], regarding the use of composite analyses with an example for heat wave-SST, highlight the importance of applying comprehensive statistical approaches before making physical inferences on apparent climate associations. Hence, it is important to employ true statistical/signal processing methods for our analysis in this study. A recent method, based on the singular spectrum analysis (SSA (SSA algorithm in this study is motivated by its usefulness in situations where the periods of seasonal or oscillatory trends are unknown; additionally, the number of such trends is not predetermined)) [53,54], is employed to extract the existing patterns within the time series by decomposing them into their principal parts. This method has not been previously used in such studies. In this way, AMOC and its components, as well as each of the average/lowest/highest winter temperatures, are decomposed into a long-term trend (slowly varying component) and seasonal/variational/oscillatory trends (periodic components—a minimum of one trend is expectable from the SSA analysis) to show the oscillations and a noise/residual signal. The steps in SSA to decompose the trends in the time series $X = (x_1, \ldots, x_N)$ with length N are as follows: 1—embedding X as mapping into K subseries of the X as lagged vectors with dimension L (L is selected as a number within $[3, N/2]$ automatically with the function *trendcomp* in MATLAB R2022b) (1) in a trajectory/embedding matrix (Henkel matrix), as columns (2); and 2—applying singular value decomposition (SVD) [55] on the trajectory matrix:

$$X_i = (x_i, \ldots, x_{i+L-1})^T, \quad 1 < L < N; 1 \leq i \leq K; K = N - L + 1 \tag{1}$$

$$X = [X_1, \ldots, X_N] = \begin{bmatrix} x_1 & \cdots & x_K \\ \vdots & \ddots & \vdots \\ x_L & \cdots & x_N \end{bmatrix} \tag{2}$$

Once the eigenvalues of matrix X are calculated, the decomposition of the time series is completed. Any separation/decomposition of times series X needs the separation of Henkel matrix X and a set of eigenvalues produced by the SVDs of each separated part. The primary focus of this study is to analyze the long-term trends observed in the temperature data and the AMOC. However, seasonal trends specifically for AMOC and its components are also presented. Furthermore, yearly averages are calculated for the temperature time series, AMOC, and the NAO index, providing insights into the anomalies within the trends. The Pearson correlation coefficient [56] is employed at lag zero and at its maximum lagged value to evaluate the similarity between the AMOC and its components. These correlation measures serve as quantitative indicators to assess the degree of association between the AMOC and its constituent elements. The potential relationships between winter AMOC and temperatures, as well as winter NAO and temperatures, are examined using two approaches. First, the yearly average of winter AMOC, NAO, and winter temperatures (spanning the entire winter season) was analyzed. Second, lagged-correlation analysis (cross-correlations) [57] is conducted between the winter climatic indices and temperatures over different time lags (in years) for different months of winter as well as for the entire winter season. The maximum correlation values are then identified. A higher positive correlation at positive lags could indicate a similarity or potential link between the winter climatic indices and winter temperatures.

4. Results

This section presents the findings related to the observations of AMOC, winter temperature analysis at selected locations, the possible connection between winter AMOC and temperatures, and the influence of other variables, such as winter NAO, on AMOC and/or temperatures.

4.1. AMOC Variations

First, the variability of the AMOC transport using the latest existing recordings is explained using two approaches. Initially, the AMOC is divided into its components, i.e., Florida current (GS transport), meridional Ekman transport, and upper mid-ocean (UMO) transport between the Bahamas and the Canary Islands, as shown in Figure 2, in terms of overturning strength (OS in Sv) versus year. GS transport (always observed positive) is based on electromagnetic cable measurements; Ekman transport (observed as sometimes negative) is based on the interaction between wind and ocean surface; while the UMO transport (always observed negative) is the vertical integral of the transport per unit depth down to the deepest northward velocity (~1100 m) on each day. Overturning transport (AMOC) is the sum of the three explained components and represents the maximum northward (positive values) transport of upper-layer waters each day.

The correlation analysis between the daily time series of AMOC and its components is conducted, and the results are presented in Figure 3. Figure 3a displays Pearson coefficients at zero lag, Figure 3b shows the corresponding p-values, and Figure 3c illustrates the maximum or minimum lagged-correlation values. All Pearson coefficients in Figure 3a have p-values that are very close or equal to zero indicating statistically significant. These p-values are smaller than the significant obtained level of 3.24×10^{-28}, corresponding to a confidence level of $100 \times (1 - 3.24 \times 10^{-28}) \sim 100\%$. Therefore, the observed correlation values are considered significant. The results reveal the highest positive similarity between

AMOC and its Ekman component as +0.69 at zero lag (with a *p*-value of zero) and as +0.692 for a one-day lead. This shows a strong direct impact of the Ekman component, which represents wind stress, in transporting heat into the AMOC during the daily periods. This finding is consistent with a previous study [44] that also confirmed this relationship for periods shorter than two months. Additionally, the maximum negative correlation (minimum) of -0.52 at zero lag and -0.552 at a one-day lag is observed between the GS and UMO. These components flow in two different directions and the negative correlation is also supported by [44] for timescales shorter than 1 year. Furthermore, Figure A1 in Appendix A presents the results using two alternative methods to calculate correlations: Kendall and Spearman. The Pearson and Spearman methods yield almost similar results, while the Kendall method shows lower correlation values in comparison.

Figure 2. Daily time series of AMOC and its components from 7 April 2004 to 10 December 2020. Positive transports correspond to northward flow, while negative values show southward flow. AMOC negative values are marked by *.

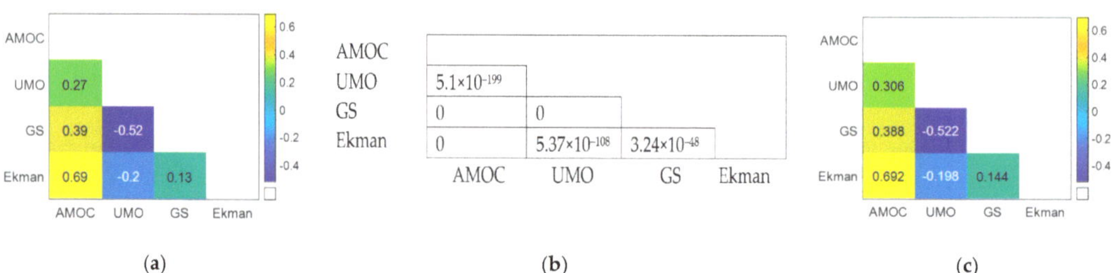

Figure 3. Correlation between daily time series of AMOC and its components: (**a**) Pearson coefficients; (**b**) *p*-values corresponded; (**c**) Minimum/Maximum Pearson lagged correlation.

In the second kind of AMOC analysis, the AMOC measurements (Figure 4a) are decomposed into three parts according to the SSA method mentioned in Section 3: a long-term trend, a seasonal trend, and the residual (noise) signal as shown in Figure 4b–d, respectively. Additionally, the analysis of critical change points (this method finds the years in which the AMOC change most significantly in terms of the mean value [53,54]; in this way, the AMOC variations can be divided into a finite number of regions (selected by the user as far as is possible by the algorithm in [53]), and the sum of the residual (squared) error from its local mean can be minimized for each region separately) identified three critical years (Figure 4a) and four different time windows (with different mean values over that period): the beginning of 2009, the end of 2009, and the spring of 2010. The minimum value of AMOC is observed in the interval 2009–2010. However, instantaneous minima are also recorded at the beginning of 2013 and 2018. The long-term trend (Figure 4b) shows a fast decrease from 2004 to 2009 from 18 to 16.5 Sv, followed by a slight decrease from 2009 to 2010, which remains almost constant from 2010 to 2012. Small decreases are observed from 2012 to 2016, and after that, AMOC changes its direction again toward an increase (i.e., it seems recovering), although some changes between 2018 and 2019 are also observed. However, the values at the end of the observation period (and of 2020) are still lower than the ones in 2004. The difference between AMOC values from 2004 to 2020 shows a general 7% decrease for 16 years.

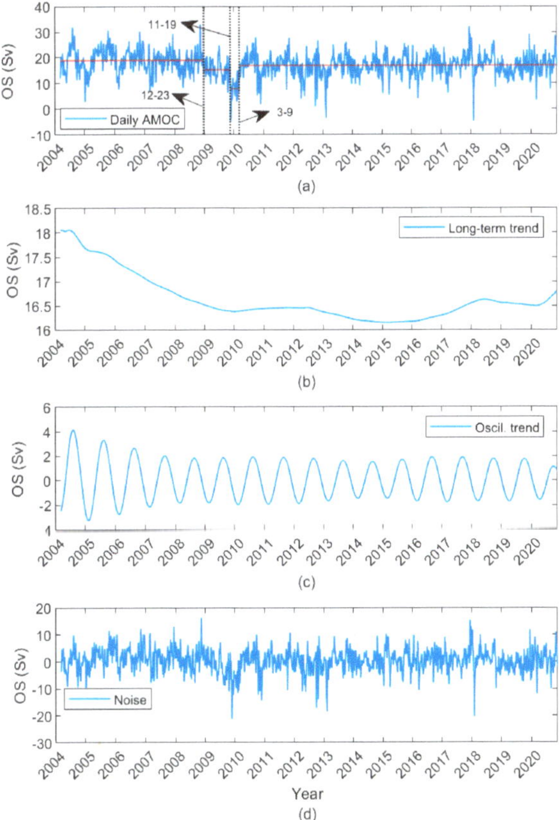

Figure 4. Overturning strength of AMOC (Sv): (**a**) Daily time series; (**b**) Long-term trend; (**c**) Oscillation trend; (**d**) Residual. The dashed vertical lines in (**a**) show the points of the year at which the mean of transport (red horizontal lines in (**a**)) changes most significantly.

If the decreasing speed remains constant, a decrease of about 20% (with low confidence) might be estimated over the first half of the 21st century. Next to [14–17], which show the weakening of AMOC at a fast rate over the 20th century or within the last 50 years, ref. [21] has predicted a 20% weakening of the AMOC during the first half of the 21st century and a stabilization and slight recovery after that. The seasonal variations in AMOC are cyclic, with a decreasing magnitude before 2009 and a constant magnitude after that. The high-frequency variations have shown a minimum value of -21 Sv and a maximum of $+15$ Sv, as shown in Figure 4d. However, the range of noise values changes mostly from -10 to $+10$ Sv, which is similar to the 10-day measurements in [44] for the period of 2004–2018.

Figure 5 shows the long-term and seasonal variations for AMOC components. The general long-term trends show GS weakening (or at least decreasing), as well as strengthening of Ekman, and strengthening of UMO in its negative direction (weakening in the positive direction). However, the behavior is different for the periods after 2017–2018. The "GS weakening in the same direction as AMOC" and the "strengthening of UMO in the opposite direction" are the main key for the AMOC weakening, despite Ekman increasing, as observed in Figures 4b and 5a,c,e. Although a high correlation is observed between daily measurements of AMOC and Ekman (Figure 3), it is clear (Figures 4b and 5a) that AMOC is governed/dominated by UMO. This result was also confirmed in [44] for time intervals longer than a year (our case, daily from 2004 to 2020). The minimum value of GS was observed around 2010. The cyclic variation for the three components also shows that the behavior before 2009–2011 is different from that after that period.

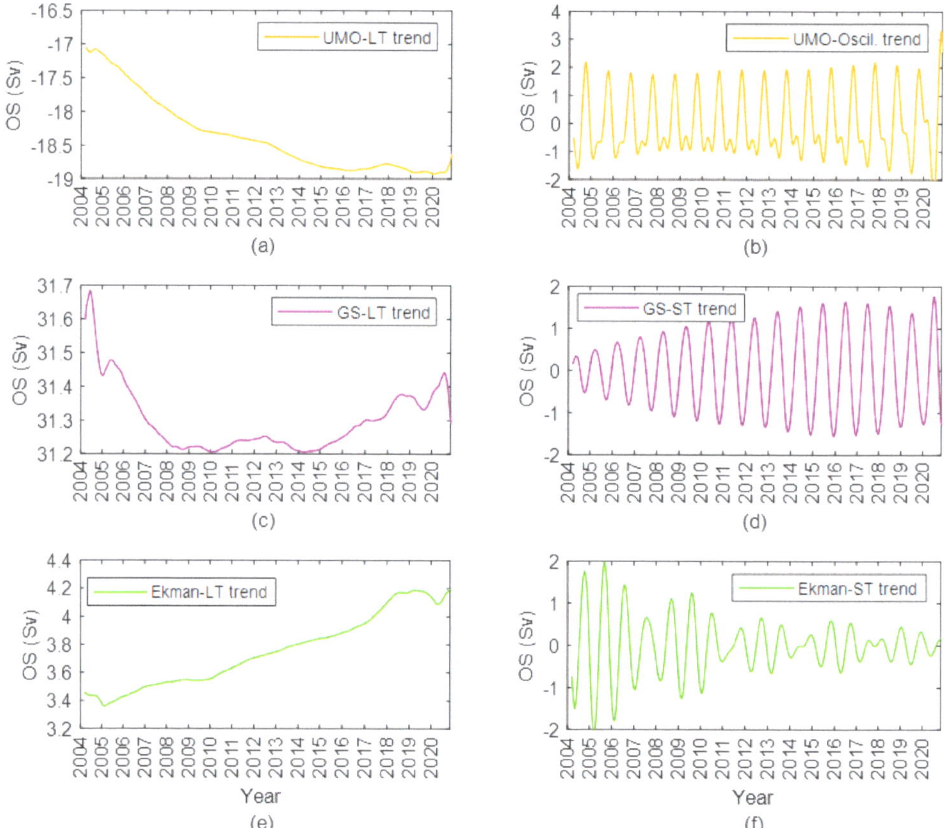

Figure 5. AMOC components in terms of long and seasonal trends: (**a,c,e**) Long-term trends for UMO, GS, and Ekman; (**b,d,f**) Seasonal trends for UMO, GS, and Ekman.

To examine the anomalies within the AMOC variations and its components, a yearly average (based on the daily time series shown in Figure 2) along with its corresponding yearly standard deviation is depicted in Figure 6. The results in Figure 6 are more similar to the results published in (Figure 4, [44]) for the AMOC and its component's interannually variability than to long-term trends in Figures 4b and 5a,c,e, in which the SSA algorithm is applied by employing SVD analysis on the trajectory matrix including lagged sub time series for each of AMOC and its components, separately. The mean and standard deviation for the different transports in the year 2020 and after that are 17 ± 4.3 Sv (AMOC), 31.3 ± 3.2 Sv (GS), 4.3 ± 3.4 Sv (Ekman), and -18.5 ± 3.7 Sv (UMO). The minimum values of AMOC are observed first for 2009–2010 and then for 2010–2011, as dashed vertical red lines in Figure 6a. The first decline period corresponds to the UMO, and the latter corresponds to Ekman. In 2019–2020, another local minimum for AMOC was also observed, corresponding to the year regarding the lowest value of GS. Another observation is the decreasing value for AMOC and UMO before 2009 and the recovery after that.

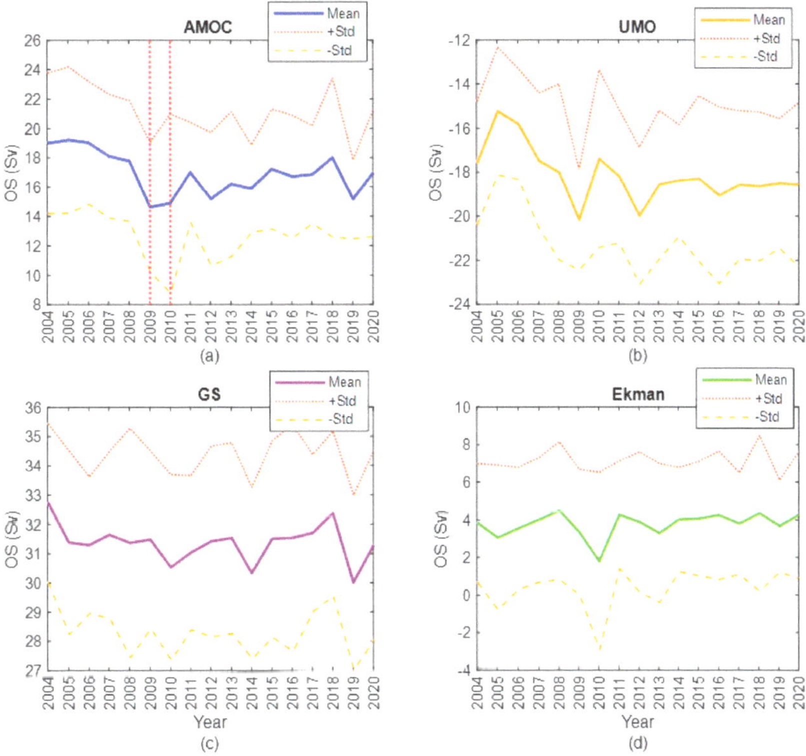

Figure 6. The annual average of AMOC and its components in (Sv): (**a**) AMOC; (**b**) UMO; (**c**) GS; (**d**) Ekman. OS values are given as mean, + standard deviation, and − standard deviation. Red dashed lines show the standard deviation in (**a**–**d**).

Note that although the long-term trend of AMOC (Figure 4b and the results presented in [44]) from 2004 to 2020 shows a small recovery after 2014, the under-study interval is from 2004 to 2020 (according to the availability of data in RAPID AMOC program for 17 years); hence, the trend looks somehow similar to averaged variations for a defined window (Figure 6a). Considering more data on AMOC and GS, which is essential in the SSA algorithm, would conclude more precise results for a long-term trend so that the recovered part is negligible. Moreover, a light decrease in AMOC is observed in 2012 and 2014, and a more severe decrease is seen in 2019, as shown in Figure 6a for the winter AMOC. The AMOC trend can be seen as an almost constant trend followed as weakened, such as the results presented in [35], in which three periods are divided for AMOC strength: before 2000 (almost constant), 2000–2020 (weakening), and 2020–2050 (forecasted weakening with more strength). We aimed to investigate and confirm whether the AMOC is weakening in a general sense based on the available data.

4.2. Winter Temperature Variations at Candidate Locations

This section provides an analysis of winter temperature variations in the selected locations using two approaches:

(a) *An investigation of anomaly detection for all winter months through a yearly average analysis.* This analysis covers the period from 2004 to the present, depending on the existing temperature data and considering the availability of AMOC data (Figure 7).

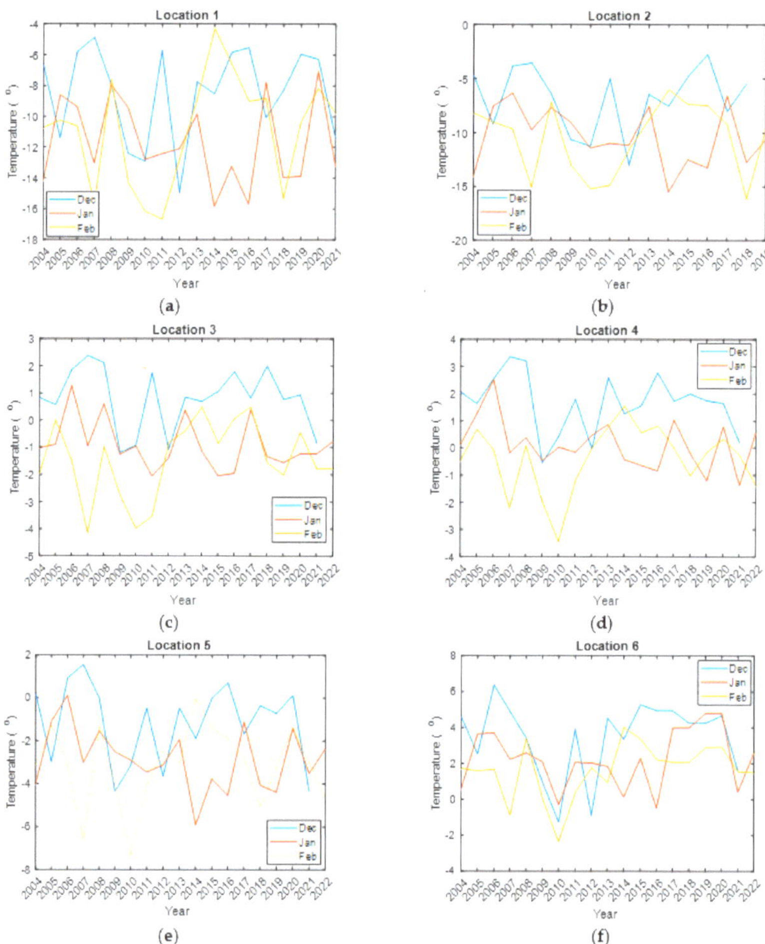

Figure 7. Yearly winter (December, January, and February) average temperatures for the selected locations 1 to 6 from 2004 to present (**a–f**).

As shown in Figure 7, clearly, the winter months in locations 1 and 2 in Sweden are much colder than the other locations in Norway. The winter temperatures over the years can be divided into three periods: when February is the coldest month (before 2012), when February is becoming warmer, some years even the warmest month (2012–2017), and when February is close to January temperatures (after 2017). It also shows that February is somehow becoming warmer over time in the selected locations. The local minima are observed in 2007, 2009, 2010, 2012, 2018, and 2019. These anomalies in minimum values are clearer in February. Looking at the yearly average of December, specifically for locations 3, 4, and 6, shows that there is a downward trend before 2010, an upward after that, and again downward till the end of the observation period.

(b) *A long-term analysis of temperature variations, focusing on daily lowest, highest, and average temperatures for all winter months.* This analysis spans a period of at least 50 years and, at most, 71 years (Figures 8–13 and Table 2). The SSA's capabilities in noise reduction, localized trend analysis, interpretability, and handling missing values and outliers (for example, the values at the connection point of two similar months over different years in the temperature time series) make it a powerful tool for extracting long-term temperature trends that may not be achievable with traditional methods.

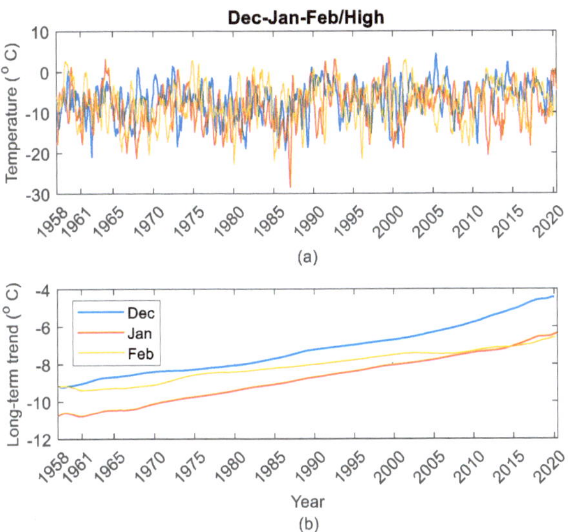

Figure 8. The daily highest temperatures (with a 10-day running averaging window) for all winter months in location 1: (**a**) Time series; (**b**) Long-term trend. Same legend as used in (**b**) apply in (**a**).

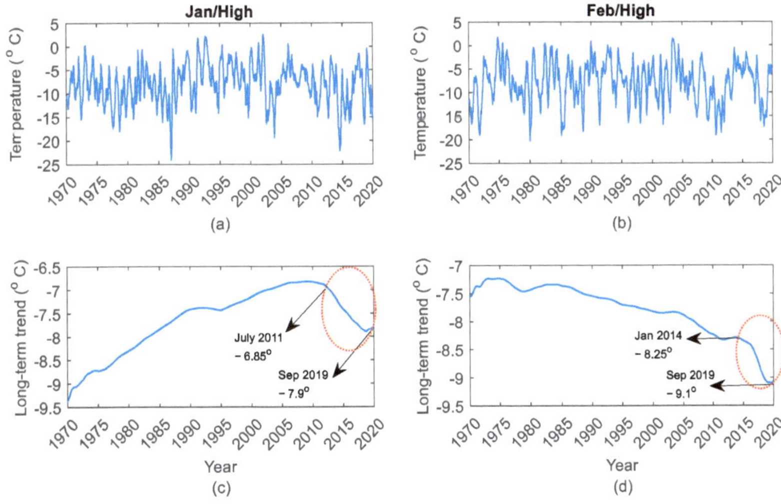

Figure 9. The daily highest temperatures (with a 10-day running averaging window) for January and February in location 2: (**a,b**) Time series; (**c,d**) Long-term trend. The arrows in (**c,d**) show the beginning and end of the downward sector in the long-term trend marked by an ellipse.

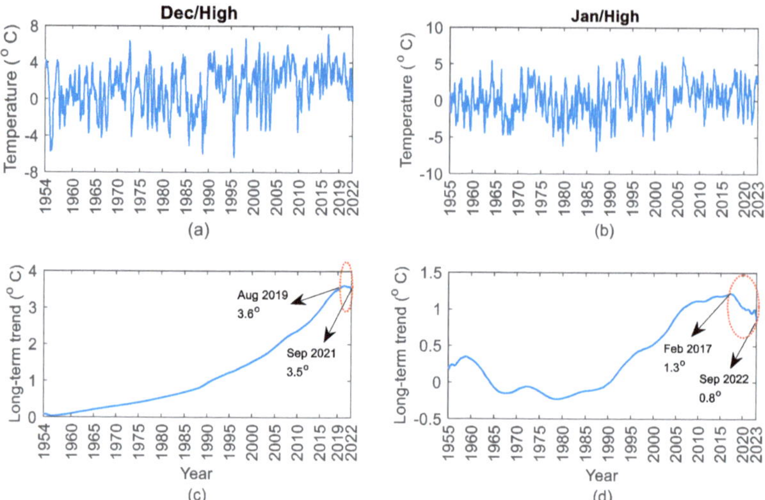

Figure 10. The daily highest temperatures (with a 10-day running averaging window) for December and January in location 3: (**a**,**b**) Time series; (**c**,**d**) Long-term trend. The arrows in (**c**,**d**) show the beginning and end of the downward sector in the long-term trend marked by an ellipse.

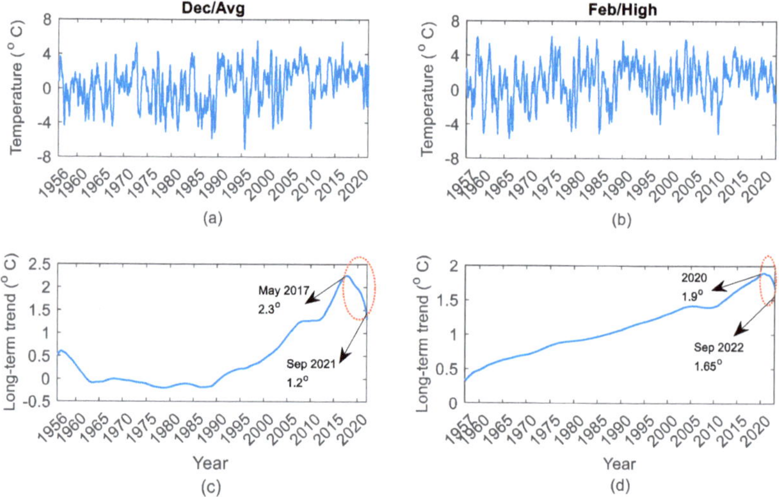

Figure 11. The daily temperatures (with a 10-day running averaging window) in location 4: (**a**,**c**) Time series and long-term trend for average temperature in December; (**b**,**d**) Time series and long-term trend for highest temperature in February. The arrows in (**c**,**d**) show the beginning and end of the downward sector in the long-term trend marked by an ellipse.

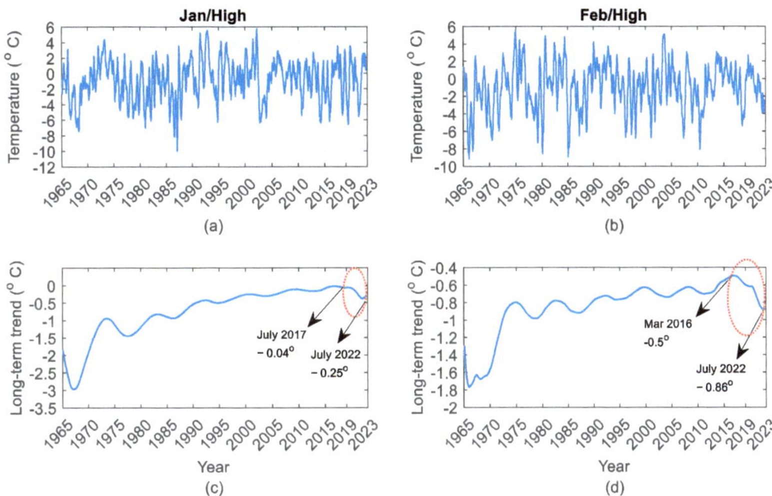

Figure 12. The daily highest temperatures (with a 10-day running averaging window) for January and February in location 5: (**a**,**b**) Time series; (**c**,**d**) Long-term trend. The arrows in (**c**,**d**) show the beginning and end of the downward sector in the long-term trend marked by an ellipse.

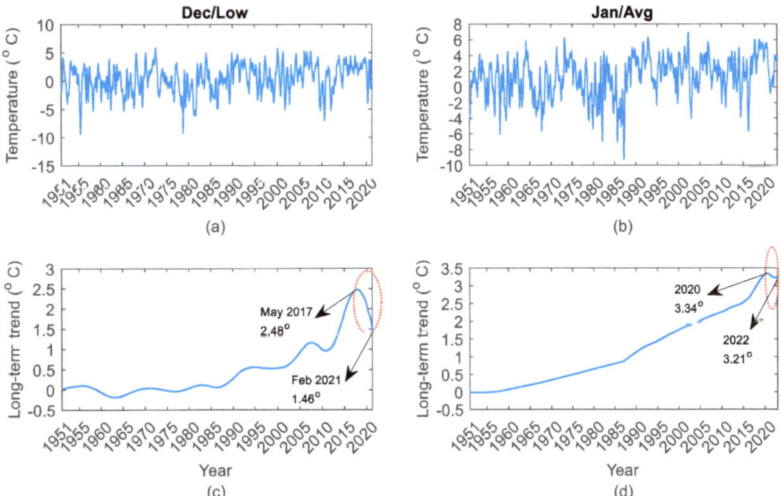

Figure 13. The daily temperatures (with a 10-day running averaging window) in location 6: (**a**,**c**) Time series and long-term trend for lowest temperature in December; (**b**,**d**) Time series and long-term trend for average temperature in January. The arrows in (**c**,**d**) show the beginning and end of the downward sector in the long-term trend marked by an ellipse.

Table 2. The obtained information in the downward temperature trends at locations 2 to 6. The colors in the column before the last one is sorted from lowest (light pink) to highest (dark green).

Location	Winter Month	Type of Temp.	From		To		Decreased Temp. (°C)	Duration of Decreased Temps. (Year)	Decreased Temp. Rate (°C/Year)	
			Temp (°C)	Year	Temp (°C)	Year			Month/Type Temp	Avg/Loc
2	Dec	High	−5	2011	−5.35	2018	0.35	8	0.044	0.090
		Low	−11.2	2011	−11.7	2019	0.5	9	0.056	
	Jan	High	−6.85	2011	−7.9	2019	1.05	9	0.117	
		Avg	−9	2011	−9.8	2019	0.8	9	0.089	
	Feb	Low	−12.6	2016	−12.93	2019	0.33	4	0.083	
		Avg	−10.4	2014	−11	2019	0.6	6	0.100	
		High	−8.25	2014	−9.1	2019	0.85	6	0.142	
3	Dec	High	3.6	2019	3.5	2021	0.1	3	0.033	0.040
	Jan	High	1.3	2017	0.8	2022	0.5	6	0.083	
	Feb	High	0.1	2000	0	2022	0.1	23	0.004	
4	Dec	Avg	2.3	2017	1.2	2021	1.1	5	0.220	0.088
	Jan	Avg	0.04	2017	−0.015	2022	0.055	6	0.009	
		High	2.25	2020	2.13	2022	0.12	3	0.040	
	Feb	High	1.9	2020	1.65	2022	0.25	3	0.083	
5	Dec	Low	−3.4	2018	−3.6	2021	0.2	4	0.050	0.059
		High	1.19	2019	0.8	2021	0.39	3	0.130	
		Avg	−1.1	2020	−1.22	2021	0.12	2	0.060	
	Jan	Low	−5	2016	−5.3	2022	0.3	7	0.043	
		High	−0.5	2016	−0.86	2022	0.36	7	0.051	
		Avg	−2.37	2016	−2.67	2022	0.3	7	0.043	
	Feb	Low	−5.6	2016	−6.06	2022	0.46	7	0.066	
		High	−0.04	2017	−0.25	2022	0.21	6	0.035	
		Avg	−2.91	2016	−3.3	2022	0.39	7	0.056	
6	Dec	Low	2.48	2017	1.46	2021	1.02	5	0.204	0.076
	Jan	Avg	3.33	2020	3.21	2022	0.12	3	0.040	
	Feb	High	3.91	2020	3.89	2022	0.02	3	0.007	
		Avg	2.36	2021	2.25	2022	0.11	2	0.055	

Figures 8–13 showcase a chosen set of temperature plots for each location, with a specific emphasis on the highest temperatures. These plots represent a selection from various potential combinations, including average, highest, and lowest temperatures, for December, January, and February, a total $6 \times 3 \times 3 = 54$ time series. The long-term trend of December/January/February highest temperatures in location 1 for 63 years from 1958 to 2020 (Figure 8b) shows an increase with a constant slope from -9.5 to -5 °C (December), -11 to -7 °C (January), and -9 to -7 °C (February). This location seems to be affected more by global warming since the long-term trend is just increasing.

Figure 9a shows the highest temperatures for January in location 2 (northwestern Sweden). The long-term trend (Figure 9c) shows the increased temperature from -9.5 (1970) to -6.85 °C first (around 2011) affected by global warming ($+2.15$ °C increasing in 42 years). Between the end of 2011 and the end of 2019, the temperature decreased from -6.85 to -7.9 °C, as indicated by an ellipse in Figure 9c (-1.05 °C decreasing in 9 years). The downward part may be attributed to specific reasons. A possible scenario is explained here. According to the results of some of the literature, such as [35], in which a broad range of AMOC variations are studied and show AMOC weakening (and also the general view of AMOC in Figure 4b, which shows weakening), the marked area, which has a faster speed of decreased temperatures than the increased temperatures before that, may show a possible role of AMOC weakening. Note that the direct impact of global warming is the warming and increasing temperatures; however, an indirect impact is on the AMOC weakening, which, by itself, has the task to transfer warm surface water. Moreover, the speed of AMOC weakening has been confirmed more often than the speed of global warming [35]. Nevertheless, these observations could not be detected by the yearly average temperatures shown in Figure 7b from 2004 to 2019. Figure 9b shows the highest temperatures for February in location 2. The long-term trend (Figure 9d) shows, first, some small variations by 2005. Between 2005 and 2010, the temperature decreased and then increased until 2014. After that, a clear decrease in temperature is seen from -8.25 °C to around -9.1 °C by the end of 2019, as indicated by an ellipse in Figure 9b. The observations for the marked period could be also detected by the yearly average temperatures shown in Figure 7b from 2014 to 2019.

Figures 10–13 show the daily temperatures for locations 3 to 6, respectively. From the first year of study to around 2017/2019 (for location 3), 2017/2020 (for locations 4 and 6), and 2016/2017 (for location 5), there is a slight (linear/nonlinear) upward long-term trend. This is followed by a downward trend, shown by an ellipse in Figures 10c,d, 11c,d, 12c,d and 13c,d.

An analysis of the results from the six locations (some of which are depicted in Figures 8–13) reveals a general long-term temperature trend indicating an increase, which can be attributed to global warming. However, it is noteworthy that 50% (27) of these various combinations of long-term trends have shown a considerable decrease in recent years. Additional details regarding the downward long-term temperature trends are presented in Table 2. The rates of decreasing temperatures, represented as slopes of linear trends, are sorted by color in the column before the last one. In this column, December exhibits the highest decreasing rate in locations 4 and 6. The average rate of change in decreasing temperatures per location is as follows: 0.09, 0.04, 0.088, 0.059, and 0.076 °C/year for locations 2 to 6, respectively. As seen, locations 2, 4, and 6 have experienced colder temperatures in recent years. In order to further examine the underlying factors contributing to the observed downward trends in long-term winter temperatures, indicating the possibility of colder winters in recent years, a thorough and detailed analysis is required. This investigation should involve studying temperature change maps and conducting analyses to establish connections between potential causes and the decrease in temperatures, as well as explaining why these effects are particularly prominent in December and vary across different locations. While multiple factors could contribute to these downward trends, Sections 4.3 and 4.4 will provide valuable insights into two significant climatic indices, namely, AMOC and NAO, as potential explanatory factors.

4.3. Possible Similarity between Winter Variations in AMOC and Temperatures of Candidate Locations

Previous research, including the study referenced as [58], has demonstrated that a 1 Sv alteration in the AMOC can lead to approximately a 0.3 °C change in SSTs for decadal-centennial changes, i.e., a 0.03 °C/(year. Sv). Moreover, these studies have also identified similarities in the variations of AMOC and SSTs. However, they did not pay attention to the regional winter temperatures affected by AMOC variations.

In a more recent investigation [59], an artificial weakening of the AMOC was conducted, reducing it by an average of 57% over 60 years, equivalent to an average of 7.55 Sv, i.e., a 1 Sv weakening for 7.94 years. The results indicated that a weakened AMOC has a cooling effect on the global near-surface air temperature in the Northern Hemisphere, with an average decrease of −1.09 °C/year to −0.14 °C/(year. Sv). The majority of cooling occurs in the Northern Hemisphere, which experiences an average temperature decrease of −2.09 °C/year to −0.26 °C/(year. Sv). Notably, the most significant changes are observed during winter, with a cooling effect of −2.42 °C to −0.3 °C/(year. Sv). These findings suggest that other factors, such as regional warming/cooling, can also influence the magnitude of cooling associated with a weakened AMOC. However, it is important to note that the analysis in the study [59] is conducted using artificially weakened AMOC simulations and climate models, rather than real observational data.

With the information derived from the previous paragraph, this section examines the possible similarity between winter variations in AMOC and winter average temperature variations in the selected locations. Figure 14 illustrates the yearly average variations in AMOC and temperatures at the locations throughout the entire winter season. This analysis covers the period from 2004 to the present, depending on the existing temperature data and considering the availability of AMOC data, which covers from April 2004. The observations from Figure 14 reveal the winter AMOC deep anomalies in 2010 and 2019. The yearly average winter temperatures across all locations indicate a synchronized response to the winter temperatures without any time lag before 2010. Notably, the light and deep minima observed in AMOC in 2007 and 2010, respectively, coincide with the winter temperature minima. From 2010 to 2016, specifically for locations 3 to 6, a slight positive lag can be observed between the AMOC and temperatures. However, starting from 2016 onwards, it appears that there is a larger time lag between smaller AMOC variations and larger variations in temperature. It is also noteworthy to pay attention to the local minima of AMOC in 2018, which is followed by a decrease in temperatures with a lag observed in locations 3 to 6.

The highest lagged correlation along with their corresponding lag values between winter AMOC and average temperatures time series spanning from 2005 to 2020, are presented in Table 3. Almost the same results were concluded for the highest and lowest temperatures. In general, all computed correlation values are higher than 0.2 for the average temperatures; however, December exhibits the highest correlation among the winter months with the AMOC variations. Specifically, locations 4, 6, and 3 demonstrate the strongest correlations (0.58, 0.52, and 0.49) compared to the other locations. The lag values increase from locations 1 and 2 to 3 and 4, and further to 5 and 6. From Table 3, the maximum lagged correlations are notably 0.49 and 0.58 at a lag of 0.065 years (approximately 24 days). For the entire winter season, the highest correlations and the corresponding lags are observed in locations 3, 4, and 6 as 0.32, 0.37, and 0.43 with the lags 28, 20, and 12 days, respectively. The lag values decrease from locations 3 to 6, while almost no lag is observed for locations 1 and 2.

The results from Figure 14 and Table 3 show (1) The possible contribution of AMOC (indirect most likely) to colder winters locally around 2010, and further, due to the positive lagged correlations between AMOC and winter temperatures. It should be noted that there is a decrease observed in 2018–2019 in Figure 14. Although there is no available information for 2020–2022, according to reference [53], there is a possibility of a decreasing trend in AMOC during that period. (2) Higher correlations between winter AMOC and temperatures (>0.49) are observed specifically in December, for locations 3, 4, and 6. (3) There is a pattern where the correlation value increases and the corresponding lag decreases from location 3 to 4 and then to 6, which are those areas along the southern part of the coastline of Norway. There is a clear pattern where the correlation value increases and the corresponding lag decreases from location 3 to 4 and then to 6.

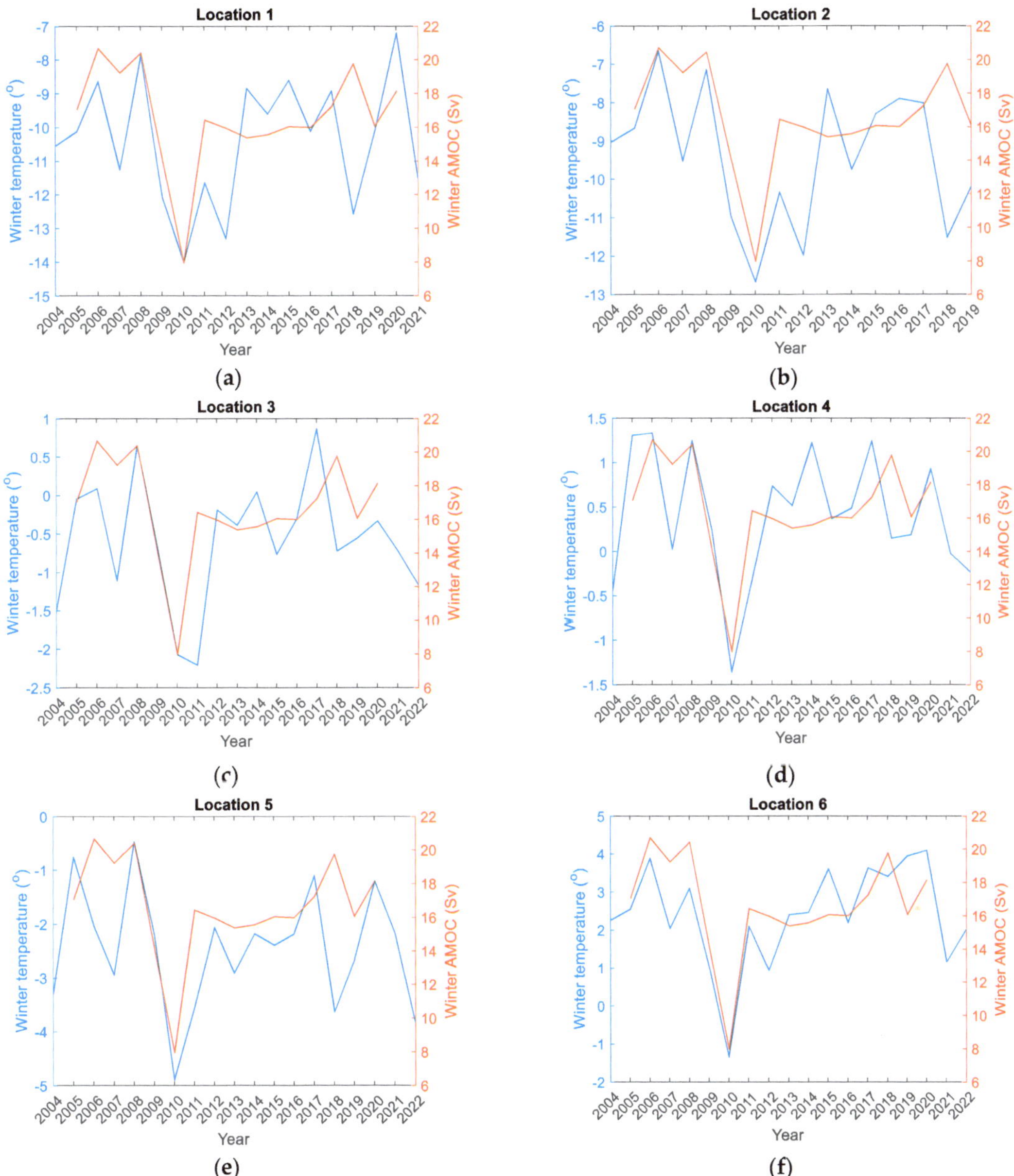

Figure 14. Yearly average variations for winter temperatures and winter AMOC in locations 1 to 6. (**a–f**) from 2004 to the latest available year of temperature data. Winter includes all three months of the season.

Table 3. Maximum lagged correlation and the corresponding lags between winter AMOC and winter average temperatures (as per separated months and whole winter) for locations 1 to 6. Colors in the columns indicate the sorted values from low (bright pink) to high (dark red).

Location	Correlation				Lag (Year)			
	Dec	Jan	Feb	Winter	Dec	Jan	Feb	Winter
1	0.44	0.24	0.3	0.25	0.032	0.871	3.150	0.011
2	0.45	0.33	0.27	0.26	0.032	0.806	3.080	0.000
3	0.49	0.32	0.29	0.32	0.065	1.129	1.381	0.078
4	0.58	0.35	0.27	0.37	0.065	1.194	2.973	0.055
5	0.47	0.2	0.3	0.28	0.097	1.000	2.973	0.044
6	0.52	0.43	0.3	0.43	0.097	0.258	0	0.033

The objective of this study is not to quantify the exact magnitude of temperature decrease resulting from changes in AMOC strength but rather to determine if there are considerably lagged correlations between them. Our focus has primarily been on the impact of AMOC weakening in low latitudes on colder winters in mid to high latitudes, even though an AMOC strengthening would be observed for high latitudes [60] linked to colder winters. According to Table 2, locations 3, 4, and 6 experienced an average temperature decrease of -0.04 °C/year, 0.088 °C/year, and 0.076 °C/year, respectively. In a simple analysis, one could consider that an AMOC weakening of 1.3 Sv, 2.6 Sv, and 2.53 Sv, based on the findings in [58], or 0.28 Sv, 0.63 Sv, and 0.54 Sv, based on the findings in [59], might have contributed to these temperature decreases. Based on the findings of this section, it can be concluded that the weakening of AMOC around 2017 played a significant role in reducing the transfer of warm surface waters. However, despite observing a moderate correlation, establishing a direct link between winter AMOC and temperatures, as well as identifying the mechanisms through which AMOC impacts temperatures on daily/monthly timescales, proves challenging. Nevertheless, there may still exist, with a lesser degree of certainty that necessitates further investigation, a potential connection between winter AMOC and winter temperatures in mid to high latitudes, as well as a potential link between winter AMOC and the downward patterns observed in Table 2. The next section will focus on examining another variable that may influence winter temperatures.

4.4. Possible Impacts of Other Variables, Particularly NAO, on the Winter AMOC and/or Temperatures

The moderate correlations observed between winter AMOC and temperatures (Section 4.3) may not necessarily imply a direct causal relationship. It is possible that a third variable, such as wind or the NAO, influences the AMOC, the temperatures, or even both. The previous study [23] explored the relationship between GS weakening, AMOC weakening, and the NAO decline in 2010, confirming the connection between AMOC weakening and NAO reduction. Additionally, it has been observed that NAO$^+$ (positive phase of NAO) strengthens the AMOC on timescales exceeding 20–30 years [12]. Research conducted on various mountain cities in Europe, Morocco, Turkey, and Lebanon in 2011 indicated that projected NAO trends could lead to increased winter modes and a decrease in the number of cold winters during the 21st century, due to the influence of global warming [61]. The direct impact of NAO$^+$ on the warm summer in 2018 was also demonstrated [13]. The spatial variability of NAO has been found to play a crucial role in regulating the European climate in addition to its temporal variability [62].

In addition to the factor of AMOC weakening, it has been observed that the AMOC exhibits a strong response to wind-driven variability, particularly by the Ekman component, which is in turn influenced by the NAO. During NAO$^+$, stronger winds over the subpolar North Atlantic increase surface heat loss to the atmosphere, promote the formation of dense water, and result in a strengthened AMOC [63]. A study conducted in [44], utilizing AMOC anomalies from the RAPID and GloSea5 datasets at 26° N, along with Atlantic indices such as NAO, examined the relationship between NAO phases and various parameters. It was found that during an NAO$^-$ (NAO$^+$) state period, there is a reduction (increase) in surface heat loss and weakening (strengthening) of winds over the subpolar North Atlantic, resulting in a weaker (strengthened) AMOC. Consequently, the transport of heat by AMOC toward the northward direction decreases (increases), leading to a cooling (warming) effect on the North Atlantic, and possibly the Norwegian Sea. This aligns with a delayed decrease (increase) in SST. Hence, a clear link between NAO, winds, AMOC, and SST is established. A study conducted by [64] examined the influence of winds on the AMOC. They proposed a fully-coupled climate mode where nudging winds poleward of 45° N, through a response from the Ekman component of AMOC, resulted in statistically insignificant trends in AMOC and SST trends in the North Atlantic. These findings were pretty consistent with the observations of AMOC from RAPID at 26.5° N. Another study [65] focused on the impact of NAO on the low-frequency variability of AMOC. The simulation results revealed that the influence of NAO varied among different models. Some models indicated less sensitivity of AMOC to NAO, while others suggested a higher sensitivity. This study also highlighted the importance of the oceanic mean state as a crucial aspect of climate change that requires improvement in models.

According to the literature investigated in this section, it is concluded that NAO could have affected the AMOC weakening. Hence, a possible reason for the cooler winters in the discussed locations could be due to NAO weakening (being in NAO$^-$ phase for a long time). Another scenario for the cooler winters is the direct impact of NAO as attached with wind changes at those locations. Therefore, the winter yearly average of the NAO index was calculated using daily values obtained from [49]. The resulting winter yearly average of the NAO index is presented in Figure 15. The data cover a similar period as Figure 14, ranging from 2004 to the latest available year of temperature data. Observations from Figure 15 reveal the presence of deep anomalies in the winter NAO index in 2010 (corresponding to AMOC anomalies) and 2021. These are the years when the NAO anomalies align with the winter temperatures across all locations. Notably, location 6 demonstrates a clearer correlation between temperatures and NAO variations compared to the other locations.

The lagged correlations between winter NAO and average temperatures time series spanning from 2005 to 2020 are shown in Figure 16. Almost the same results were concluded for the highest and lowest temperatures. Table 4 presents the highest lagged correlations between the winter NAO index and winter average temperatures, along with their corresponding lag values. The analysis is conducted both on a monthly basis and for the entire winter period spanning from 2005 to 2020, which corresponds to the same period as the AMOC-temperature analysis presented in Table 3. Overall, all computed correlation values for average temperatures are higher than 0.28. Among the winter months, December shows the highest correlation with winter NAO variations. However, it is worth noting that the correlation values between the NAO index and temperatures are higher than those reported in Table 3 regarding the possible link between winter AMOC and temperatures. Among the locations, the weakest correlation is observed in location 1 in northern Sweden, with a value of 0.57. Location 3 in the northernmost part of Norway exhibits a slightly higher correlation of 0.58. As we move to locations 4, 2, 5 (Locations 2 and 5 are in almost the same latitude positions), and 6, the correlation values increase to 0.6, 0.62, and 0.66, respectively. To illustrate this, Figure 17, as a sample, depicts the daily variations in the NAO index and temperature in December for locations 5 and 6, in which, on most days, the temperature follows the NAO with or without lag. For the entire winter, the highest correlations and the corresponding lags are observed in locations 2, 4, and 6 as 0.42, 0.42, and 0.55 with the lags 16, 12, and 20 days, respectively.

The results obtained from Figures 15 and 16, and Table 4 provide insights into the possible link between winter NAO and temperatures; the following conclusions are drawn:

(1)   There is a strong relationship between winter NAO and temperatures, particularly during December, for locations 2, 4, 5, and 6. These locations, situated closer to the coastal areas of Norway rather than the northernmost regions, exhibit correlation

values exceeding 0.6. Such values suggest a potentially significant influence of NAO on the temperature patterns observed in these locations.

(2) Among locations 2, 4, 5, and 6, there is a higher probability of having colder winters influenced by the NAO for locations 2, 4, and 5. This conclusion is supported by the following observations: (i) the correlation values between the winter NAO index and winter temperatures are greater than 0.42; (ii) the highest average rates of temperature decrease in Table 2 support this pattern for locations 2, 4, and 6.

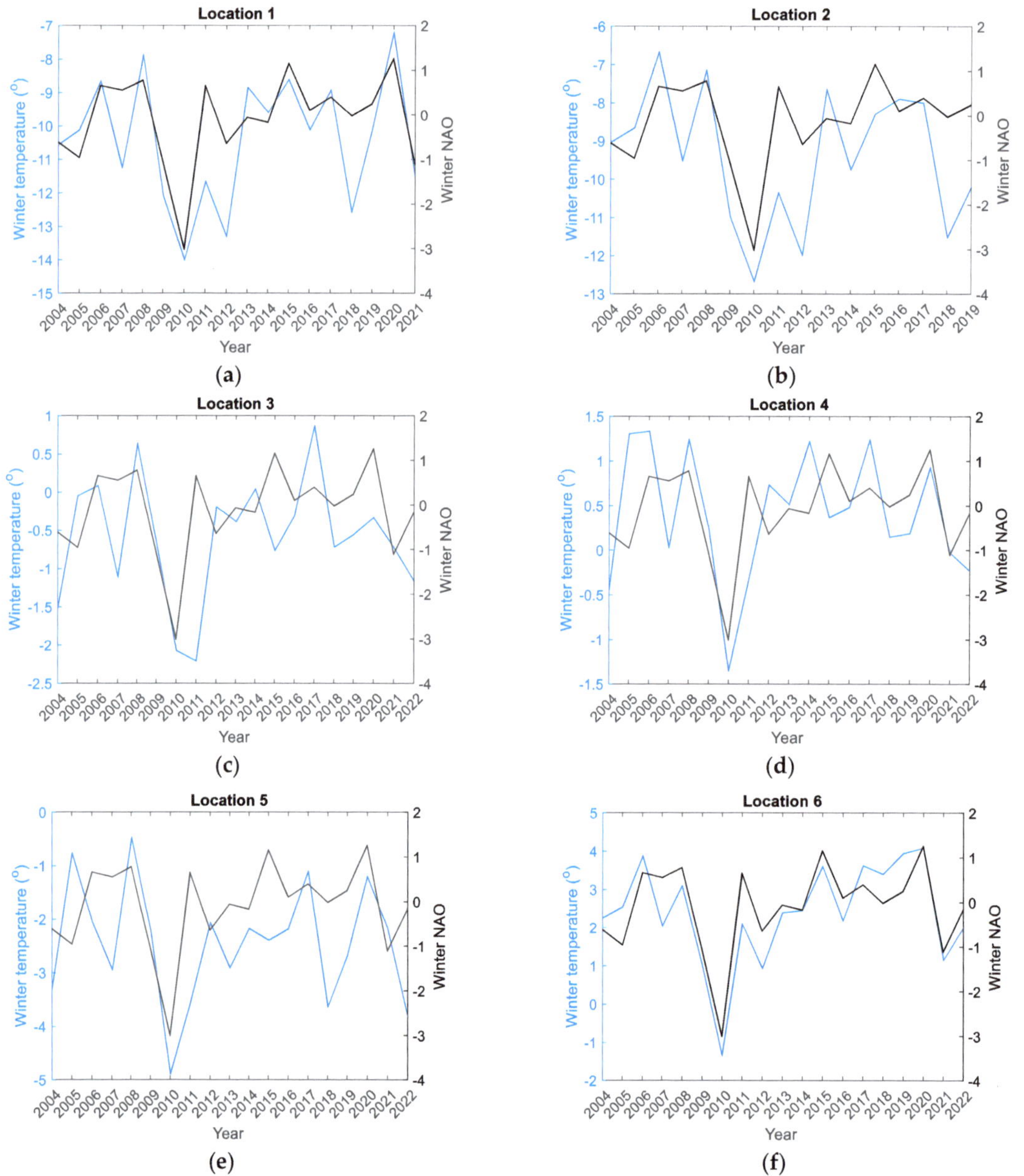

Figure 15. Yearly average variations for winter temperatures and winter NAO index in locations 1 to 6: (**a**–**f**) from 2004 to the latest available year of temperature data. Winter includes all three months of the season.

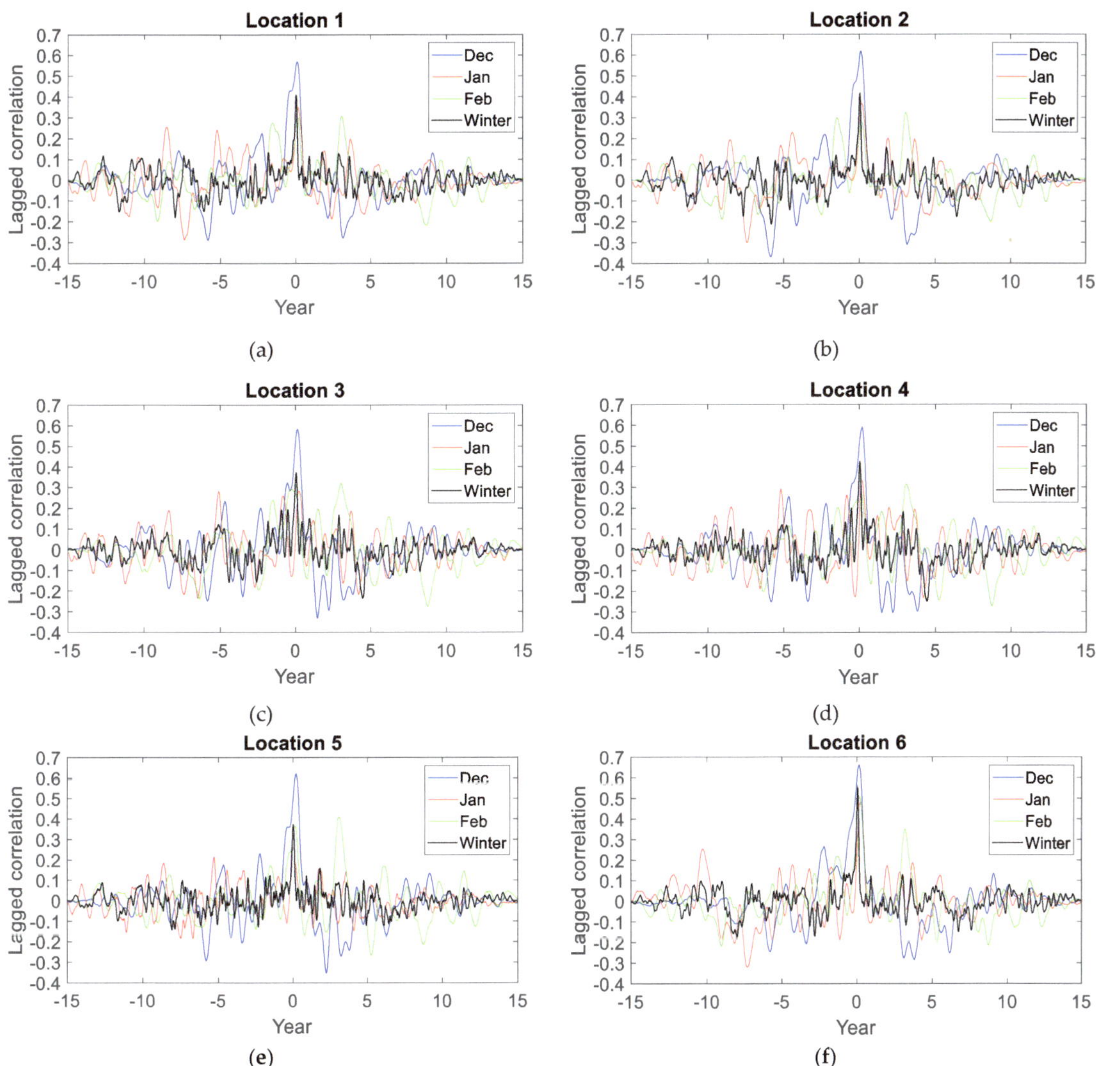

Figure 16. Lagged correlations versus lag (per year) between winter NAO and winter average temperatures (as per separated months and whole winter) for locations 1 to 6 from 2005 to 2020: (**a**–**f**). Positive correlations at positive lags indicate that the NAO strengthening/weakening leads to an increasing/decreasing temperature.

In addition to the previously mentioned factors, it is important to consider other atmospheric parameters such as atmospheric pressure, humidity, solar radiation, and wind when analyzing the temperatures over those locations. In particular, local wind speed variations, influenced by factors like the NAO or regional storm activities, could potentially contribute to the observed temperature reductions in those locations. Researchers [66] examined the connection between the winter NAO and wind climate in Norway from 1920 to 2010. The findings indicated a strong correlation between NAO$^+$ and a higher occurrence of southwest winds from the southwest parts (such as location 6 in our study), as well as a decrease in the frequencies of northeast winds (such as location 3 in our study). However, there was no significant relationship found between the wind climate and the NAO in the northernmost part of the country (such as locations 2, 4, and 5). Therefore, based on this research, colder temperatures experienced in location 6 might also be attributed to the increased wind patterns in the area. It is also worth noting that wind simulations for certain cities in southwestern Norway (near location 6 in our study), between the 1990s and 2050s, forecast higher possible temperatures [67] around our considered period, which could mitigate the impact of local winds on observed temperatures. These findings show that further investigation and research are necessary to fully understand and explore the factors like winds, and this is an ongoing endeavor for the authors of this study.

Table 4. Maximum lagged correlation and the corresponding lags between winter NAO index and winter average temperatures (as per separated months and whole winter) for locations 1 to 6. Colors in the columns indicate the sorted values from low (bright pink) to high (dark red).

Location	Correlation				Lag (Year)			
	Dec	Jan	Feb	Winter	Dec	Jan	Feb	Winter
1	0.57	0.35	0.31	0.40	0.097	0.129	3.044	0.044
2	0.62	0.37	0.33	0.42	0.065	0.129	3.044	0.044
3	0.58	0.28	0.32	0.37	0.097	0.226	2.973	0.044
4	0.60	0.33	0.32	0.42	0.161	0.129	3.044	0.033
5	0.62	0.32	0.41	0.37	0.194	0	3.044	0.011
6	0.66	0.48	0.51	0.55	0.129	0.194	0.142	0.055

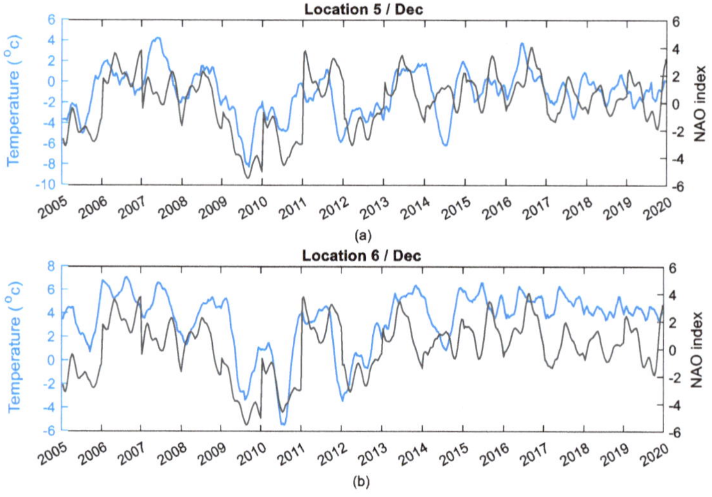

Figure 17. The daily variations in NAO (black line) and daily average temperature (with a 10-day running averaging window) in December from 2005 to 2020 for (**a**) Location 5 and (**b**) Location 6.

5. Cold Winter Impact on the Electrical Power System's Aspects

The results from Section 4.3 demonstrate a less likely relationship between the weakening of AMOC (caused by the indirect effects of climate change, specifically global warming (the direct impact of global warming is evident in the warmer weather patterns we have observed (Figure 8, for instance); however, there is also an indirect impact of global warming with a possible delay, which can weaken the AMOC; this weakening can lead to a weaker transfer of warm surface water toward the north, resulting in colder weather [59] conditions in the affected regions)) and the potential occurrence of colder winters in northwest Sweden and Norway. Additionally, there is a higher level of confidence regarding the presence of these colder winters in the coastal areas of Norway. The findings from Section 4.4 also highlight the more likely role of NAO in influencing the winter temperatures directly.

Therefore, this section focuses on analyzing the potential impact of cold winters, whether caused by AMOC, NAO, or other climate factors, on the operation of electrical power and energy systems in Norway and Sweden. The analysis encompasses aspects such as electricity generation, consumption, and the security of the electrical grids. Potential risks are identified from the perspective of ensuring a reliable electricity supply and the resilience of the power grid. Moreover, potential avenues for future research in this area are discussed.

5.1. Colder Winters and Electricity Generation

According to the public data available in [68], the installed generation capacities in 2022 were as follows: hydro 82.59% (33.36 GW) and wind 12.64% (5.1 GW) for Norway, and hydro 37.3% (16.3 GW), nuclear 19.79% (6.9 GW), and wind 27.69% (12.1 GW) for Sweden. The dominating investments in both countries are in wind power units [69]. In Norway, hydropower units generated 90% (134.4 TWh), and the generation of wind power units covered 6.4% of the total electricity generation in 2020 [69]. Electricity generation in Sweden amounted to 161 TWh, of which 29% was from nuclear, 45% from hydro, 17% from wind, and 8% from combustion-based power units in 2020 [70,71]. Electricity generation from solar power units in both countries is becoming increasingly important; however, it is still negligible.

The operation of hydro generation units during winter months highly depends on the capacity of reservoirs since water inflow is generally very low [69]. Fortunately, water inflow in Norway has shown an increasing trend during the past 60 years, and the increase is relatively the largest during the winter [72]. However, colder winters might reduce the reservoir capacities due to the possibility of ice formation, which is different for high-head and low-head hydropower units [73]. Therefore, further research on the impact of colder winters on electricity generation from hydropower units in Norway is needed. Such research is also essential because Norwegian reservoirs are likely to mitigate the intermittent generation of wind power units [74] and support the lack of energy during the winter months in Europe. Also, the potential risk of reduced electricity generation due to the shutdown of Sweden's nuclear power units should be considered.

Furthermore, a possible reduction in electricity generation in Norway and Sweden will also impact the neighboring countries. The results of modeling the multi-national impacts of Finland's closure of coal-fired generation and Sweden's decrease in nuclear generation showed reduced import possibilities, increased electricity prices, and the expected rise of the EU CO_2 allowance prices in the Baltic countries [75]. In Nordic countries, CO_2 intensity is expected to decrease due to the planned structural changes in the energy systems. However, short-term (2009–2010) and long-term (until 2030) hour-by-hour analyses of marginal electricity generation show that the highest CO_2 intensity is from October to March, especially in Finland [76].

5.2. Colder Winters and Electricity Consumption

The annual electricity consumption per person in Nordic countries, especially in Sweden, is one of the highest in the world [77]. In Sweden, the residential and service sector uses the most electricity, followed by the industrial sector and the transport sector [71]. According to public data available in [78,79], electricity consumption in both countries has been increasing in recent years, as shown in Figure 18, where the average trend is 1.64 TWh/Year for Sweden and 1.19 TWh/Year for Norway. Considering this increasing trend in electricity consumption, it can be seen that the local maxima for Norway are around 2010, 2018, and 2021, which are in concert with the local minima of the yearly winter temperatures, as seen in Figure 7c,d.

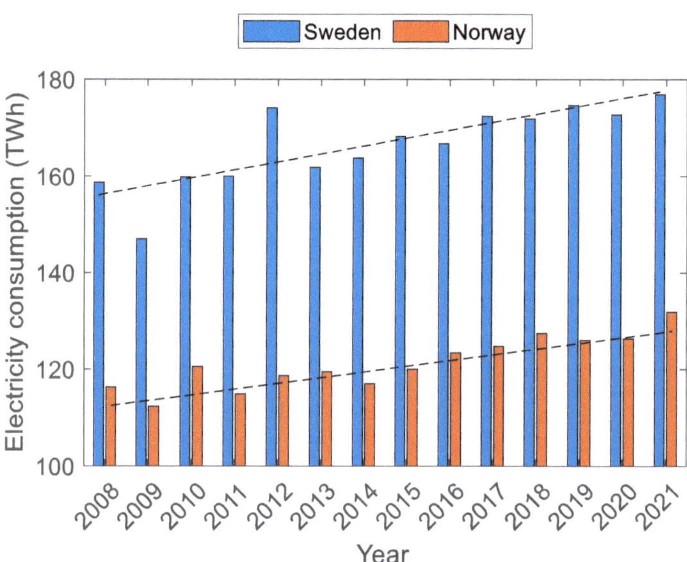

Figure 18. Electricity consumption in Sweden and Norway from 2008 to 2021.

Norway faces an energy deficit almost annually, with the winter season being particularly crucial. This is due to the predominant use of electric heaters in most residential buildings, increased demand from the industrial sector, and the growing need for charging electric cars [74]. Almost 20% of heating comes from electricity [80,81]; consequently, colder winters will directly increase electricity consumption, mainly because people will spend more time at home. Therefore, energy savings should be prioritized in the retrofitting of buildings, even though the investments may not be profitable, as concluded for studied Swedish cities [82]. One promising technology that can be used for multi-family houses is a PV/thermal system combined with a ground-source heat pump system [83]. Energy efficiency should also be increased through the integrated electricity and heating sectors of municipal energy systems, as proposed in [84] for a case study of Piteå in northern Sweden. Another possible source of the increased electricity consumption could be the hydrogen-based steelmaking technology, also known as HYBRIT [85]. In Sweden, HYBRIT requires approximately 10% of electricity generation, which is possible only when electricity exports are reduced [86]. Fortunately, hydrogen storage has the potential to provide balancing services to the power grid.

5.3. Colder Winters, More Likely Storms, and Security of Electricity Supply

As discussed in the previous sections, colder winters in Norway and Sweden generally increase energy consumption and decrease energy generation. In Norway, the electrical energy balance in the winter of 2002/2003 was especially critical due to the limited transmission capacity of power lines between the neighboring regions [74]. The security of electricity supply in Norway was also in focus in the winters of 2009/2010 and 2010/2011 (the years in which AMOC and NAO showed local minima values); however, high prices encouraged lower consumption, higher production, and increased imports of electricity [69]. In Sweden, cold winter events already require an increase in the balancing capacity of the power system, which is needed due to the intermittent generation of wind and solar power units [87,88]. Thus, future research must address the critical question of investments in new storage capacities and equipment for increasing and controlling electrical energy exchange between neighboring regions. Another consideration that might impact the power system operation regarding stability is the amount of inertial energy. From 2017 to 2020, the total inertial energy in Nordic countries decreased by almost 10% [89]. However, the amount of total inertial energy is higher in winter, while the inertial energy of hydropower even started to increase in 2019. Nevertheless, the increased investments in wind power units and the possible shutdown of nuclear power units in Sweden might also reduce the total inertial energy in winter.

More intense winters could result in more storms battering Europe. This, with a weak scenario, could be a further consequence of AMOC weakening [59]. While there may be limited scientific research on this specific aspect, some ecosystem scientists have mentioned it, as reported in sources such as the Guardian [90] or the ClearIAS [91] websites. A study conducted by [92] further supports the notion of increasing storms during the negative phase of the NAO. The major event in Nordic countries was the 2005 Gudrun storm (the year that one of the NAO declines appeared, as reported in Figure 17), causing economic damage to the electric power service, calculated to be around EUR 3 billion [77]. With important evidence, another winter storm in 2011 (the year that AMOC and NAO declines appeared) caused significant disruption in Norway because the high winds brought trees down on power lines [69]. Furthermore, researchers in [93] showed for 30 cities in Sweden that uncertainties in renewable energy potential and demand could lead to a drop in power supply reliability (up to 16%) due to extreme weather events. Such extreme weather events inevitably result in the operation of protection relays to disconnect the faulty elements (power lines, power transformers, and generation units). In order to enhance the resilience of the power grid [94], several measures should be considered, such as the implementation of wide-area monitoring systems in the transmission grid [95–99], smart and closed-loop operation of the distribution grid [100–105], as well as the installation of power quality monitoring and mitigation systems in order to check the impact on disturbances such as RMS voltages (daily or in short time intervals) [106,107].

6. Conclusions

This study aimed to investigate the winter temperatures in Norway and northern Sweden over a period ranging from 50 to 71 years. Six locations were selected, including two in Sweden (1 and 2) and four in Norway (3 to 6). The analysis utilized the SSA algorithm to examine the temperature's long-term trends. The overall long-term trend indicated an increase, which could be attributed to global warming. However, when considering different combinations of highest, lowest, and average temperatures for December, January, and February, 50% of the variations showed a significant decrease in recent years. The average rate of decreasing temperatures was observed as: 0.09, 0.04, 0.088, 0.059, and 0.076 °C/year for locations 2 to 6, respectively, in which locations 2, 4, and 6 experienced colder temperatures, particularly in December, in recent years. The time series of AMOC, a significant climate index, was analyzed from 2004 through to 2020, and the results showed that the values were rarely negative, implying a net flow southward. A maximum positive correlation was observed between AMOC and the Ekman component, showing a direct impact of this component on the AMOC transports. The long-term trend of AMOC measurements presented a 7% general decrease over 17 years, which would lead to an approximate 20% decrease/slowdown forecasted over the first half of the 21st century. However, more data on AMOC would result in more precise results for the AMOC long-term trend concluded from the SSA algorithm. Calculating yearly average values of AMOC transfer variations and its components also showed an anomaly (local minima) during 2009–2010 for all, in 2014 for GS, and in 2019 for both GS and AMOC.

Secondly, the potential similarity between winter AMOC variations and winter temperatures in the six selected locations at mid to high latitudes was investigated. This analysis involved examining the yearly average of winter AMOC and temperatures as well as calculating the lagged correlations between them. The results revealed (1) The possible contribution of AMOC (indirect most likely) to colder winters was realized, particularly around 2010, and further, due to the positive lagged correlations between AMOC and winter temperatures. (2) Higher correlations between winter AMOC and December temperatures (>0.49) were observed specifically in December, for locations 3, 4, and 6. Moreover, higher correlation values were observed for locations 3, 4, and 6. (3) There was a clear pattern where the correlation value increases and the corresponding lag decreases from location 3 to 4 and then to 6, which are those areas along the southern part of the coastline of Norway.

Thirdly, the potential link between another significant climate factor, the NAO, and winter temperatures across the six selected locations was investigated. Similar to the AMOC-temperature analysis, we conducted the same analysis to assess the relationship between the winter NAO index and temperatures, and the results yielded that (1) There is a strong association between the winter NAO and temperatures, specifically in December, for locations 2, 4, 5, and 6, which are situated closer to the coastal areas of Norway but not the northernmost regions. The correlation values between the winter NAO index and December temperatures exceed 0.6, indicating a possible significant influence of NAO on these locations. (2) Among locations 2, 4, 5, and 6, there is a higher probability of experiencing colder winters impacted by the NAO for locations 2, 4, and 5. This conclusion is supported by the following observations: (i) correlation values between the winter NAO index and winter temperatures surpass 0.42; (ii) the highest average rates of temperature decrease were observed earlier for locations 2, 4, and 6.

Fourthly, we examined the impact of colder winters on various aspects of electrical power and energy systems such as electricity generation, electricity consumption, and the security of supply in Sweden and Norway. It was concluded that (1) Colder winters have the potential to reduce reservoir capacities in Norway due to the possibility of ice formation in hydropower units. (2) Reduced electricity generation in Sweden's winters could shut down the nuclear power units. (3) A possible reduction in electricity generation in Norway and Sweden will also impact the neighboring countries. (4) Colder winters directly increase electricity consumption as the demand for electrical heaters in residual buildings rises. Additionally, increased demand is observed in the industrial sector and for charging electric vehicles. (5) A notable example is the winter of 2010, during which a decline in AMOC, NAO, and winter temperatures coincided with increased electricity consumption in Norway. (6) Winter storms, particularly in colder winters, can pose challenges to the resilience and security of power grids, potentially leading to disruptions in the supply of electricity.

In general, our study reveals several important findings. The cities located near the borders of Norway exhibit an overall upward temperature trend that can be followed with a downward trend. Although there was a moderate correlation, specifically for December, between AMOC and temperatures, there has not been clear evidence of a direct impact of AMOC on the winter temperatures on daily/monthly timescales. Considering the NAO variations, in detail, highlighted that the temperatures in December can be impacted directly from NAO, attached with stronger lagged correlations, albeit to varying degrees across different sites. While we did not specifically examine the AMOC-NAO connection in this study, based on the existing literature it might be concluded that NAO could impact both winter temperatures and AMOC. Understanding the interplay between these climate factors is crucial for comprehending temperature variations. To explain the reasons behind the observed downward temperature trends in most locations and subsequent colder winters in recent years, a detailed investigation is needed. The investigation must consider the maps of the temperature changes and analysis to support the links between the reasons and downward temperatures, and explanations as to why it affects particularly December and some locations differently. While there are many potential reasons for these downtrend trends, some possible scenarios could be the weakening of the climatic indices investigated in this study. Colder winters in Norway and Sweden, whether influenced by AMOC, NAO, or other factors, pose challenges for electrical power and energy systems. Researchers must address the challenges of balancing between generation and consumption as well as ensuring the resilience of power grids, which might be crucial in winter, and it is not a good idea to wait and experience such cold winters unprepared. Finally, it is important to note that the climatic indices of AMOC/NAO are complex and variable systems, and there is still considerable uncertainty surrounding their extent, so further research should focus on improving our understanding of these climate phenomena and their possible role in winter climate patterns. We recommend that future studies employ more robust and physically based methods to estimate the colder winters and the phenomena impacting them, moving beyond the statistical/signal processing approaches used in this study. In particular, it would be beneficial to investigate the influence of winter winds in greater detail across the study locations. Additionally, incorporating another Atlantic index, i.e., Atlantic multidecadal variability (AMV), could provide valuable insights. Expanding the analysis to include more locations across Sweden and Norway, and creating a comprehensive correlation–location map would also enhance our understanding of regional variations.

Author Contributions: Conceptualization, Y.M.; methodology, Y.M.; software, Y.M.; validation, Y.M., A.P., B.P., S.M.M. and D.K.; formal analysis, Y.M. and A.P.; investigation, Y.M., A.P., B.P., S.M.M. and D.K.; resources, Y.M.; data curation, Y.M.; writing—original draft preparation, Y.M.; writing—review and editing, Y.M., A.P., B.P., S.M.M. and D.K.; visualization, Y.M., A.P., B.P., S.M.M. and D.K.; supervision, D.K.; project administration, D.K. All authors have read and agreed to the published version of the manuscript.

Funding: This research was funded by the Kempe Foundation (*Kempestiftelserna*) (https://www.kempe.com/), grant (funding) number JCK22-0025. Data from the RAPID AMOC monitoring project are also funded by the Natural Environment Research Council.

Data Availability Statement: Data regarding the temperature at the selected locations are freely available from https://seklima.met.no/observations, accessed on 1 February 2023 and https://www.smhi.se/en/theme/climate-centre. Data from the RAPID AMOC monitoring are also freely accessible from www.rapid.ac.uk/rapidmoc, accessed on 26 January 2023. The website https://ftp.cpc.ncep.noaa.gov/cwlinks/ provides free access to the daily NAO index monitoring data, accessed on 15 May 2023.

Conflicts of Interest: The authors declare no conflict of interest. The funders had no role in the design of the study; in the collection, analyses, or interpretation of data; in the writing of the manuscript; or in the decision to publish the results.

Appendix A

Figure A1 gives the correlation coefficient and corresponding *p*-values using Kendal and Spearman methods.

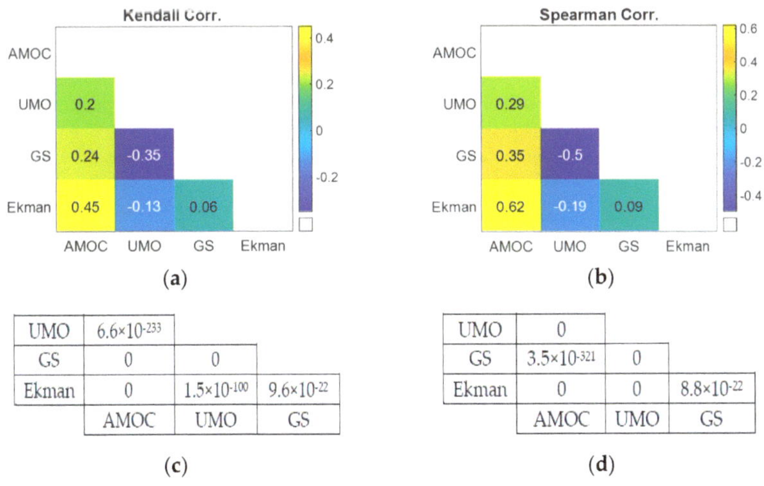

	AMOC	UMO	GS
UMO	6.6×10^{-233}		
GS	0	0	
Ekman	0	1.5×10^{-100}	9.6×10^{-22}

(c)

	AMOC	UMO	GS
UMO	0		
GS	3.5×10^{-321}	0	
Ekman	0	0	8.8×10^{-22}

(d)

Figure A1. Coefficient correlations and corresponding *p*-values: (**a**,**c**) Kendal; (**b**,**d**) Spearman.

Figure A2 shows an uncompressed version of the temperature plotted in Figure 10a to give an understanding of smoothed jumps between each month over different years.

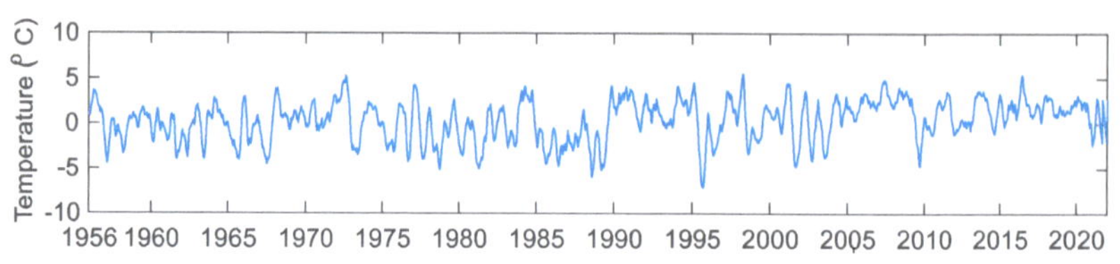

Figure A2. An uncompressed version of the temperature plotted in Figure 10a.

References

1. O'Hare, G. Updating Our Understanding of Climate Change in the North Atlantic: The Role of Global Warming and the Gulf Stream. *Geography* **2011**, *96*, 5–15. [CrossRef]
2. Lozier, M.S. Overturning in the North Atlantic. *Ann. Rev. Mar. Sci.* **2011**, *4*, 291–315. [CrossRef]
3. Willis, J.K. Can in Situ Floats and Satellite Altimeters Detect Long-Term Changes in Atlantic Ocean Overturning? *Geophys. Res. Lett.* **2010**, *37*. [CrossRef]
4. Xu, X.; Chassignet, E.P.; Johns, W.E.; Schmitz, W.J.; Metzger, E.J. Intraseasonal to Interannual Variability of the Atlantic Meridional Overturning Circulation from Eddy-Resolving Simulations and Observations. *J. Geophys. Res.* **2014**, *119*, 5140–5159. [CrossRef]
5. Johns, W.E.; Baringer, M.O.; Beal, L.M.; Cunningham, S.A.; Kanzow, T.; Bryden, H.L.; Hirschi, J.J.M.; Marotzke, J.; Meinen, C.S.; Shaw, B.; et al. Continuous, Array-Based Estimates of Atlantic Ocean Heat Transport at 26.5° N. *J. Clim.* **2011**, *24*, 2429–2449. [CrossRef]
6. McDonagh, E.L.; King, B.A.; Bryden, H.L.; Courtois, P.; Szuts, Z.; Baringer, M.; Cunningham, S.A.; Atkinson, C.; McCarthy, G. Continuous Estimate of Atlantic Oceanic Freshwater Flux at 26.5° N. *J. Clim.* **2015**, *28*, 8888–8906. [CrossRef]
7. Talley, L.D.; Feely, R.A.; Sloyan, B.M.; Wanninkhof, R.; Baringer, M.O.; Bullister, J.L.; Carlson, C.A.; Doney, S.C.; Fine, R.A.; Firing, E.; et al. Changes in Ocean Heat, Carbon Content, and Ventilation: A Review of the First Decade of GO-SHIP Global Repeat Hydrography. *Ann. Rev. Mar. Sci.* **2016**, *8*, 185–215. [CrossRef]
8. Seidov, D.; Mishonov, A.; Reagan, J.; Parsons, R. Multidecadal Variability and Climate Shift in the North Atlantic Ocean. *Geophys. Res. Lett.* **2017**, *44*, 4985–4993. [CrossRef]
9. Lavoie, D.; Lambert, N.; Gilbert, D. Projections of Future Trends in Biogeochemical Conditions in the Northwest Atlantic Using CMIP5 Earth System Models. *Atmosphere-Ocean* **2019**, *57*, 18–40. [CrossRef]
10. Portis, D.H.; Walsh, J.E.; El Hamly, M.; Lamb, P.J. Seasonality of the North Atlantic Oscillation. *J. Clim.* **2001**, *14*, 2069–2078. [CrossRef]
11. Yang, H.; Lohmann, G.; Wei, W.; Dima, M.; Ionita, M.; Liu, J. Intensification and Poleward Shift of Subtropical Western Boundary Currents in a Warming Climate. *J. Geophys. Res. Ocean.* **2016**, *121*, 4928–4945. [CrossRef]
12. Delworth, T.L.; Zeng, F. The Impact of the North Atlantic Oscillation on Climate through Its Influence on the Atlantic Meridional Overturning Circulation. *J. Clim.* **2016**, *29*, 941–962. [CrossRef]
13. Li, M.; Yao, Y.; Simmonds, I.; Luo, D.; Zhong, L.; Chen, X. Collaborative Impact of the Nao and Atmospheric Blocking on European Heatwaves, with a Focus on the Hot Summer of 2018. *Environ. Res. Lett.* **2020**, *15*, 114003. [CrossRef]
14. Buckley, M.W.; Marshall, J. Observations, Inferences, and Mechanisms of the Atlantic Meridional Overturning Circulation: A Review. *Rev. Geophys.* **2016**, *54*, 5–63. [CrossRef]
15. Rahmstorf, S.; Box, J.E.; Feulner, G.; Mann, M.E.; Robinson, A.; Rutherford, S.; Schaffernicht, E.J. Exceptional Twentieth-Century Slowdown in Atlantic Ocean Overturning Circulation. *Nat. Clim. Chang.* **2015**, *5*, 475–480. [CrossRef]
16. Smeed, D.A.; McCarthy, G.D.; Cunningham, S.A.; Frajka-Williams, E.; Rayner, D.; Johns, W.E.; Meinen, C.S.; Baringer, M.O.; Moat, B.I.; Duchez, A.; et al. Observed Decline of the Atlantic Meridional Overturning Circulation 2004–2012. *Ocean Sci.* **2014**, *10*, 29–38. [CrossRef]
17. IPCC. *CLIMATE CHANGE 2013 Climate Change 2013*; IPCC: Geneva, Switzerland, 2013; ISBN 9781107661820.
18. Longworth, H.R.; Bryden, H.L.; Baringer, M.O. Historical Variability in Atlantic Meridional Baroclinic Transport at 26.5° N from Boundary Dynamic Height Observations. *Deep. Res. Part II Top. Stud. Oceanogr.* **2011**, *58*, 1754–1767. [CrossRef]
19. Bryden, H.L.; Longworth, H.R.; Cunningham, S.A. Slowing of the Atlantic Meridional Overturning Circulation at 25 Degrees N. *Nature* **2005**, *438*, 655–657. [CrossRef]
20. Kanzow, T.; Cunningham, S.A.; Johns, W.E.; Hirschi, J.J.M.; Marotzke, J.; Baringer, M.O.; Meinen, C.S.; Chidichimo, M.P.; Atkinson, C.; Beal, L.M.; et al. Seasonal Variability of the Atlantic Meridional Overturning Circulation at 26.5° N. *J. Clim.* **2010**, *23*, 5678–5698. [CrossRef]
21. Jewett, L.; Romanou, A. Ocean acidification and other ocean changes. In *Climate Science Special Report: Fourth National Climate Assessment*; Wuebbles, D.J., Fahey, D.W., Hibbard, K.A., Dokken, D.J., Stewart, B.C., Maycock, T.K., Eds.; U.S. Global Change Research Program: Washington, DC, USA, 2017; Volume I, pp. 364–392. [CrossRef]

22. Weijer, W.; Cheng, W.; Garuba, O.A.; Hu, A.; Nadiga, B.T. CMIP6 Models Predict Significant 21st Century Decline of the Atlantic Meridional Overturning Circulation. *Geophys. Res. Lett.* **2020**, *47*, e2019GL086075. [CrossRef]
23. Zhang, W.-Z.; Chai, F.; Xue, H.; Oey, L.-Y. Remote Sensing Linear Trends of the Gulf Stream from 1993 to 2016. *Ocean Dyn.* **2020**, *70*. [CrossRef]
24. Andres, M.; Donohue, K.A.; Toole, J.M. The Gulf Stream's Path and Time-Averaged Velocity Structure and Transport at 68.5° W and 70.3° W. *Deep Sea Res. Part I Oceanogr. Res. Pap.* **2020**, *156*, 103179. [CrossRef]
25. Fox-Kemper, B. Ocean, Cryosphere and Sea Level Change. In Proceedings of the AGU Fall Meeting Abstracts, New Orleans, LA, USA, 13–17 December 2021; Volume 2021, p. U13B-09.
26. Yang, H.; Chen, G.; Tang, Q.; Hess, P. Quantifying Isentropic Stratosphere-Troposphere Exchange of Ozone. *J. Geophys. Res. Atmos.* **2016**, *121*, 3372–3387. [CrossRef]
27. Chen, C.; Wang, G.; Xie, S.-P.; Liu, W. Why Does Global Warming Weaken the Gulf Stream but Intensify the Kuroshio? *J. Clim.* **2019**, *32*, 7437–7451. [CrossRef]
28. Caesar, L.; Rahmstorf, S.; Robinson, A.G.; Feulner, G.; Saba, V.S. Observed Fingerprint of a Weakening Atlantic Ocean Overturning Circulation. *Nature* **2017**, *556*, 191–196. [CrossRef] [PubMed]
29. Ortega, P.; Robson, J.; Sutton, R.; Andrews, M. Mechanisms of Decadal Variability in the Labrador Sea and the Wider North Atlantic in a High-Resolution Climate Model. *Clim. Dyn.* **2017**, *49*, 2625–2647. [CrossRef]
30. Belonenko, T.; Morozova, L.; Gordeeva, S. Key to the Atlantic Gates of the Arctic. *Russ. J. Earth Sci.* **2022**, *22*, 2004. [CrossRef]
31. Caesar, L.; McCarthy, G.; Thornalley, D.; Cahill, N.; Rahmstorf, S. Current Atlantic Meridional Overturning Circulation Weakest in Last Millennium. *Nat. Geosci.* **2021**, *14*, 1–3. [CrossRef]
32. Biastoch, A.; Schwarzkopf, F.U.; Getzlaff, K.; Rühs, S.; Martin, T.; Scheinert, M.; Schulzki, T.; Handmann, P.; Hummels, R.; Böning, C.W. Regional Imprints of Changes in the Atlantic Meridional Overturning Circulation in the Eddy-Rich Ocean Model VIKING20X. *Ocean Sci.* **2021**, *17*, 1177–1211. [CrossRef]
33. University of South Florida (USF Innovation). Melting Greenland ice Sheet may Affect Global Ocean Circulation, Future Climate: University of South Florida and International Scientists Find Influx of Freshwater Could Disrupt the Atlantic Meridional Overturning Circulation, an Important Component of Global Ocean Circulation. ScienceDaily, 22 January 2016. Available online: www.sciencedaily.com/releases/2016/01/160122122629.htm (accessed on 24 February 2023).
34. Ciemer, C.; Winkelmann, R.; Kurths, J.; Boers, N. Impact of an AMOC Weakening on the Stability of the Southern Amazon Rainforest. *Eur. Phys. J. Spec. Top.* **2021**, *230*, 3065–3073. [CrossRef]
35. Liu, W.; Fedorov, A.V.; Xie, S.-P.; Hu, S. Climate Impacts of a Weakened Atlantic Meridional Overturning Circulation in a Warming Climate. *Sci. Adv.* **2020**, *6*, eaaz4876. [CrossRef]
36. Wang, H.; Zuo, Z.; Qiao, L.; Zhang, K.; Sun, C.; Xiao, D.; Bu, L. Frequency of Winter Temperature Extremes over Northern Eurasia Dominated by AMOC. *npj Clim. Atmos. Sci.* **2021**, *5*, 84. [CrossRef]
37. Smeed, D.; Wood, R.; Cunningham, S.; Mccarthy, G.; Kuhlbrodt, T.; Office, M.; Centre, H. Impacts of Climate Change on the Atlantic Heat Conveyor. *MCCP Sci. Rev.* **2013**, *4*, 49–59. [CrossRef]
38. Buchan, J.; Hirschi, J.J.M.; Blaker, A.T.; Sinha, B. North Atlantic SST Anomalies and the Cold North European Weather Events of Winter 2009/10 and December 2010. *Mon. Weather Rev.* **2014**, *142*, 922–932. [CrossRef]
39. Bryden, H.L.; King, B.A.; McCarthy, G.D.; McDonagh, E.L. Impact of a 30% Reduction in Atlantic Meridional Overturning during 2009–2010. *Ocean Sci.* **2014**, *10*, 683–691. [CrossRef]
40. Munday, A.J.; Hunt, J.B. University of Southampton. *Tribology* **1970**, *3*, 106–107. [CrossRef]
41. Bellomo, K.; Meccia, V.; D'Agostino, R.; Fabiano, F.; von Hardenberg, J.; Corti, S. The Climate Impacts of an Abrupt AMOC Weakening on the European Winters. In Proceedings of the EGU General Assembly 2022, Vienna, Austria, 23–27 May 2022. EGU22-1023. [CrossRef]
42. RAPID: Monitoring the Atlantic Meridional Overturning Circulation at 26.5 N since 2004. Available online: https://rapid.ac.uk/rapidmoc/overview.php (accessed on 26 January 2023).
43. Duchez, A.; Courtois, P.; Harris, E.; Josey, S.A.; Kanzow, T.; Marsh, R.; Smeed, D.A.; Hirschi, J.J.M. Potential for Seasonal Prediction of Atlantic Sea Surface Temperatures Using the RAPID Array at 26° N. *Clim. Dyn.* **2016**, *46*, 3351–3370. [CrossRef]
44. Moat, B.I.; Smeed, D.A.; Frajka-Williams, E.; Desbruyères, D.G.; Beaulieu, C.; Johns, W.E.; Rayner, D.; Sanchez-Franks, A.; Baringer, M.O.; Volkov, D.; et al. Pending Recovery in the Strength of the Meridional Overturning Circulation at 26° N. *Ocean Sci.* **2020**, *16*, 863–874. [CrossRef]
45. Wanner, H.; Brönnimann, S.; Casty, C.; Gyalistras, D.; Luterbacher, J.; Schmutz, C.; Stephenson, D.B.; Xoplaki, E. North Atlantic Oscillation–Concepts And Studies. *Surv. Geophys.* **2001**, *22*, 321–381. [CrossRef]
46. Stephenson, D.B.; Wanner, H.; Brönnimann, S.; Luterbacher, J. The History of Scientific Research on the North Atlantic Oscillation. In *The North Atlantic Oscillation: Climatic Significance and Environmental Impact*; American Geophysical Union (AGU): Washington, DC, USA, 2003; pp. 37–50. ISBN 9781118669037.
47. Norsk Klimaservicesenter. Available online: https://seklima.met.no/observations (accessed on 26 January 2023).
48. Swedish National Knowledge Centre for Climate Change Adaptation | SMHI. Available online: https://www.smhi.se/en/theme/climate-centre (accessed on 26 January 2023).
49. Monitoring North Atlantic Oscillation Since 1950. Available online: Https://Ftp.Cpc.Ncep.Noaa.Gov/Cwlinks/ (accessed on 15 May 2023).

50. Sinha, B.; Smeed, D.A.; McCarthy, G.; Moat, B.I.; Josey, S.A.; Hirschi, J.J.-M.; Frajka-Williams, E.; Blaker, A.T.; Rayner, D.; Madec, G. The Accuracy of Estimates of the Overturning Circulation from Basin-Wide Mooring Arrays. *Prog. Oceanogr.* **2018**, *160*, 101–123. [CrossRef]

51. McCarthy, G.D.; Brown, P.J.; Flagg, C.N.; Goni, G.; Houpert, L.; Hughes, C.W.; Hummels, R.; Inall, M.; Jochumsen, K.; Larsen, K.M.H.; et al. Sustainable Observations of the AMOC: Methodology and Technology. *Rev. Geophys.* **2020**, *58*, e2019RG000654. [CrossRef]

52. Boschat, G.; Simmonds, I.; Purich, A.; Cowan, T.; Pezza, A.B. On the Use of Composite Analyses to Form Physical Hypotheses: An Example from Heat Wave-SST Associations. *Sci. Rep.* **2016**, *6*, 29599. [CrossRef] [PubMed]

53. Deng, C. Time Series Decomposition Using Singular Spectrum Analysis Part of the Longitudinal Data Analysis and Time Series Commons. Master's Thesis, East Tennessee State University, Johnson City, TN, USA, 2014.

54. Golyandina, N. Particularities and Commonalities of Singular Spectrum Analysis as a Method of Time Series Analysis and Signal Processing. *WIREs Comput. Stat.* **2020**, *12*, e1487. [CrossRef]

55. Miraftabzadeh, S.M.; Longo, M.; Brenna, M.; Pasetti, M. Data-Driven Model for PV Power Generation Patterns Extraction via Unsupervised Machine Learning Methods. In Proceedings of the 2022 North American Power Symposium (NAPS), Salt Lake City, UT, USA, 9–11 October 2022; pp. 1–5.

56. Miraftabzadeh, S.M.; Longo, M.; Brenna, M. Knowledge Extraction from PV Power Generation with Deep Learning Autoencoder and Clustering-Based Algorithms. *IEEE Access* **2023**, *11*, 69227–69240. [CrossRef]

57. Li, H. *Spectral Analysis of Signals [Book Review]*; IEEE: Piscataway, NJ, USA, 2008; Volume 24, ISBN 0131139568.

58. Muir, L.C.; Fedorov, A.V. How the AMOC Affects Ocean Temperatures on Decadal to Centennial Timescales: The North Atlantic versus an Interhemispheric Seesaw. *Clim. Dyn.* **2015**, *45*, 151–160. [CrossRef]

59. Bellomo, K.; Agostino, R.D.; Fabiano, F.; Larson, S.M.; Corti, S. Impacts of a Weakened AMOC on Precipitation over the Euro-Atlantic Region in the EC-Earth3 Climate Model. *Clim. Dyn.* **2022**. [CrossRef]

60. Chiang, J.C.H.; Cheng, W.; Kim, W.M.; Kim, S. Untangling the Relationship Between AMOC Variability and North Atlantic Upper-Ocean Temperature and Salinity. *Geophys. Res. Lett.* **2021**, *48*, e2021GL093496. [CrossRef]

61. López-Moreno, J.I.; Vicente-Serrano, S.M.; Morán-Tejeda, E.; Lorenzo-Lacruz, J.; Kenawy, A.; Beniston, M. Effects of the North Atlantic Oscillation (NAO) on Combined Temperature and Precipitation Winter Modes in the Mediterranean Mountains: Observed Relationships and Projections for the 21st Century. *Glob. Planet. Chang.* **2011**, *77*, 62–76. [CrossRef]

62. Rousi, E.; Rust, H.W.; Ulbrich, U.; Anagnostopoulou, C. Implications of Winter NAO Flavors on Present and Future European Climate. *Climate* **2020**, *8*, 13. [CrossRef]

63. Jackson, L.C.; Biastoch, A.; Buckley, M.W.; Desbruyères, D.G.; Frajka-Williams, E.; Moat, B.; Robson, J. The Evolution of the North Atlantic Meridional Overturning Circulation since 1980. *Nat. Rev. Earth Environ.* **2022**, *3*, 241–254. [CrossRef]

64. Roach, L.A.; Blanchard-Wrigglesworth, E.; Ragen, S.; Cheng, W.; Armour, K.C.; Bitz, C.M. The Impact of Winds on AMOC in a Fully-Coupled Climate Model. *Geophys. Res. Lett.* **2022**, *49*, 1–10. [CrossRef]

65. Kim, H.J.; An, S.I.; Park, J.H.; Sung, M.K.; Kim, D.; Choi, Y.; Kim, J.S. North Atlantic Oscillation Impact on the Atlantic Meridional Overturning Circulation Shaped by the Mean State. *npj Clim. Atmos. Sci.* **2023**, *6*. [CrossRef]

66. Iversen, E.C.; Burningham, H. Relationship between NAO and Wind Climate over Norway. *Clim. Res.* **2015**, *63*, 115–134. [CrossRef]

67. Xu, Y. Estimates of Changes in Surface Wind and Temperature Extremes in Southwestern Norway Using Dynamical Downscaling Method under Future Climate. *Weather Clim. Extrem.* **2019**, *26*, 100234. [CrossRef]

68. ENTSO-E Transparency Platform. Available online: https://transparency.entsoe.eu/ (accessed on 24 February 2023).

69. Energifaktanorge.No-Fakta Om Norsk Energi Og Vannressurser-Energifakta Norge. Available online: https://energifaktanorge.no/ (accessed on 24 February 2023).

70. An Overview of Energy in Sweden 2022 Now Available. Available online: https://www.energimyndigheten.se/en/news/2022/an-overview-of-energy-in-sweden-2022-now-available/ (accessed on 24 February 2023).

71. Quarterly Statistics-Swedish Wind Energy Association. Available online: https://swedishwindenergy.com/statistics (accessed on 24 February 2023).

72. Haddeland, I.; Hole, J.; Holmqvist, E.; Koestler, V.; Sidelnikova, M.; Veie, C.A.; Wold, M. Effects of Climate on Renewable Energy Sources and Electricity Supply in Norway. *Renew. Energy* **2022**, *196*, 625–637. [CrossRef]

73. Heggenes, J.; Stickler, M.; Alfredsen, K.; Brittain, J.E.; Adeva-Bustos, A.; Huusko, A. Hydropower-Driven Thermal Changes, Biological Responses and Mitigating Measures in Northern River Systems. *River Res. Appl.* **2021**, *37*, 743–765. [CrossRef]

74. François, B.; Martino, S.; Tøfte, L.S.; Hingray, B.; Mo, B.; Creutin, J.-D. Effects of Increased Wind Power Generation on Mid-Norway's Energy Balance under Climate Change: A Market Based Approach. *Energies* **2017**, *10*, 227. [CrossRef]

75. Farsaei, A.; Syri, S.; Olkkonen, V.; Khosravi, A. Unintended Consequences of National Climate Policy on International Electricity Markets—Case Finland's Ban on Coal-Fired Generation. *Energies* **2020**, *13*, 1930. [CrossRef]

76. Olkkonen, V.; Syri, S. Spatial and Temporal Variations of Marginal Electricity Generation: The Case of the Finnish, Nordic, and European Energy Systems up to 2030. *J. Clean. Prod.* **2016**, *126*, 515–525. [CrossRef]

77. Gündüz, N.; Küfeoğlu, S.; Lehtonen, M. Impacts of Natural Disasters on Swedish Electric Power Policy: A Case Study. *Sustainability* **2017**, *9*, 230. [CrossRef]

78. Available online: https://www.statista.com/Statistics/1027078/Electricity-Consumption-in-Sweden/ (accessed on 5 May 2023).

79. Available online: https://www.statista.com/Statistics/1024933/Net-Consumption-of-Electricity-in-Norway/ (accessed on 25 January 2023).

80. Energimyndigheten. Available online: https://www.energimyndigheten.se/ (accessed on 26 January 2023).

81. Renewable Energy Production in Norway. Available online: https://www.regjeringen.no/en/topics/energy/renewable-energy/renewable-energy-production-in-norway/id2343462/ (accessed on 11 May 2016).

82. Mata, É.; Wanemark, J.; Nik, V.M.; Sasic Kalagasidis, A. Economic Feasibility of Building Retrofitting Mitigation Potentials: Climate Change Uncertainties for Swedish Cities. *Appl. Energy* **2019**, *242*, 1022–1035. [CrossRef]

83. Sommerfeldt, N.; Madani, H. In-Depth Techno-Economic Analysis of PV/Thermal plus Ground Source Heat Pump Systems for Multi-Family Houses in a Heating Dominated Climate. *Sol. Energy* **2019**, *190*, 44–62. [CrossRef]

84. Fischer, R.; Elfgren, E.; Toffolo, A. Towards Optimal Sustainable Energy Systems in Nordic Municipalities. *Energies* **2020**, *13*, 290. [CrossRef]

85. Hybrit. Available online: https://www.hybritdevelopment.se/en/ (accessed on 24 February 2023).

86. Öhman, A.; Karakaya, E.; Urban, F. Enabling the Transition to a Fossil-Free Steel Sector: The Conditions for Technology Transfer for Hydrogen-Based Steelmaking in Europe. *Energy Res. Soc. Sci.* **2022**, *84*, 102384. [CrossRef]

87. Höltinger, S.; Mikovits, C.; Schmidt, J.; Baumgartner, J.; Arheimer, B.; Lindström, G.; Wetterlund, E. The Impact of Climatic Extreme Events on the Feasibility of Fully Renewable Power Systems: A Case Study for Sweden. *Energy* **2019**, *178*, 695–713. [CrossRef]

88. Miraftabzadeh, S.M.; Longo, M. High-Resolution PV Power Prediction Model Based on the Deep Learning and Attention Mechanism. *Sustain. Energy Grids Networks* **2023**, *34*, 101025. [CrossRef]

89. Graham, J.; Heylen, E.; Bian, Y.; Teng, F. Benchmarking Explanatory Models for Inertia Forecasting Using Public Data of the Nordic Area. In Proceedings of the 2022 17th International Conference on Probabilistic Methods Applied to Power Systems (PMAPS), Lisbon, Portugal, 12–15 June 2022; pp. 1–6.

90. Atlantic Ocean Circulation at Weakest in a Millennium, Say Scientists | Climate Crisis | The Guardian. Available online: https://www.theguardian.com/environment/2021/feb/25/atlantic-ocean-circulation-at-weakest-in-a-millennium-say-scientists (accessed on 26 January 2023).

91. Available online: https://www.clearias.com/atlantic-meridional-overturning-circulation/ (accessed on 20 December 2022).

92. Chartrand, J.; Pausata, F.S.R. Impacts of the North Atlantic Oscillation on Winter Precipitations and Storm Track Variability in Southeast Canada and the Northeast United States. *Weather Clim. Dyn.* **2020**, *1*, 731–744. [CrossRef]

93. Perera, A.T.D.; Nik, V.M.; Chen, D.; Scartezzini, J.-L.; Hong, T. Quantifying the Impacts of Climate Change and Extreme Climate Events on Energy Systems. *Nat. Energy* **2020**, *5*, 150–159. [CrossRef]

94. de Morais, S.G.; Resener, M.; Ramos, J.S.M.; Mohammadi, Y. Estabilidade Transitória de Sistemas de Distribuição Ativos Com Recursos Energéticos Distribuídos: Um Estudo de Caso. *Secr. SBA* **2020**, *2*, 1. [CrossRef]

95. Mohammadi, Y.; Moradi, M.H.; Chouhy Leborgne, R. Employing Instantaneous Positive Sequence Symmetrical Components for Voltage Sag Source Relative Location. *Electr. Power Syst. Res.* **2017**, *151*, 186–196. [CrossRef]

96. Mohammadi, Y.; Salarpour, A.; Chouhy Leborgne, R. Comprehensive Strategy for Classification of Voltage Sags Source Location Using Optimal Feature Selection Applied to Support Vector Machine and Ensemble Techniques. *Int. J. Electr. Power Energy Syst.* **2021**, *124*, 106363. [CrossRef]

97. Shazdeh, S.; Colpîra, H.; Bevrani, H. A PMU-Based Back-up Protection Scheme for Fault Detection Considering Uncertainties. *Int. J. Electr. Power Energy Syst.* **2023**, *145*, 108592. [CrossRef]

98. Liu, T.; Tang, Z.; Huang, Y.; Xu, L.; Yang, Y. Online Prediction and Control of Post-Fault Transient Stability Based on PMU Measurements and Multi-Task Learning. *Front. Energy Res.* **2023**, *10*, 1084295. [CrossRef]

99. Alqudah, M.; Kezunovic, M.; Obradovic, Z. Automated Power System Fault Prediction and Precursor Discovery Using Multi-Modal Data. *IEEE Access* **2023**, *11*, 7283–7296. [CrossRef]

100. Mohammadi, Y.; Leborgne, R.C. A New Approach for Voltage Sag Source Relative Location in Active Distribution Systems with the Presence of Inverter-Based Distributed Generations. *Electr. Power Syst. Res.* **2020**, *182*, 106222. [CrossRef]

101. Moradi, M.H.; Mohammadi, Y.; Hoseyni Tayyebi, M. A Novel Method to Locate the Voltage Sag Source: A Case Study in the Brazilian Power Network (Mato Grosso). *Prz. Elektrotechniczny* **2012**, *88*, 112–115.

102. Mohammadi, Y.; Leborgne, R.C.; Polajžer, B. Modified Methods for Voltage-Sag Source Detection Using Transient Periods. *Electr. Power Syst. Res.* **2022**, *207*, 107857. [CrossRef]

103. Mohammadi, Y.; Miraftabzadeh, S.M.; Bollen, M.H.J.; Longo, M. Voltage-Sag Source Detection: Developing Supervised Methods and Proposing a New Unsupervised Learning. *Sustain. Energy, Grids Networks* **2022**, *32*, 100855. [CrossRef]

104. Štumbrger, G.; Rošer, M.; Pintarič, M.; Polajžer, B. Permanent Closed-Loop Operation as a Measure for Improving Power Supply Reliability in a Rural Medium Voltage Distribution Network. *Renew. Energy Power Qual. J.* **2021**, *19*, 523–527. [CrossRef]

105. Mohammadi, Y.; Leborgne, R.C. Improved DR and CBM Methods for Finding Relative Location of Voltage Sag Source at the PCC of Distributed Energy Resources. *Int. J. Electr. Power Energy Syst.* **2020**, *117*, 105664. [CrossRef]

106. Mohammadi, Y.; Miraftabzadeh, S.M.; Bollen, M.H.J.; Longo, M. An Unsupervised Learning Schema for Seeking Patterns in Rms Voltage Variations at the Sub-10-Minute Time Scale. *Sustain. Energy Grids Networks* **2022**, *31*, 100773. [CrossRef]
107. Mohammadi, Y.; Mahdi Miraftabzadeh, S.; Bollen, M.H.J.; Longo, M. Seeking Patterns in Rms Voltage Variations at the Sub-10-Minute Scale from Multiple Locations via Unsupervised Learning and Patterns' Post-Processing. *Int. J. Electr. Power Energy Syst.* **2022**, *143*, 108516. [CrossRef]

Review

Review: The Energy Implications of Averting Climate Change Catastrophe

Patrick Moriarty [1,*] and Damon Honnery [2]

[1] Department of Design, Monash University, Caulfield Campus, P.O. Box 197, Caulfield East, VIC 3145, Australia

[2] Department of Mechanical and Aerospace Engineering, Monash University, Clayton Campus, P.O. Box 31, Clayton, VIC 3800, Australia; damon.honnery@monash.edu

[*] Correspondence: patrick.moriarty@monash.edu; Tel.: +61-39-903-2584

Abstract: Conventional methods of climate change (CC) mitigation have not 'bent the curve' of steadily rising annual anthropic CO_2 emissions or atmospheric concentrations of greenhouse gases. This study reviews the present position and likely future of such methods, using the recently published literature with a global context. It particularly looks at how fast they could be implemented, given the limited time available for avoiding catastrophic CC (CCC). This study then critically examines solar geoengineering, an approach often viewed as complementary to conventional mitigation. Next, this review introduces equity considerations and shows how these even further shorten the available time for effective action for CC mitigation. The main findings are as follows. Conventional mitigation approaches would be implemented too slowly to be of much help in avoiding CCC, partly because some suggested technologies are infeasible, while others are either of limited technical potential or, like wind and solar energy, cannot be introduced fast enough. Due to these problems, solar geoengineering is increasingly advocated for as a quick-acting and effective solution. However, it could have serious side effects, and, given that there would be winners and losers at the international level as well as at the more regional level, political opposition may make it very difficult to implement. The conclusion is that global energy consumption itself must be rapidly reduced to avoid catastrophic climate change, which requires strong policy support.

Keywords: climate change; climate equity; energy equity; energy reductions; fossil fuels; global sustainability; policy changes; renewable energy; technological optimism

Citation: Moriarty, P.; Honnery, D. Review: The Energy Implications of Averting Climate Change Catastrophe. *Energies* **2023**, *16*, 6178. https://doi.org/10.3390/en16176178

Academic Editor: Eugen Rusu

Received: 8 July 2023
Revised: 13 August 2023
Accepted: 23 August 2023
Published: 25 August 2023

1. Introduction

Interest in climate change (CC) and means of CC mitigation is at an all-time high. According to the Scopus database, a total of over 426,000 papers have so far been published with the term 'climate change' in either the title, abstract, or keywords. In 2022, the figure was over 48,000, more than double the 2016 number. However, this vast number of reviewed papers has not led to any reduction in carbon emissions. On the contrary, CO_2 emissions from energy and industry rose from 21.3 gigatonnes (Gt) in 1990, the year of the first Intergovernmental Panel on Climate Change (IPCC) report, to 33.9 Gt in 2021 [1]. This was paralleled by the rise in overall greenhouse gas (GHG) emissions, which reached an estimated 59.0 ± 6.6 $GtCO_2$ equivalent ($GtCO_2$-eq) in 2019 [2].

A further factor to consider is that Earth faces several other environmental challenges in addition to CC [3]. Steffen et al. [4] originally identified nine planetary boundaries, including CC, for which the crossing of any could prove catastrophic. These other global problems include the deterioration of the ocean environment and ongoing acidification, biodiversity loss, and air, water, and land pollution, especially by plastics [5–9]. Also, as Crist et al. [10] warned, the world's present population, let alone projected further increases [11], make achieving a sustainable future Earth even more difficult.

What is novel in this paper is the stress on the crucial importance of the time factor in assessing the feasibility of the various possible responses to CC and its interaction with equity considerations. For CC mitigation, the important factor is not the ultimate potential for each proposal, but whether it can be effectively deployed in time to avert not only catastrophic CC [12] but also the other challenges to global sustainability. Synergistic interactions among these various threats can potentially further shorten the time we have available for effective action to avoid crossing a given threshold [13] Or, as the IPCC [2] put it, we can have high confidence that 'Climatic and non-climatic risks will increasingly interact, creating compound and cascading risks that are more complex and difficult to manage'. A full discussion of this time dimension is lacking in virtually all of the many studies that address CC mitigation.

In Section 2, the frequency of published papers in Scopus on various possible approaches to dealing with climate change, as well as the approach used for article selection in this paper, are presented and discussed. Section 3 stresses the crucial importance of timing: can any of these proposed solutions make a real difference in the crucial next decade or two? Section 4, in turn, examines the various conventional mitigation methods from this time-based viewpoint. In Section 5, solar geoengineering (SG) is considered as an alternative to the slow shift to low-carbon fuels, but it has known and possibly unknown serious risks. Section 6 examines the complex questions of equity in income, energy use, and CO_2 emissions in both low- and high-income countries. Finally, Section 7 discusses all these methods and finds that none of them, singly or together, can affect the reduction in climate forcing that is needed over the critical next couple of decades. The overall conclusion is that the changes needed—including energy reductions and the need to sustain biodiversity and cut land, air, and water pollution—necessitate the end of global economic growth.

2. Materials and Methods

Figure 1 uses the Scopus database to show how annual publications on various methods of CC mitigation have changed over the years. Although not an energy-related CC mitigation approach, SG was included since it is regarded by many as an alternative (or at least a complement) to conventional mitigation approaches [14]. It is evident that interest in bioenergy with carbon capture and storage (BECCS), SG (also called solar radiation management (SRM)), and direct air capture (DAC) only took off after around 2010. In contrast, the more general term carbon dioxide removal (CDR), which includes BECCS and DAC as well as methods like reforestation, enhanced weathering (EW), and soil carbon sequestration, has had many annual publications for decades. As mentioned, the term 'climate change' returned a total of over 426,000 papers, with annual numbers beginning to rise sharply in the late 1980s. Over the same period, a further 70,400 papers included the term 'global warming' in place of 'climate change', lifting the combined total to almost half a million articles.

In this paper, the various approaches to avoiding CCC are critically discussed. The emphasis on papers selected for discussion in general meet two criteria. First, they preferably should be global in scope, since CC is a global problem; local solutions may not be feasible elsewhere and could even be globally counterproductive. Second, given the progress in both understanding the nature of CC and the assessment of the viability of proposed mitigation solutions, very recent papers were preferred over older ones. The IPCC's sixth assessment reports [15,16], particularly its 2023 Synthesis Report [2], were relied on for the science of global warming and the up-to-date surveys of mitigation methods. The annual publications by BP [1] and the International Energy Agency (IEA) [17] were used for global and national energy statistics.

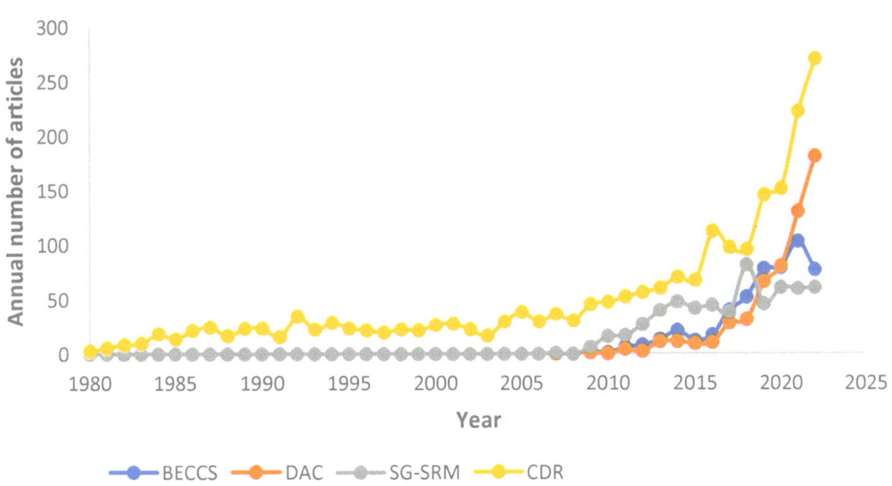

Figure 1. Plot of the annual number of annual publications in the Scopus database with the terms 'BECCS', 'DAC', 'SG OR SRM', and CDR in the title, abstract, or keywords, from 1980 to 2022.

3. Importance of Timing for Low-Carbon Energy

A complication for CC mitigation is the short time left for effective action. Already, the world is experiencing a spate of record-breaking extreme weather events—floods, heat waves, droughts, and wildfires [18,19]. Both their severity and frequency are anticipated to rise in a non-linear manner as the temperature rises; the increase from 1.0 to 1.5 °C can be expected to produce more damage than the previous increase from 0.5 to 1.0 °C, just as this latter rise was more damaging than that from 0 to 0.5 °C [2]. This does not mean that, when the mean global temperature surpasses 1.5 or 2.0 °C above pre-industrial levels, we should give up all attempts at mitigation. Even a 3.0 °C rise, while disastrous in its effects, is much less severe than a 4 °C increase [2].

Different mitigation methods not only have different average costs and potentials but also have different time frames for their implementation. For all forms of renewable energy (RE) except bioenergy, lifetime energy input costs are dominated by energy for construction, as the annual operating energy costs are small. Due to this, the rate of introduction of new RE is important, as formalised in dynamic energy analysis (DEA).

Capellán-Pérez et al. [20] examined the consequences of a complete global shift to 100% RE for electricity by 2060. Their modelled results showed that the average energy return on investment (EROI) would fall from its current value of about 12 to about 3 by 2050 and would then stabilise at about 5. The authors pointed out that these low values are well below those thought needed to maintain a (growth-oriented) industrial economy. The reason for these low EROI values is that much of the output from the RE plants is needed to build new RE plants, limiting the amount of energy available to run the rest of the economy. From another angle, Fizaine and Court [21] argued that, for the US, 'growth is only possible if its primary energy system has at least a minimum EROI of approximately 11:1'. The conclusion is that if the aim is to keep industrial economies going, DEA/EROI considerations show that the rate of uptake of RE for electricity—and for primary energy generally—must be curtailed.

A further factor that could slow down the rate of non-carbon energy sources is that, in many OECD countries, electricity production is falling [1]. This is the case for major economies such as Germany and the UK, where usage peaked a decade or more ago [1]. With falling demand, there is less need for new electricity power capacity of any type, which again hinders growth in RE electricity in these countries. Although the output of wind and solar is increasing in both Europe and the Organization for Economic Cooperation and

Development (OECD) overall [1], further growth in RE may be dictated by the replacement rate of ageing generation infrastructure rather than growth in demand. For total commercial primary energy consumption, the decrease is even more pronounced, with the OECD overall and, especially, European Union (EU) countries, experiencing a peak around 2007 [1]. Table 1 shows the change in the share of total primary energy and total low carbon primary energy of the OECD and non-OECD over the period 2011–2022.

Table 1. OECD and non-OECD share of primary energy and low carbon fuel primary energy for 2011 and 2022.

Energy Type\Year	2011	2022
OECD primary energy share (%)	46.5	38.8
Non-OECD primary energy share (%)	53.5	61.2
OECD low carbon primary energy share (%)	61.1	48.1
Non-OECD low carbon primary energy share (%)	38.9	51.9

Source [1].

If CO_2 was a short-lived gas in the atmosphere—with, say, an atmospheric lifetime of only one year—then any reduction in annual emissions would also reduce atmospheric CO_2 concentrations. The problem is, of course, that CO_2 has a very long atmospheric lifetime. Although the exact figure is disputed (see, e.g., [22,23]), full recovery to its pre-industrial atmospheric levels could take centuries. It follows that most of the CO_2 the world has emitted since the 1950s will still be present over the crucial next few decades.

The multiple challenges to sustainability discussed in the Introduction complicate the search for timely CC mitigation solutions in two ways, which adds to the urgency of a rapid response to ongoing CC. Climate change—and how we respond to it—affects other environmental problems such as biodiversity loss [6]. More generally, various global limits can act synergistically, lowering a threshold and, thus, the time available for effective action to avoid crossing a given threshold [13]. Unfortunately, it is not possible to give any dates for when the various approaches would be able to play a dominant role in CC mitigation. Only for RE are estimates available for various scenarios, with the IEA [24] forecasting RE as just over half of global primary energy by 2050 in their Announced Pledges Scenario (APS).

4. Assessment of Conventional Approaches

In 1990, the IPCC Intergovernmental Panel on Climate Change (IPCC) released its first report. At that time, the conventional methods for mitigation could have provided a feasible solution. These approaches include greatly increased use of the various forms of renewable energy (RE); nuclear power; increased energy efficiency; and CDR, both by biological and mechanical means. But, as is shown here, these solutions, even taken together, cannot give the world much relief from climate change. The reasons include the following, with one or more applicable to each approach:

- They cannot deliver major CC mitigation in a timely manner;
- Their mitigation potential is too small;
- Feedback effects reduce their mitigation potential;
- Political opposition limits their deployment at scale;
- Their expansion conflicts with other important aims.

The authors discussed the difficulties facing these various approaches in previous publications (see, e.g., [25,26]). Hence, in this section, emphasis is placed on the first of these points: how rapidly could each of these reduce global climate forcing?

4.1. Non-Fossil Fuel Energy Sources

Solar and wind energy are not only the fastest-growing RE sources [24] but also those with the greatest expansion potential. Nevertheless, DEA indicates that their rate of growth could be limited if sufficient energy is available for the non-energy sectors of the economy [19]. As already discussed, the rate at which their output can grow is governed by their EROI [27]. A characteristic of all RE sources except biomass is that nearly all energy inputs—for materials mining and processing, for construction, for access roads, and for transmission and distribution power lines—must be made upfront, before any energy output can be obtained. The maintenance energy costs are relatively minor. Only dismantling and site cleanup energy costs must be postponed until the plant's end of life.

If the EROI for wind and solar energy is high, then only minor energy inputs are needed, and the net energy available for the non-energy economy sectors is high. But if it is low—below a value of about 5–10—an 'energy cliff' [27] is encountered, such that input energy costs are significant, and DEA analysis is needed. The problem is that the EROI values for wind and, especially, for photovoltaic (PV) systems are strongly contested (see, e.g., [28–30]), with some researchers giving very high values for PV and others giving very low values.

The key explanation for this divergence is the inclusion or otherwise of important input costs, especially those termed Ecosystem Maintenance Energy (ESME) costs [31]. These include the energy costs of avoiding pollution from the mining of the often-scarce materials needed for wind and PV energy systems. All too often, such mining in tropical African countries and elsewhere ignores the local pollution that is generated. Even when tailing dams are constructed, they often fail [32]. This suggests that the input energy costs for RE electricity systems (which have much higher materials input per gigawatt (GW) of capacity than fossil fuel (FF) plants [33]) are often significantly under-estimated, which means that their EROI values are inflated. Lower EROI values also mean that emission savings are also lower than expected. Further, while adding energy storage systems such as batteries to smooth the supply in RE networks can recover curtailed energy, they ultimately act to reduce EROI and come with considerable ESME costs [31].

Hydro, bioenergy, and geothermal electricity are expected to exhibit only slow growth in all the IEA [24] scenarios, and together are several times smaller than wind and solar combined. Despite their minor potential, it is still useful to look at their GHG emissions profile over time. Tropical hydro systems emit high levels of CO_2 and methane gas over their early years of operation. Geothermal plants also emit CO_2, and only achieve carbon balance after several centuries [34]. For bioenergy plantations, Sterman et al. [35] stressed that many decades are needed for regrowth, so the CO_2 drawdown from plantations would not be available in the coming decades. The development of RE projects can impact not only local biodiversity [36] but also many globally significant biodiversity areas [37], even beyond the area occupied by the RE plant [38].

For hydro, bioenergy, and geothermal electricity, time considerations show that over the early years of operation, GHG reductions are far less than expected. A further complication for hydropower is that ongoing CC could change river flows and their timing, leading to faster reservoir siltation rates, all of which could reduce lifetime TWh and, thus, EROI. Glacier loss in the Himalayas could initially lead to higher hydro potential, though there would be decreased potential as glaciers shrink. There is also increased risk to Himalayan hydropower projects from 'Glacial Lake Outburst Floods' [39]. For bioenergy, competition for food could push bioenergy production out of prime farmland (such as is used in the US for corn ethanol production), again lowering EROI, because of increased need for water and fertiliser inputs.

Nuclear energy's share of global electricity production is expected to fall further, having peaked at 14.6% in 2006—well before the 2011 Fukushima accident—before falling to 9.8% in 2012 [1]. There are several reasons for this market share decline. Nuclear plants take a long time to plan and build, particularly compared with wind or PV solar farms. This is especially true for plants in the major OECD countries, where political opposition led to moratoriums on new plants and long construction times for plants being built in a number of countries. A related point is that many plants are nearing the end of their service lives, so closures would hinder net nuclear output growth, even if new plants are built. The end result is that nuclear power is most unlikely to play more than a minor role in energy production over the coming decades [25].

The IEA [24] presented three future energy scenarios and gave the expected contribution of all RE sources, as well as nuclear energy, to the global primary energy supply up to 2050. Table 2 shows their percentage contributions in 2010 and 2021 and the expected values in 2030, 2040, and 2050 for the APS. This scenario is actually an optimistic one, given that the world is not on track to reach this target. Even so, less than half of global energy in 2040 is projected to come from non-carbon sources. In the IEA's back-casting exercise to see what would be needed for the 'Net Zero Emissions by 2050' (NZE 2050) scenario, RE and nuclear energy together would still provide less than 40% of global primary energy in 2030.

Table 2. Share of RE and nuclear energy in global primary energy in 2010, 2020, and 2021 and the EIA's APS scenario for 2030, 2040, and 2050.

Energy Type\Year	2010	2020	2021	2030	2040	2050
RE (all types) (%)	8.3	11.7	11.9	23.8	38.2	50.7
RE (all types) EJ	45	69	74	141	239	319
Nuclear energy (%)	5.6	4.9	4.8	6.1	7.8	8.9
Nuclear energy EJ	30	29	30	39	49	56

Source [24].

4.2. Carbon Dioxide Removal (CDR)

Carbon dioxide removal can take many forms, both biological and mechanical. Biological approaches include reforestation and sequestration in soils and a technology untried at scale, bioenergy with carbon capture and storage (BECCS). These various approaches are described in detail in [3]. Their climate mitigation potential over the next two decades appears minor. Carbon capture and storage (CCS), needed for CO_2 capture from FF power stations, as well as for BECCS and DAC, despite its discussion for three decades, presently sequesters only a few tens of millions of tonnes of CO_2, compared with the tens of billions needed to be a major CC mitigation solution. It is also more costly than other approaches [40]. CO_2 utilisation is attracting increased attention but is presently insignificant. Table 3 gives the values for the GtC emissions avoided for each of the listed scenarios for 2030, 2040, and 2050. Even in these optimistic scenarios, in 2030 only 0.4–1.2 Gt of CO_2 would be captured, compared with the nearly 23 Gt still released in the IEA NZE 2050 scenario.

Bastin et al. [41] calculated that a global tree planting program could sequester a total of 205 Gt of CO_2, largely by increasing soil carbon and afforestation in grasslands and scrublands. Veldman et al. [42], in their critique of the Bastin et al. paper [41], claimed that their estimate was too high by a factor of five and that a more realistic—but still useful—value was 42 Gt. This far lower estimate was partly caused by over-estimating soil organic carbon gains, failing to account for warming from boreal forests because of reduced albedo, and neglecting existing human use of savannas, grasslands, and shrublands. In an earlier review, Boysen et al. [43] argued that such global terrestrial carbon fixation could only counteract business-as-usual warming at the expense of nearly all natural ecosystems.

Figure 2 outlines the various technical approaches that can be taken to reduce CO_2 emissions into the atmosphere. The already-discussed low-carbon energy sources (RE and nuclear), while already well-established, continue to benefit from technical improvements (e.g., in solar PV cell efficiency), whereas CO_2 removal methods are in their infancy or are yet to be attempted. A key advantage for CDR is that it enables the present fossil fuel economy to continue—at least until readily exploitable reserves of FF, particularly oil, are depleted, with the likely consequence of the delayed implementation of low carbon alternatives. Aside from the moral hazard attached to this approach, the question of when 'peak oil' will occur is unclear, with some arguing that there are only a few years left before it occurs (e.g., [44]), while others argue that it will occur decades in the future (e.g., [45]). A recent view is that the question is irrelevant, since 'peak demand' would come well before 'peak supply' [46]; but, if SG is adopted, peak oil could be a limiting factor.

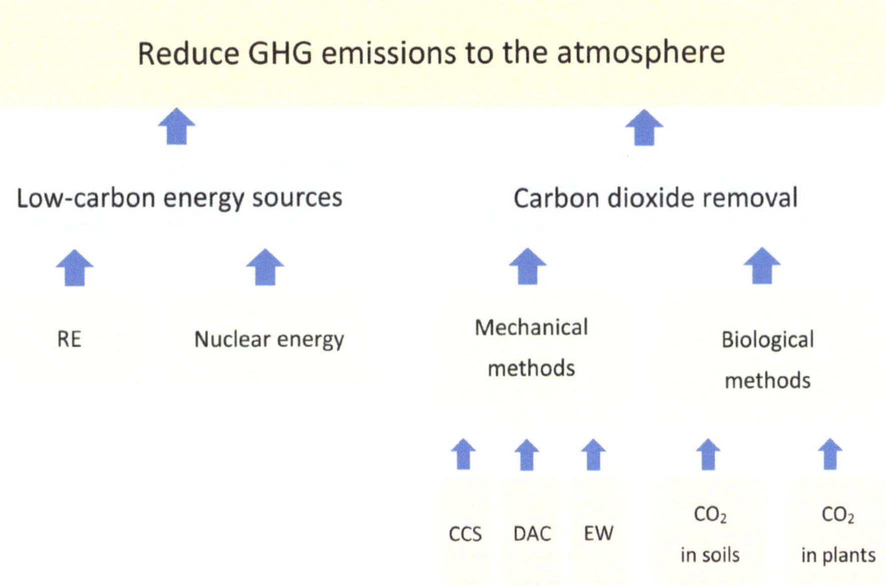

Figure 2. Diagram of conventional approaches for reducing GHG emissions into the atmosphere.

Table 3. CCUS (including BECCS and DAC) in various zero emissions scenarios by 2050 (in annual Gt of CO_2 avoided).

Scenario	2021	2030	2040	2050
IEA 2022 (NZE)	0.04	1.22	4.42	6.23
BP 2023	0.04	NA	NA	6.05
DNV 2022	0.04	0.4	3.6	5.8

Sources [24,47,48].

4.3. Energy Efficiency

The theoretical potential for energy efficiency improvements is large [49,50], but several obstacles stand in the way of rapid efficiency gains, even though energy efficiency is likely the cheapest method of CC mitigation. One obstacle to rapid change is the existence of the large and still-growing generating capacity of FF power stations, as well as a large and still growing global vehicle fleet [51]. Most efficiency improvement methods rely on new equipment replacing inefficient old equipment.

Energy savings from efficiency improvements are also reduced by the well-known energy rebound effect [52]—the lower fuel costs of (say) vehicle operation can induce extra travel. Furthermore, the desire for private vehicles in countries with low present ownership levels tends to swamp any efficiency gains. The deep energy/carbon reductions from efficiency gains are also offset by the widespread introduction of new energy-using equipment or practices, such as ride-on lawnmowers, mechanical hedge clippers, and leaf blowers for gardens. A recent innovation, Bitcoin mining, is very energy intensive; a 2023 study found that its global electricity use exceeded that of many countries, including Norway [53]. Another example is bottled water, which is first collected from the source and then distributed from bottling plants in small trucks, replacing the far more energy-efficient tap water.

In the case of vehicular transport, three developments negate efficiency gains. The first is the desire for faster travel—time efficiency (speed) can conflict with energy efficiency. Hence, public transport is replaced by car travel, and aircraft dominate long-distance travel.

The second development is the increase in non-propulsion energy needs in vehicles, for entertainment, driver aids, and environmental control. The third is the global shift to larger sports utility vehicles replacing cars. In the US in 2021, such vehicles formed 77% of all four-wheel private vehicle sales, at USD 14.57 million [54]. Rapid reductions in GHGs from energy efficiency improvements seem unlikely in a market-based global economy.

5. Solar Geoengineering: Impact on Low-Carbon Energy

The above discussion shows that none of the conventional methods for CC mitigation look capable of delivering major reductions in carbon emissions any time soon, let alone reducing atmospheric levels of GHGs, unless strong supporting policies are introduced. To be clear: conventional approaches without the strong policy support needed have failed so far, as shown by rising annual GHG emissions, as discussed in the Introduction. Thus, early advocates envisaged SG as a way of completely counteracting climate forcing without the need to change either global energy consumption or the energy mix.

It is acknowledged, however, that deploying SG to counter (say) a doubling of atmospheric CO_2 ppm compared with the pre-industrial value of around 280 ppm could lead to unacceptable side effects, worsening climate impacts (such as precipitation decreases) in some regions [14] in an already water-stressed world [55]. Instead, it is proposed that SG be used to counteract perhaps 50% of global warming [14].

In its most discussed form, SG involves the annual placement of sulphate aerosols in the lower stratosphere to increase Earth's albedo. In order to offset half the climate forcing from anthropogenic CC, a radiative forcing of about -2 W/m^2 is needed. This could be achieved by the annual placement of 12 Mt of sulphur into the lower stratosphere, perhaps using airplanes. Annual costs were estimated as anywhere between USD 20 and 200 billion [56].

One possible important effect of SG (and also all CDR methods) is that it could discourage the uptake of low-carbon sources of energy. Proponents for SG claim that it is far cheaper for a given reduction in climate forcing than low-carbon energy and, further, can be rapidly implemented in a year or two [57,58]. It can also be rapidly terminated should the side effects prove unacceptable. However, Trisos et al. [59] warned of the 'potentially dangerous consequences for biodiversity of solar geoengineering implementation and termination'.

Above all (again, like CDR), it enables the continuation of the fossil fuel economy, which has strong support from industry and FF exporting economies such as Organization of the Petroleum Exporting Countries (OPEC) countries [57].

Another problem with CO_2 atmospheric emissions into the atmosphere is that oceans absorb 25–30% of this CO_2, where it causes ocean acidification (OA), a serious threat to ocean ecosystems [8]. This OA would continue unabated under SG, though not if, for example, RE is used for CC mitigation, since atmospheric CO_2 emissions are avoided. For a fair comparison, the monetary, energy, and environmental costs of countering OA must be included for SG. Slaked lime is one option for such ocean alkalinity enhancement (OAE). Slaked lime could be relatively cheaply spread by freight ships, but its unavoidably high local concentrations could have serious adverse effects on ocean ecosystems [60].

Aircraft would enable more uniform spreading, but Gentile et al. [61] found that depending on aircraft height and dispersal time, aircraft energy use would involve a 28–77% energy penalty, with the cost per tonne of CO_2 neutralised between USD 31 and 1920. Since each extra molecule of ocean CO_2 must be neutralised, the quantities involved are very large; Fakhraee et al. [62] found that 6–30 Gt of CaO or MgO would be needed annually, depending on the assumptions made. In summary, when the need for OAE is factored in, SG may well be more expensive than more conventional options and would also entail additional ecological risks, which are still poorly understood.

Due to these serious problems, many scientists even opposed further research into SG [63–65]. In the words of McGuire [64], it is simply 'the wrong answer to the wrong question'

One localised form of SG—painting urban roofs and pavements with a high albedo coating—avoids the freeloader effect that bedevils meaningful reduction efforts, since the benefits solely accrue for the urban residents, and, further, does not need any new technology. Although its radiative forcing is negligible at -0.01 W/m^2 and, hence, is useless for global cooling, it is being implemented and has no adverse effects elsewhere. In contrast, any globally effective SG initiatives would have potentially serious ecological effects, as well as facing political opposition from nations perceiving themselves to be disadvantaged.

6. Global Equity in Energy and Climate Change Impacts

So far, this review—in line with the great majority of papers on energy—did not factor in equity considerations. As an editorial in *Nature*, referring to the enormously influential 2009 paper by Rockström et al. [66], 'A safe operating space for humanity', put it: 'A gap in the original concept was that it lacked environmental justice and equity—it needed to take into account the fact that everyone, especially the most vulnerable, has an absolute right to water, food, energy and health, alongside the right to a clean environment' [67]. The original 2009 paper found that three of the nine planetary boundaries had been crossed. However, the authors now list eight boundaries, namely 'climate, natural ecosystem area, ecosystem functional integrity, surface water, groundwater, nitrogen, phosphorus and aerosols', and, when equity considerations are factored in, they argue that seven of these thresholds have already been crossed [68]. Why is there an increase in the number of planetary thresholds considered to have been breached? The answer lies in the fact that different geographical regions and even different groups of people in the same region, for instance an urban area, can experience the impacts of climate change very differently.

Equity has many aspects, and the ones relevant to energy use and subsequent GHG emissions include income, energy use, and CO_2 emissions' distribution, both at the national and household levels. Chancel and Piketty [69] examined world income distribution over the past century. They found that at the international level inequality was falling, but at the household level it was increasing. Energy inequality is also still high, even at the international level, particularly if only commercial fuels are considered [17]. Kartha et al. [70] showed that CO_2 emissions are very unequally distributed among the world's households. The top 10% of households accounted for 49% of emissions in 2015, with the bottom 50% only emitting 7%. When split into sectorial emissions, the poorest 50% of the world's population emitted less than 20% of the total GHG emissions from transport and energy but an almost equal share from agriculture. On an average per capita basis, IEA statistics [1] show that emissions from the highest emitting country are 200 times those of the lowest. As the IPCC [2] stated, 'Vulnerable communities who have historically contributed the least to current climate change are disproportionately affected'.

Another form of inequity is revealed when the cumulative emissions of CO_2 are considered. Since CO_2 is a long-lived gas in the atmosphere, cumulative as well as annual emissions are important. In 1965, OECD countries accounted for 68.8% of global CO_2 emissions from fossil fuel use and industry. By 2022, the much-enlarged list of OECD countries accounted for only 33.7% of such CO_2 emissions, although the average emissions per capita in OECD countries was almost twice as high as the global average [1]. But, when cumulative emissions are considered, 55.6% of all energy-related emissions since 1850 have been from OECD countries, with the figure dropping to 45.9% when all GHG emissions are considered [71].

6.1. Inequality in Low-Income Countries, Especially in the Tropics

Low-income countries, particularly those in tropical Africa, Asia, and South America, are anticipated to experience the negative effects of CC both earlier and more severely than high-income countries. There are several reasons for this difference:

- Tropical ecosystems are near their upper thermal limit, so rising temperatures could exceed optimum plant germination temperature or even exceed the upper limit for germination [72]. (Further, [73] argued that many tropical ecosystems have adapted to a narrow temperature range, although Sentinella et al. [72] dispute this claim.) Thus, temperature rises could have adverse consequences for agriculture. In contrast, in more temperate climates, rising temperatures shift more species closer to their optimum germination temperature [73].

- Even for similar extreme weather events like floods or droughts, the risks for low-income communities and households are much higher than in wealthier countries, as poorer communities have fewer resources, both material and administrative, for coping and recovery and tend to lose a bigger share of their wealth. Even worse, a vicious cycle can occur between losses from disasters—whatever the cause—and poverty: '(...) poverty is a major driver of people's vulnerability to natural disasters, which in turn increase poverty in a measurable and significant way' [74]. Cappelli et al. [75] even argued for a vicious cycle that 'keeps some countries stuck in a disasters-inequality trap'.
- Further, there are significant differences in human mortality from extreme weather events depending on the level of vulnerability. As the IPCC [2] noted, 'Between 2010 and 2020, human mortality from floods, droughts and storms was 15 times higher in highly vulnerable regions, compared to regions with very low vulnerability'.

The question to ask here is how already-adopted conventional policies for CC mitigation—and proposals such as SG—affect the prospects for more equality in an unequal world. The example of traditional biomass fuel is instructive. A possible conflict exists between the 'simplistic' desire of many CC mitigation advocates for low-income countries to move directly to RE and forego FFs. As Ramachandran [76] argued, for cooking meals in places like India, FFs such as liquid petroleum gas (LPG) should greatly reduce the damaging health effects of particulate pollution that occur with traditional biomass fuels. Vital health concerns can and should sometimes override CC mitigation.

An important example illustrates the difficulties involved in trying to balance CC mitigation and equity. One heavily favoured adaptation to rising global temperatures and heat waves conflicts with CC mitigation efforts: the use of air conditioner (A/C) units. Globally, A/C numbers have very closely followed an exponential curve since at least 1990, and in 2021 they numbered over two billion. If this exponential growth pattern persists, the IEA [77] forecasts this figure to rise to over 5.5 billion units by 2050, with especially large increases in A/C units expected for both China and India. Even as early as 2016, A/C units consumed 10% of global electricity or more than 2000 terawatt hr (TWh) [78]. However, solar electricity output, with its peak during the hottest hours, is well-matched to provide power for A/C units.

There is no easy solution to this dilemma. The need for A/C units for most but not all countries is evident from the work of Raymond et al. [79], who documented how, in some regions of the world, wet bulb temperatures on occasion exceed 35 °C, which marks the upper physiological limit of human tolerance. Humans can become acclimatised to lower temperatures [79], but, beyond 35 °C wet bulb temperatures, A/C appears to be the only solution. Even so, a mixture of acclimatisation and A/C could be used, with A/C only used for higher temperatures and not for room temperatures above 20 °C. As Hanna and Tait [80] argued, both 'behavioral and technological adaptations' are necessary for adaptation to rising global warming.

Although 90.4% of the global population had access to electricity in 2020, households without electricity were heavily concentrated in tropical African countries [81]. Furthermore, it is important that electricity companies are publicly owned, as energy in the hands of private companies is not a guarantee of access for everyone. Residents of many such countries are still mainly engaged in agriculture, requiring prolonged periods of being outside. Most of their fuel is still from traditional biomass, which also requires much time outside for its collection. Further, at present, apart from sleeping, many other human activities take place outside the house [82]. So, even if electricity was available, and the cost of A/C units and power consumption could be afforded, it may not help such tropical residents avoid life-threatening temperatures.

6.2. Inequality in High-Income Countries

A few years ago, it could be argued that although poor countries would be the first to experience the full brunt of CC, high-income countries such as those in the core OECD would not experience much adverse change until global temperatures reached 3 °C above pre-industrial levels [83]. We now know better, as evidenced by the record-breaking heat waves in Europe [84] and the forest fires in California [85].

It is important to consider equity problems, not only between high- and low-income countries but also within high-income countries as well, as shown for the US by Polonik et al. [86]. Large cities often exhibit a pronounced urban heat island (UHI) effect. The UHI effect has several contributory factors, including heat release from vehicles, buildings, etc.; the 'canyon effect' of tall buildings blocking back radiation from escaping; and reduced evapotranspiration from paved surfaces [87]. Chakraborty et al. [88], based on a study of the distribution of the UHI effect and income in 25 cities around the globe, found that the UHI effect—together with its deleterious health effects—disproportionately affected low-income groups. The main reason was that low-income areas in cities tend to have a much smaller area given to parks and vegetation—and, conversely, a higher share of paved areas—which reduces evapotranspiration from their surfaces. The risks in all countries from extreme temperatures are higher for urban dwellers [89].

7. Discussion and Conclusions

The discussion above shows that the technical solutions for mitigating climate change have so far not been successful. Further, given the limited time we have to avoid extremely disruptive CC, these methods, even together, can only be a complementary approach to tackling CC over the next decade or so. This conclusion has even more force when inequality—of incomes, energy use, and climate change damages—are factored into CC mitigation policies. As already discussed, Gupta et al. [68] and Rockström et al. [90] argued on equity grounds that no further temperature increase should be allowed—even a 1.5 °C rise is too high.

The limitations of this review mainly arise from the extreme uncertainty surrounding how the future climate will evolve, both regionally and globally. Witze [84] summed up this uncertainty as follows: 'Unprecedented temperatures are coming faster and more furiously than researchers expected, raising questions about what to anticipate in the future'. This, in turn, is partly the result of uncertainty about whether (and when) the world's nations will implement policies that seriously tackle CC. Another uncertainty is the possibility of some breakthrough technology that can quickly mitigate CC. However, given the multiple environmental problems we face, experience shows that any innovation could well exacerbate these other risks to our future.

What options are left for avoiding CCC, given the failure of existing and proposed approaches? The only approach is a rapid reduction in GHG emissions, not only by low-carbon or CDR methods but also by rapid reductions in energy use itself, initially in high-energy-use nations. The response to the COVID-19 crisis in the form of stringent lockdowns and the resulting emissions reductions indicates the importance of strong policies [91]. This conclusion is at odds with the continued growth in global energy use, as forecast by various government and energy organisations [2,24,47,48,51]. In a previous review, the authors [3] detailed the possible policy changes that are needed to support RE introduction and energy and GHG reductions.

As shown, large energy efficiency improvements cannot be expected in the context of continuing global economic growth. Jason Hickel and colleagues [92] stressed the urgent need for what is lacking from the IPCC and in other official documents: CC mitigation scenarios that do not assume the continuation of global economic growth. Such global economic 'degrowth' would not be uniform, in that reductions would first need to apply to the OECD and other high-income countries—or, even better, high-income households in every country.

In a later paper, Hickel and colleagues [93] gave some ideas for how such degrowth could be achieved in high-income countries, mainly by focusing more on satisfying human needs. In particular, they advocated for cutting production in sectors such as animal products, private transport, aviation, and fast fashion and ending the planned obsolescence of goods. They also advocated for providing high-quality public health care, housing, and education, so human welfare can be improved with low resource use. At the same time, equity demands some growth in low-income countries—or households. Here, the UN's Sustainable Development Goals (SDGs) [94] could be used as a starting point in meeting basic human needs.

In an earlier paper, the authors showed how large reductions in GHG emissions are possible, particularly in agriculture worldwide, with crop pests being a key problem in low-income countries and food waste in high-income nations. Also, for passenger transport, especially in high-mobility countries, large GHG reductions are possible by shifting the emphasis from vehicular mobility to access and by promoting non-motorised modes of transport and public transport [3].

Deep emission reductions from a rapid reduction in FF use will prove very difficult to politically implement in high-income countries, and there is no guarantee of success. In fact, the model results of van Ruijven and colleagues [95] indicated that energy use would strongly grow until 2050. Although most energy growth would come from assumed economic growth, the changing climate led to further energy growth of 11–58%, depending on the scenario.

This review identifies a number of shortcomings and gaps in the published literature. A vital one is a better idea as to how the climate—especially the frequency, duration, and severity of extreme weather events—will respond to further increases in atmospheric GHGs. More work is also needed to produce realistic costs for the various options and for when they could be deployed.

Although the majority of the population in OECD countries thinks that CC is a serious problem, one that needs to be urgently addressed, this support may be predicated on there being a relatively painless solution like a massive shift to low-carbon fuels or the use of CDR, particularly if it is promoted as a means of providing more time for deploying low-carbon technologies. As this review argues, such technological optimism is likely unwarranted, so fundamental social and political changes are needed. But, to echo the words of UK's former prime minister, Margaret Thatcher: 'There is no alternative'.

Author Contributions: Conceptualisation, P.M. and D.H.; methodology, P.M. and D.H.; investigation, P.M.; data curation, D.H.; writing—original draft preparation, P.M.; writing—review and editing, D.H.; visualisation, D.H.; project administration, P.M. All authors have read and agreed to the published version of the manuscript.

Funding: This research received no external funding.

Data Availability Statement: All data used in this paper are from sources such as BP, the IEA, OPEC, Scopus, the IPCC, etc. These sources all allow their data to be freely used if they are cited—as they are in this review. All tables are the authors' own creation from these data.

Conflicts of Interest: The authors declare no conflict of interest.

Nomenclature

A/C	air conditioner
APC	Announced Pledges Scenario
BECCS	bioenergy with carbon capture and storage
CC	climate change
CCC	catastrophic climate change
CCS	carbon capture and storage
CDR	carbon dioxide removal

CO_2	carbon dioxide
CO_2-eq	carbon dioxide equivalent
DAC	direct air capture
DEA	dynamic energy analysis
EIA	Energy Information Administration
EJ	exajoule (1018 joules)
EROI	energy return on investment
ESME	Ecosystem Maintenance Energy
ESS	Earth System Science
EU	European Union
EW	enhanced weathering
FF	fossil fuels
GHG	greenhouse gas
Gt	gigatonne = 1^9 tonnes
GW	gigawatt (10^9 watts)
IEA	International Energy Agency
IPCC	Intergovernmental Panel on Climate Change
Mt	megatonne (10^6 tonnes)
OA	ocean acidification
OAE	ocean alkalinity enhancement
OECD	Organization for Economic Cooperation and Development
OPEC	Organization of the Petroleum Exporting Countries
ppm	parts per million (atmospheric)
PV	photovoltaic
RE	renewable energy
SDG	Sustainable Development Goal
SG	solar geoengineering
SRM	solar radiation management
t CO_2/cap	tonnes of CO_2 per capita
TWh	terawatt hours (10^{12} watt hrs)
USD	US dollars
UNEP	United Nations Environment Program

References

1. Energy Institute. Statistical Review of World Energy 2023, 72nd ed. Available online: https://www.energyinst.org/statistical-review (accessed on 26 March 2023).
2. Intergovernmental Panel on Climate Change (IPCC). Synthesis Report of the IPCC Sixth Assessment Report (AR6): Summary for Policymakers. Available online: https://report.ipcc.ch/ar6syr/pdf/IPCC_AR6_SYR_SPM.pdf (accessed on 21 March 2023).
3. Moriarty, P.; Honnery, D. Review: Renewable energy in an increasingly uncertain future. *Appl. Sci.* **2023**, *13*, 388. [CrossRef]
4. Steffen, W.; Richardson, K.; Rockström, J.; Cornell, S.E.; Fetzer, I.; Bennett, E.M.; Biggs, R.; Carpenter, S.R.; de Vries, W.; de Wit, C.A.; et al. Planetary boundaries: Guiding human development on a changing planet. *Science* **2015**, *347*, 1259855. [CrossRef] [PubMed]
5. Bradshaw, C.J.A.; Ehrlich, P.R.; Beattie, A.; Ceballos, G.; Crist, E.; Diamond, J.; Dirzo, R.; Ehrlich, A.H.; Harte, J.; Harte, M.E.; et al. Underestimating the challenges of avoiding a ghastly future. *Front. Conserv. Sci.* **2021**, *1*, 615419. [CrossRef]
6. Brodie, J.F.; Watson, J.E.M. Human responses to climate change will likely determine the fate of biodiversity. *Proc. Natl. Acad. Sci. USA* **2023**, *120*, e2205512120. [CrossRef]
7. Dirzo, R.; Ceballos, G.; Ehrlich, P.R. Circling the drain: The extinction crisis and the future of humanity. *Philos. Trans. R. Soc. B* **2022**, *377*, 20210378. [CrossRef]
8. Georgian, S.; Hameed, S.; Morgan, L.; Amon, D.J.; Sumaila, U.R.; Johns, D.; Ripple, W.J. Scientists' warning of an imperiled ocean. *Biol. Conserv.* **2022**, *272*, 109595. [CrossRef]
9. World Economic Forum. Plastic Pollution Is a Public Health Crisis. How Do We Reduce Plastic Waste? 2022. Available online: https://www.weforum.org/agenda/2022/07/plastic-pollution-ocean-circular-economy/ (accessed on 3 June 2023).
10. Crist, E.; Ripple, W.J.; Ehrlich, P.R.; Rees, W.E.; Wolf, C. Scientists' warning on population. *Sci. Total Environ.* **2022**, *845*, 157166. [CrossRef]
11. United Nations (UN). World Population Prospects 2022. 2022. (Also Earlier UN Forecasts). Available online: https://population.un.org/wpp/ (accessed on 12 April 2023).
12. Moriarty, P.; Honnery, D. The risk of catastrophic climate change: Future energy implications. *Futures* **2021**, *128*, 102728. [CrossRef]
13. Lade, S.J.; Steffen, W.; de Vries, W.; Carpenter, S.R.; Donges, J.F.; Gerten, D.; Hoff, H.; Newbold, T.; Richardson, K.; Rockström, J. Human impacts on planetary boundaries amplified by Earth system interactions. *Nat. Sustain.* **2020**, *3*, 119–128. [CrossRef]

14. Irvine, P.J.; Keith, D.W. Halving warming with stratospheric aerosol geoengineering moderates policy-relevant climate hazards. *Environ. Res. Lett.* **2020**, *15*, 044011. [CrossRef]
15. Intergovernmental Panel on Climate Change (IPCC). Climate Change 2022: Mitigation of Climate Change. (Also, Earlier Reports). 2022. Available online: https://www.ipcc.ch/report/ar6/wg3/ (accessed on 1 July 2023).
16. Intergovernmental Panel on Climate Change (IPCC). *Climate Change 2022: Impacts, Adaptation and Vulnerability*; CUP: Cambridge, UK, 2022. [CrossRef]
17. International Energy Agency (IEA). *Key World Energy Statistics 2021*; (Also. Earlier Editions); IEA/OECD: Paris, France, 2021. Available online: https://www.iea.org/reports/key-world-energy-statistics-2021 (accessed on 15 April 2023).
18. Valavanidis, A. Extreme Weather Events Exacerbated by the Global Impact of Climate Change. Available online: Chem-tox-ecotox.org/ScientificReviews (accessed on 28 May 2023).
19. Vaughan, A. Is the climate becoming too extreme to predict? *New Sci.* **2021**, *251*, 11. [CrossRef]
20. Capellán-Pérez, I.; de Castro, C.; González, L.J.M. Dynamic Energy Return on Energy Investment (EROI) and material requirements in scenarios of global transition to renewable energies. *Energy Strategy Rev.* **2019**, *26*, 100399. [CrossRef]
21. Fizaine, F.; Court, V. Energy expenditure, economic growth, and the minimum EROI of society. *Energy Policy* **2016**, *95*, 172–186. [CrossRef]
22. Archer, D.; Eby, M.; Brovkin, V.; Ridgwell, A.; Cao, L.; Mikolajewicz, U.; Caldeira, K.; Matsumoto, K.; Munhoven, G.; Montenegro, A.; et al. Atmospheric lifetime of fossil fuel carbon dioxide. *Annu. Rev. Earth Planet. Sci.* **2009**, *37*, 117–134. [CrossRef]
23. Schwartz, S.E. Observation based budget and lifetime of excess atmospheric carbon dioxide. *Atmos. Chem. Phys.* **2021**, *preprint*. Available online: https://acp.copernicus.org/preprints/acp-2021-924/ (accessed on 20 April 2023).
24. International Energy Agency (IEA). *World Energy Outlook 2022*; IEA/OECD: Paris, France, 2022. Available online: https://www.iea.org/topics/world-energy-outlook (accessed on 19 April 2023).
25. Moriarty, P. Global nuclear energy: An uncertain future. *AIMS Energy* **2021**, *9*, 1027–1042. [CrossRef]
26. Moriarty, P.; Honnery, D. The limits of renewable energy. *AIMS Energy* **2021**, *9*, 812–829. [CrossRef]
27. Hall, C.A.S. Will EROI be the primary determinant of our economic future? The view of the natural scientist versus the economist. *Joule* **2017**, *1*, 635–638. [CrossRef]
28. Moriarty, P.; Honnery, D. Feasibility of a 100% global renewable energy system. *Energies* **2020**, *13*, 5543. [CrossRef]
29. Ferroni, F.; Hopkirk, R.J. Energy Return on Energy Invested (ERoEI) for photovoltaic solar systems in regions of moderate insolation. *Energy Policy* **2016**, *94*, 336–344. [CrossRef]
30. Raugei, M.; Sgouridis, S.; Murphy, D.; Fthenakis, V.; Frischknecht, R.; Breyer, C.; Bardi, U.; Barnhart, C.; Buckley, A.; Carbajales-Dale, M.; et al. Energy Return on Energy Invested (ERoEI) for photovoltaic solar systems in region of moderate insolation: A comprehensive response. *Energy Policy* **2017**, *102*, 377–384. [CrossRef]
31. Daaboul, J.; Moriarty, P.; Honnery, D. Net green energy potential of solar photovoltaic and wind energy generation systems. *J. Clean. Prod.* **2023**, *415*, 137806. [CrossRef]
32. Halabi, A.L.M.; Siacara, A.T.; Sakano, V.K.; Pileggi, R.G.; Futai, M.M. Tailings dam failures: A historical analysis of the risk. *J. Fail. Anal. Prev.* **2022**, *22*, 464–477. [CrossRef]
33. Mills, M.P. Mines, Minerals, and "Green" Energy: A Reality Check. Manhattan Institute Report. July 2020. Available online: http://www.gogreencanada.ca/green_energy_reality_check.pdf (accessed on 3 May 2023).
34. O'Sullivan, M.; Gravatt, M.; Popineau, J.; O' Sullivan, J.; Mannington, W.; McDowell, J. Carbon dioxide emissions from geothermal power plants. *Renew. Energy* **2021**, *175*, 990–1000. [CrossRef]
35. Sterman, J.D.; Siegel, L.; Rooney-Varga, J.N. Does replacing coal with wood lower CO_2 emissions? Dynamic lifecycle analysis of wood bioenergy. *Environ. Res. Lett.* **2018**, *13*, 015007. [CrossRef]
36. Voigt, C.C.; Straka, T.M.; Fritze, M. Producing wind energy at the cost of biodiversity: A stakeholder view on a green-green dilemma. *J. Renew. Sustain. Energy* **2019**, *11*, 063303. [CrossRef]
37. Rehbein, J.A.; Watson, J.E.M.; Lane, J.L.; Sonter, L.J.; Venter, O.; Atkinson, S.C.; Allan, J.R. Renewable energy development threatens many globally important biodiversity areas. *Glob. Chang. Biol.* **2020**, *26*, 3040–3051. [CrossRef]
38. Niebuhr, B.B.; Sant'Ana, D.; Panzacchi, M.; van Moorter, B.; Sandström, P.; Ronaldo, G.; Morato, R.G.; Skarin, A. Renewable energy infrastructure impacts biodiversity beyond the area it occupies. *Proc. Natl. Acad. Sci. USA* **2022**, *119*, e2208815119. [CrossRef]
39. Wasti, A.; Ray, P.; Wi, S.; Folch, C.; Ubierna, M.; Karki, P. Climate change and the hydropower sector: A global review. *WIREs Clim. Chang.* **2022**, *13*, e757. [CrossRef]
40. Schmelz, W.J.; Hochman, G.; Miller, K.G. Total cost of carbon capture and storage implemented at a regional scale: Northeastern and midwestern United States. *Interface Focus* **2020**, *10*, 20190065. [CrossRef]
41. Bastin, J.-F.; Finegold, Y.; Garcia, C.; Mollicone, D.; Rezende, M.; Routh, D.; Sacande, M.; Sparrow, B.; Sparrow, C.M.; Zohner, T.W. The global tree restoration potential. *Science* **2019**, *365*, 76–79. Available online: https://www.science.org/doi/10.1126/science.abc8905 (accessed on 10 October 2022). [CrossRef]
42. Veldman, J.W.; Aleman, J.C.; Alvarado, S.T.; Anderson, T.M.; Archibald, S.; Bond, W.J.; Boutton, T.W.; Buchmann, N.; Buisson, E.; Canadell, J.G.; et al. Comment on "The global tree restoration potential". *Science* **2019**, *366*, eaay7976. [CrossRef] [PubMed]

43. Boysen, L.R.; Lucht, W.; Gerten, D.; Heck, V.; Lenton, T.M.; Schellnhuber, H.J. The limits to global-warming mitigation by terrestrial carbon removal. *Earth's Future* **2017**, *5*, 463–474. [CrossRef]
44. Bentley, R. Colin Campbell, oil exploration geologist and key proponent of 'Peak Oil'. *Biophys. Econ. Sustain.* **2023**, *8*, 3. [CrossRef]
45. Deming, D.M. King Hubbert and the rise and fall of peak oil theory. *AAPG Bull.* **2023**, *107*, 851–861. [CrossRef]
46. Halttunen, K.; Slade, R.; Staffell, I. What if we never run out of oil? From certainty of "peak oil" to "peak demand". *Energy Res. Soc. Sci.* **2022**, *85*, 102407. [CrossRef]
47. BP. *BP Energy Outlook 2023 Edition*; BP: London, UK, 2023. Available online: https://www.bp.com/content/dam/bp/business-sites/en/global/corporate/pdfs/energy-economics/energy-outlook/bp-energy-outlook-2023.pdf (accessed on 10 June 2023).
48. DNV. Energy Transition Outlook 2022: Executive Summary. 2022. Available online: https://www.dnv.com/energy-transition-outlook/index.html (accessed on 10 June 2023).
49. Lovins, A.B. How big is the energy efficiency resource? *Environ. Res. Lett.* **2018**, *13*, 090401. [CrossRef]
50. Lovins, A. Reframing automotive fuel efficiency. *SAE Int. J. Sustain. Transp. Energy Environ. Policy* **2020**, *1*, 59–84. [CrossRef]
51. Organization of the Petroleum Exporting Countries (OPEC). *OPEC World Oil Outlook*; OPEC: Vienna, Austria, 2021. Available online: http://www.opec.org (accessed on 23 October 2022).
52. Steren, A.; Rubin, O.D.; Rosenzweig, S. Energy-efficiency policies targeting consumers may not save energy in the long run: A rebound effect that cannot be ignored. *Energy Res. Soc. Sci.* **2022**, *90*, 102600. [CrossRef]
53. Huestis, S. Cryptocurrency's Energy Consumption Problem. 2023. Available online: https://rmi.org/cryptocurrencys-energy-consumption-problem/#:~:text=Bitcoin%20alone%20is%20estimated%20to,fuel%20used%20by%20US%20railroads (accessed on 12 April 2023).
54. Davis, S.C.; Boundy, R.G. Transportation Energy Data Book, Edition 40. ORNL/TM-2022/2376. Available online: https://tedb.ornl.gov/wp-content/uploads/2022/03/TEDB_Ed_40.pdf (accessed on 18 May 2023).
55. Naddaf, M. The world faces a water crisis—4 powerful charts show how. *Nature* **2023**, *615*, 774–775. [CrossRef]
56. Robock, A. Benefits and risks of stratospheric solar radiation management for climate intervention (geoengineering). *Bridge* **2020**, *50*, 59–67. Available online: http://climate.envsci.rutgers.edu/pdf/RobockBridge.pdf (accessed on 14 October 2022).
57. Moriarty, P.; Honnery, D. Renewable energy and energy reductions or solar geoengineering for climate change mitigation? *Energies* **2022**, *15*, 7315. [CrossRef]
58. Royal Society. *Geoengineering the Climate: Science, Governance and Uncertainty*; Royal Society: London, UK, 2009. Available online: https://royalsociety.org/topics-policy/publications/2009/geoengineering-climate/ (accessed on 8 March 2021).
59. Trisos, C.H.; Amatulli, G.; Gurevitch, J.; Robock, A.; Xia, L.; Zambri, B. Potentially dangerous consequences for biodiversity of solar geoengineering implementation and termination. *Nat. Ecol. Evol.* **2018**, *2*, 475–482. [CrossRef] [PubMed]
60. Vaughan, A. Engineering the oceans. *New Sci.* **2022**, *255*, 46–49. [CrossRef]
61. Gentile, E.; Tarantola, F.; Lockley, A.; Vivian, C.; Caserini, S. Use of aircraft in ocean alkalinity enhancement. *Sci. Total Environ.* **2022**, *822*, 153484. [CrossRef]
62. Fakhraee, M.; Li, Z.; Planavsky, N.J.; Reinhard, C.T. Environmental impacts and carbon capture potential of ocean alkalinity enhancement. *Res. Sq.* **2022**, preprint. [CrossRef]
63. Open Letter: We Call for an International Non-Use Agreement on Solar Geoengineering. 2022. Available online: https://www.solargeoeng.org/non-use-agreement/open-letter/ (accessed on 10 August 2023).
64. McGuire, B. Hacking the Earth. What could go wrong with geoengineering? *Responsible Sci. J.* **2021**, *3*, 18–19.
65. Biermann, F.; Oomen, J.; Gupta, A.; Ali, S.H.; Conca, K.; Hajer, M.A.; Kashwan, P.; Kotzé, L.J.; Leach, M.; Messner, D.; et al. Solar geoengineering: The case for an international non-use agreement. *WIREs Clim. Chang.* **2022**, *13*, e754. [CrossRef]
66. Rockström, J.; Steffen, W.; Noone, K.; Persson, A.; Chapin, F.S., III; Lambin, E.F.; Lenton, T.M.; Scheffer, M.; Folke, C.; Schellnhuber, H.J.; et al. A safe operating space for humanity. *Nature* **2009**, *461*, 472–475. [CrossRef]
67. A measure for environmental justice. *Nature* **2023**, *618*, 7. Available online: https://www.nature.com/articles/d41586-023-01749-9 (accessed on 4 August 2023).
68. Gupta, J.; Liverman, D.; Prodani, K.; Aldunce, P.; Bai, X.; Broadgate, W.; Ciobanu, D.; Gifford, L.; Gordon, C.; Hurlbert, M.; et al. Earth system justice needed to identify and live within Earth system boundaries. *Nat. Sustain.* **2023**, *6*, 630–638. [CrossRef]
69. Chancel, L.; Piketty, P. Global Income Inequality, 1820–2020: The Persistence and Mutation of Extreme Inequality. 2021. Available online: https://halshs.archives-ouvertes.fr/halshs-03321887ffhalshs-03321887 (accessed on 23 May 2023).
70. Kartha, S.; Kemp-Benedict, E.; Ghosh, E.; Nazareth, A.; Gore, T. *The Carbon Inequality Era. Joint Research Report*; Stockholm Environment Institute: Stockholm, Sweden, 2020.
71. Jones, M.W.; Peters, G.P.; Gasser, G.; Andrew, R.M.; Schwingshackl, C.; Gütschow, J.; Houghton, R.A.; Friedlingstein, P.; Pongratz, J.; Le Quéré, C. National contributions to climate change due to historical emissions of carbon dioxide, methane, and nitrous oxide since 1850. *Sci. Data* **2023**, *10*, 155. [CrossRef]
72. Sentinella, A.T.; Warton, D.I.; Sherwin, W.B.; Offord, C.A.; Moles, A.T. Tropical plants do not have narrower temperature tolerances, but are more at risk from warming because they are close to their upper thermal limits. *Glob. Ecol. Biogeogr.* **2020**, *29*, 1387–1398. [CrossRef]
73. Perez, T.M.; Stroud, J.T.; Feeley, K.J. Thermal trouble in the tropics. *Science* **2016**, *351*, 1392–1393. [CrossRef] [PubMed]
74. Hallegatte, S.; Vogt-Schilb, A.; Rozenberg, J.; Bangalore, M.; Beaudet, C. From poverty to disaster and back: A review of the literature. *Econ. Disasters Clim. Chang.* **2020**, *4*, 223–247. [CrossRef]

75. Cappelli, F.; Costantini, V.; Consoli, D. The trap of climate change-induced "natural" disasters and inequality. *Glob. Environ. Chang.* **2021**, *70*, 102329. [CrossRef]

76. Ramachandran, V. Blanket bans on fossil fuels hurt women. *Nature* **2022**, *607*, 9. [CrossRef]

77. International Energy Agency (IEA). *Global Air Conditioner Stock, 1990–2050*; IEA: Paris, France, 2022. Available online: https://www.iea.org/data-and-statistics/charts/global-air-conditioner-stock-1990-2050 (accessed on 2 March 2023).

78. International Energy Agency (IEA). *The Future of Cooling: Opportunities for Energy Efficient Air Conditioning*; OECD/IEA: Paris, France, 2018. Available online: https://www.iea.org/reports/the-future-of-cooling (accessed on 2 February 2023).

79. Raymond, C.; Matthews, T.; Horton, R.M. The emergence of heat and humidity too severe for human tolerance. *Sci. Adv.* **2020**, *6*, eaaw1838. [CrossRef]

80. Hanna, E.G.; Tait, P.W. Limitations to thermoregulation and acclimatization challenge human adaptation to global warming. *Int. J. Environ. Res. Public Health* **2015**, *12*, 8034–8074. [CrossRef]

81. World Bank (2023) Access to Electricity (% of Population). Available online: https://data.worldbank.org/indicator/EG.ELC.ACCS.ZS (accessed on 3 May 2023).

82. Mezue, K.; Edwards, P.; Nsofor, I.; Goha, A.; Anya, I.; Madu, K.; Baugh, D.; Nunura, F.; Gaulton, G.; Madu, E. Sub-Saharan Africa tackles COVID-19: Challenges and opportunities. *Ethn. Dis.* **2020**, *30*, 693–694. [CrossRef]

83. Schiermeier, Q. Telltale warming likely to hit poorer countries first. *Nature* **2018**, *556*, 415–416. [CrossRef] [PubMed]

84. Witze, A. Extreme heatwaves: Surprising lessons from the record warmth. *Nature* **2022**, *608*, 464–465. [CrossRef] [PubMed]

85. Turco, M.; Abatzoglou, J.A.; Herrera, S.; Zhuang, Y.; Jerez, S.; Lucas, D.D. Anthropogenic climate change impacts exacerbate summer forest fires in California. *Proc. Natl. Acad. Sci. USA* **2023**, *120*, e2213815120. [CrossRef] [PubMed]

86. Polonik, P.; Ricke, K.; Reese, S.; Burney, J. Air quality equity in US climate policy. *Proc. Natl. Acad. Sci. USA* **2023**, *120*, e2217124120. [CrossRef]

87. Levermore, G.; Parkinson, J.; Lee, K.; Laycock, P.; Lindley, S. The increasing trend of the urban heat island intensity. *Urban Clim.* **2018**, *24*, 360–368. [CrossRef]

88. Chakraborty, T.; Hsu, A.; Manya, D.; Sheriff, G. Disproportionately higher exposure to urban heat in lower-income neighborhoods: A multi-city perspective. *Environ. Res. Lett.* **2019**, *14*, 105003. [CrossRef]

89. Cities must protect people from extreme heat. *Nature* **2021**, *595*, 331–332. [CrossRef]

90. Rockström, J.; Gupta, J.; Qin, D.; Lade, S.J.; Abrams, J.F.; Andersen, L.S.; McKay, D.I.L.; Bai, X.; Bala, G.; Bunn, S.E.; et al. Safe and just Earth system boundaries. *Nature* **2023**, *619*, 102–111. [CrossRef]

91. Mazon, J.; Pino, D.; Vinyoles, M. Is declaring a climate emergency enough to stop global warming? Learning from the COVID-19 pandemic. *Front. Clim.* **2022**, *4*, 848587. [CrossRef]

92. Hickel, J.; Brockway, P.; Kallis, G.; Keyßer, L.; Lenzen, M.; Slameršak, A.; Steinberger, J.; Ürge-Vorsatz, D. Urgent need for post-growth climate mitigation scenarios. *Nat. Energy* **2021**, *6*, 766–768. [CrossRef]

93. Hickel, J.; Kallis, G.; Jackson, T.; O'Neill, D.W.; Schor, J.B.; Steinberger, J.; Victor, P.A.; Ürge-Vorsatz, D. Degrowth can work—Here's how science can help. *Nature* **2022**, *612*, 400–403. [CrossRef] [PubMed]

94. United Nations (UN). The Sustainable Development Goals Report. 2020. Available online: https://unstats.un.org/sdgs/report/2020/The-Sustainable-Development-Goals-Report-2020.pdf (accessed on 3 May 2023).

95. Van Ruijven, B.J.; De Cian, E.; Wing, I.S. Amplification of future energy demand growth due to climate change. *Nat. Commun.* **2019**, *10*, 2762. [CrossRef] [PubMed]

Review

The Role of the Energy Use in Buildings in Front of Climate Change: Reviewing a System's Challenging Future

George Ekonomou and Angeliki N. Menegaki *

Department of Economic and Regional Development, Agricultural University of Athens, Amfissa Campus, 331 00 Amfissa, Greece; grg.ekonomou@gmail.com
* Correspondence: amenegaki@aua.gr

Abstract: Energy keeps the global economy alive, while also being extensively exposed to various climate change impacts. In this context, severe business competition (e.g., the building sector) and the unwise use of natural resources and ecosystem services (e.g., fossil fuel energy sources) seem to sharpen the relevant effects of climate change. Indicatively, contemporary issues at the interface of building energy performance and environmental quality levels include consequences from global warming, the increasing release of carbon dioxide to peak electrical loads, power grids, and building planning, and energy demand and supply issues. In light of such concerns, the present review paper attempts to disclose the multifaceted and multidisciplinary character of building energy use at the interface of the economy, the environment, and society against climate change. This review highlights energy efficiency concepts, production, distribution, consumption patterns, and relevant technological improvements. Interestingly, the reviewed contributions in the relevant literature reveal the need and necessity to alter the energy mix and relevant energy use issues. These include developments in climate-proof and effective systems regarding climate change impacts and shocks. Practical implications indicate that the sustainable development goals for clean energy and climate action should be followed if we wish to bring a sustainable future closer and faster to our reality.

Keywords: climate change; building energy use; energy efficiency; sustainable future

Citation: Ekonomou, G.; Menegaki, A.N. The Role of the Energy Use in Buildings in Front of Climate Change: Reviewing a System's Challenging Future. *Energies* **2023**, *16*, 6308. https://doi.org/10.3390/en16176308

Academic Editors: Boštjan Polajžer, Davood Khodadad, Younes Mohammadi and Aleksey Paltsev

Received: 24 July 2023
Revised: 26 August 2023
Accepted: 26 August 2023
Published: 30 August 2023

1. Introduction

The sustainable use of resources or efficient allocation can lead to low-performance rates of natural systems (e.g., overconsumption, overexploitation). These issues correlate widely with unstructured, unplanned, and intense economic activities and human intervention (e.g., built environment, land coverage). As a result, already severe climate change conditions are sharpening. In turn, this situation reflects the availability (e.g., supply and demand perspective), sustainability (e.g., fossil fuels or renewables), and quality status of the provided ecosystem services—for instance, provisioning services, such as energy. Conventional energy is derived from scarce resources, and governments should use it conservatively and efficiently [1]. The targeted outcome supports the never-ending pursuit of optimizing resource exploitation within the limits set by natural dynamics and socio-economic forces. Supportively, ref. [2] claims that investigating the relationship between environmental indicators and macroeconomic variables is highly important to foster relevant policies like, for instance, fiscal policies on CO_2 emissions.

Energy systems emphasize the concept of green buildings offering an engineering and science base [3]. Moreover, the interdependencies of energy systems and building constraints (e.g., engineering, planning, design, carbon footprints) are crucial to achieving carbon-neutral building energy systems throughout their lifecycle [4]. Buildings are considered the most significant energy-saving space in the world, and they remain suitable fields to apply technologies for emission reduction [5]. Ref. [6] notes that the building stock relies primarily on energy generated by fossil fuels for heating and cooling purposes.

From this perspective, seeing how climate change interrelates with building energy use and efficiency would be beneficial. Both affect energy systems extensively, which in turn interrelate with environmental quality. This is a great opportunity to review and understand how the concept of 'buildings' affects the transition toward sustainable development. The present study aims to thoroughly review high-impact research efforts that discuss the impacts of climate change and building energy-related issues at the interface. The current review is led by the following research question: "What are the effects of climate change on buildings' power and energy systems concerning existing research?" To accomplish this study purpose, we process an integrative review process to meet the purpose of this study. Ref. [7] claims that integrative reviews interrelate to varied data sources, enhancing a holistic understanding of the topic in question and confronting the complexity inherent in scientific research.

Energy resource availability seems to be one of the most critical research issues, especially about building sectors' concerns. Correspondingly, the challenge lies in recognizing high-leverage interventions, such as today's decisions on future building energy trends, might create fundamental changes for improving energy systems. These significant concerns stimulated our research to explore relevant literature and gather inputs and insights across science in light of a better future. What are the effects of climate change on energy production, distribution, and consumption related to building end-use demand? What are the prospects? How might these changes affect economic growth and welfare status? Are there any established linkages and causalities? These are challenging questions in academia and business, which pursue pathways to optimize resources and processes to 'build' sustainability within the economic system. For instance, this becomes evident in high-energy demand sectors (e.g., the building sector) and relevant consumption patterns (e.g., end-use needs).

However, these considerations do not reproduce significant progress to achieve the desired balance between socioeconomic and nature dynamics, even though relevant literature has stressed the significance of energy efficiency in high-leverage industries. An integrative review of climate change impacts on building energy-related issues has yet to be processed.

The structure of the paper is the following: The Methodology section presents the theoretical background of the review process followed in this study. The next section includes the data extraction process. The following two sections concentrate on the building energy review process and the results of this study. The Recommendation section focuses on the gathered information, gaps, and future research perspectives. Finally, the last section concludes the results.

2. Methodology

By perceiving the challenge of exploiting natural resources sustainably, this study broadly reviews a series of selected published studies that discuss the climate change impacts on power and energy systems.

Whether systematic or integrative, literature reviews offer a way of summarizing individual research studies and other types of articles. Thus, these reviews integrate current topic knowledge [8]. It should be mentioned that the main difference between the systematic and the integrative review process is that the former concerns experimental study trials. The latter considers both non-experimental and experimental studies.

The present study processes an integrative review to gather and summarize previous research efforts on power and energy systems. This process allows the researcher to understand the issues of interest more deeply and thoroughly. Supportively, the integrative review approach includes a wide range of methodologies. For instance, experimental and non-experimental research, theoretical or empirical, and qualitative or quantitative studies offer great applicability for multiple research fields. Interestingly, ref. [9] asserts that such a process aims to define concepts, review theories and 'gaps,' contribute to the literature, and analyze methodologies adopted to describe research issues. In this framework, such an approach is suitable for a scope broadly related to a phenomenon or the research field of interest [7]. An integrative review process provides opportunities to incorporate findings and analysis of information into decision-making processes.

According to ref. [10], researchers adopt the integrative literature review since it exceeds merely analyzing and synthesizing research findings or primary studies [11]. They also argue that this process allows for integrating qualitative and quantitative data, opinions, discussion papers, and policy documents. This process adds sources of scientific information, creating a more comprehensive understanding of the specific phenomenon under research [12–14]. The integrative review provides a challenge to integrate existing knowledge from various communities of practice and recommend future initiatives for research [15].

Furthermore, as ref. [7] reports, little attention has been paid to efforts combining empirical and theoretical reports. The integrative review process widely considers this issue. As a review method, it also increases its potential to turn primary research methods into a more significant part of evidence-based practice initiatives. Consequently, the value of this process highlights its broadest character and enhances rigor. In this perspective, an integrative review process comprises five steps: problem identification, literature search, data evaluation, data analysis, and presentation of findings. Such an approach facilitates a researcher's review effort to integrate concepts, theories, evidence, and methodologies for the topic in question [16].

Integrative literature reviews are suitable to address mature research fields and topics or new, developing, emerging scientific issues as a research topic matures and the interest in the literature increases. Consequently, the relevant knowledge base is expanding and growing for this particular topic [17].

When processed in a detailed, well-organized, and thoughtful manner, many benefits derive from an integrative review process. For instance, strength evaluation of the reviewed studies' evidence, gap identification, research opportunities for further research efforts, integration (bridging) of relevant areas in a scientific domain, identification of core issues in science, generation of a research question, identification of conceptual and theoretical frameworks, and exploration of all successful methods used from researchers to reach results [18]. Consequently, in practical terms, it is an inclusive way to summarize various types of evidence justified by many methodologies, whereas it delivers a wider scientific view of the topic [11].

Receiving background knowledge from a sizable body of reviewed studies can lead current research efforts to define the scope and extent of a research topic [19]. Multiple types of data sources permit synthesizing the findings and identifying the main topic under review, enabling authors to develop a new understanding of the topic [20]. Figure 1 presents the steps to obtain the final number of reviewed publications (flowchart).

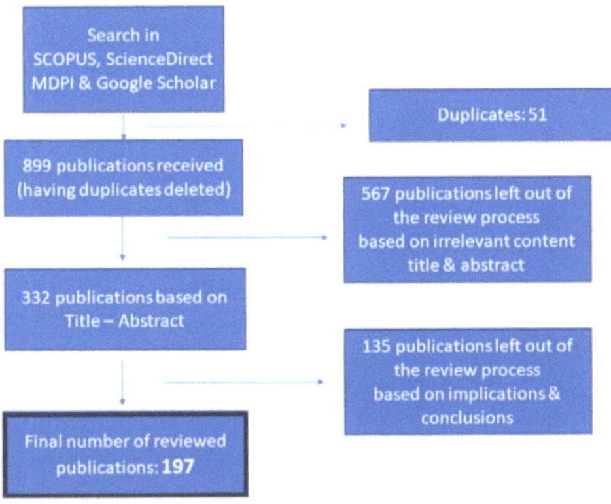

Figure 1. Steps to process this study's integrative review.

3. Data

This study's integrative review process was structured based on reliable and accredited publications within the scientific community. Search terms included energy systems, energy consumption, energy production, energy distribution, energy efficiency, energy growth nexus, power systems, electrical power, climate change impacts, climate adaptation, climate mitigation, renewable energy, and energy mix. Three of the most popular, acknowledged, and dependable databases were used to retrieve published studies relevant to the purpose of the present work: Scopus, Science Direct, and MDPI. The literature review was extended by searching the Google search engine and relevant Google Scholar data to find peer-reviewed articles published in journals indexed in the abovementioned databases. After receiving results, papers were screened for duplicates or slight relevance with the subject of interest. Essentially, publications were excluded if the study's primary purpose was not aligned with the impacts of climate change on buildings' power and energy systems. Then, studies were evaluated based on the Abstract and eligibility criteria. Criteria for keeping the studies for further review were the explicit purpose of the study, conclusions, and specific theoretical and practical implications based on test results or contributions. Another criterion was the novelty of methodologies used to support their scientific argument. Each study was thoroughly read and then listed based on the classification needs of this integrative review process. Particularly, the inclusion criteria for proceeding further with the review process were: a well-defined and visibly justified contribution to the relevant literature (e.g., research gap); the paper should have undertaken a blind peer-review process before getting published; the year of publication (e.g., studies published after the year 2000); robustness and reliability of methodology adopted; and language restrictions (e.g., written in the English language). Our data extraction purpose was to focus on and carefully analyze studies that have made acknowledged contributions to the relevant literature. Additionally, studies should have meticulously progressed relevant research efforts concerning climate change and its impacts on power and energy systems.

The review process included a variety of methods, materials, and tools used in scientific approaches from different viewpoints. These methods should highlight the multifaceted and interdisciplinary nature of the research subject. Diversity in methodology and variations in research results were identified during the process. A comprehensive analysis of the studies was made to classify points of relevance to the present effort. Then, comparisons with similar papers on the same research field took place. Next, determination of trends and tendencies in the literature was implemented. Last, the integration and summation of the significant findings related to the thematic field of the present review process was completed.

This followed data extraction process of the present study remains very constructive in retrieving each study's desired vital points and research results. To increase the reliability of this work, the data extraction process was carefully made and double-checked by both authors to overcome mistakes due to data entry errors and potential misinterpretations of concepts and methodologies of reviewed published studies.

The data extraction process allowed us to receive 197 publications. These selected publications were divided into seven categories based on their thematic field. Table 1 presents the number of retrieved studies based on the thematic area. Table 2 shows the number of reviewed studies based on the year of publication. Figure 2 illustrates a spider chart for data from Table 1, whereas Figure 3 concerns an additional spider chart from Table 2. Figure 4 shows the total number of reviewed studies. Furthermore, a trend analysis has been added (Figure 5) considering the number of publications per reference year to have a complete picture of the received results.

Table 1. Thematic fields and reviewed studies.

Building Materials	Data Analysis (e.g., Econometric Analysis)	Methods and Technology (e.g., Benchmarking Methods, Smart Technology	Model Optimization (e.g., Building Energy Systems)	Occupants (e.g., Behavior)	Policy (e.g., Policy Plans, Frameworks)	Simulations and Scenarios (e.g., Heating, Cooling, Energy Use)
11	40	35	35	27	11	38

Table 2. Reviewed studies based on year of publication.

2000	2001	2002	2003	2004	2005	2006	2007	2008	2009
1	2	4	0	4	0	1	4	5	2
2010	2011	2012	2013	2014	2015	2016	2017	2018	2019
5	6	19	8	10	12	9	9	9	8
2020	2021	2022	2023						
14	8	24	33						

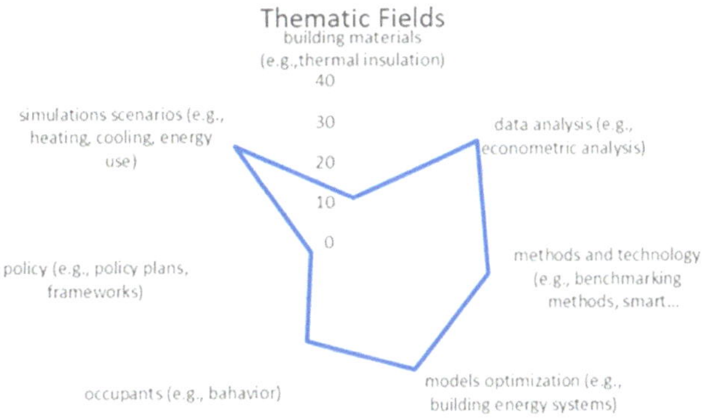

Figure 2. Spider chart for thematic fields and reviewed studies.

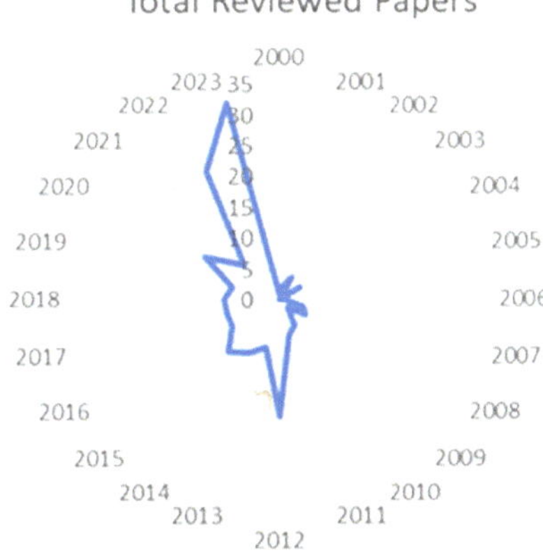

Figure 3. Spider chart for reviewed studies based on year of publication.

Figure 4. Chart for the total number of reviewed papers.

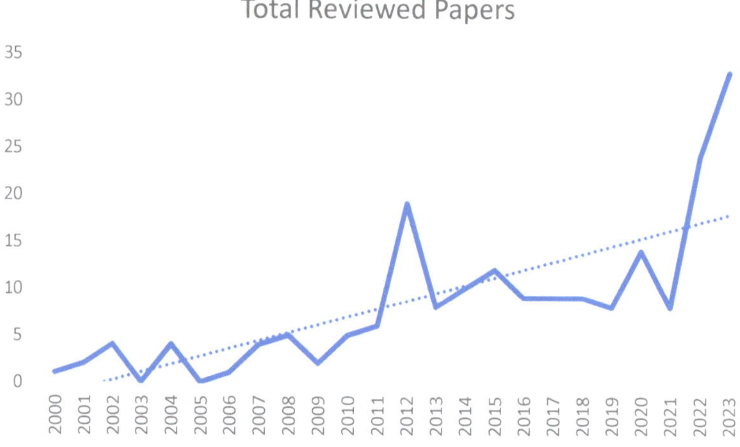

Figure 5. Trend analysis for the number of publications and reference year. The left axis indicates the number of reviewed publications. The horizontal axis indicates reference years.

4. Buildings and Energy

There is a growing interest in energy use and consumption and its environmental implications. This is mainly due to fossil fuel use, over time, rapidly and gradually, as the core energy source and the related greenhouse gas emissions (GHGs) and carbon dioxide ("CO_2") releases. This situation results in raising the global temperature to a great extent. Buildings contribute largely to energy-related emissions [21]. Therefore, the role of buildings (e.g., residential, non-residential) and their lifespan in this process (e.g., energy demand, energy-related emissions, emissions footprint) are considered fundamental. Ref. [22] mentions that the energy supply side should be able to cover future energy demands. In turn, energy demand varies based on various factors. As indicated in ref. [23], the critical determinants behind the building energy service demand vary according to different trends in the socio-economic system, technological factors, behavioral aspects, climate issues [24–27], and numerous electrification pathways. One key issue for limiting energy consumption regarding demand reduction concerns improving building stocks [28]. Buildings concretely represent the energy used in various processes (e.g., mining, processing, manufacturing, and transporting building materials) and the energy consumed in constructing and decommissioning the buildings [21]. Given the long lifetime of buildings, estimated at 50 years, it is significant to review their response to climate change throughout the years. The future perspectives on energy consumption (e.g., heating and cooling) should be considered. Not surprisingly, the issue of energy use in buildings can be incorporated into well-structured and organized mitigation and adaptation measures against climate change. Interestingly, this seems to be a high-impact issue related to weather conditions, climate zones, and energy efficiency. Energy efficiency concerns technological advancements and smart energy systems that use less energy to produce the same or better outcomes and tasks. It was calculated that in 2002, buildings globally accounted for about 33% of the world's GHGs [29]. Recent estimations indicate that buildings still cause 36% of the European Union's energy-related GHGs [30]. This issue in the building sector is currently at the top of the agenda signifying its importance in reaching the European Union's energy and climate objectives for 2030 and 2050. Specifically, ref. [31] clearly states that from 2028 new public sector buildings will be zero-emission. Additionally, from 2030 all new buildings will be zero-emission buildings. The agreement launched a new energy category for buildings, "A0", concerning energy performance certificates indicating zero-emission buildings. The final target is to activate renovations, move forward to a gradual phase-out of the worst-performing buildings, and improve profoundly regarding the national building stock. This means better and more energy-efficient buildings will result in a decarbonized building stock by 2050. Furthermore, these targets are expressed thoroughly in ref. [32] for improving the well-being of people and a net-zero age. Interestingly, ref. [33] conducted a study concerning the Building Renovation Passport (BRP) concept in terms of definitions and content (structure) to offer useful building-related documentation.

Given such worries, more sustainable investments will become a reality (e.g., buildings with eco-friendly materials and advanced energy systems). People (e.g., entrepreneurs and individuals) will make more informed decisions regarding energy-saving and cost-saving options (e.g., heating, cooling, and running appliances and devices). According to ref. [34], it is an imperative need to improve the energy intensity per square meter concerning the building sector by 30% by 2030 to stay consistent with the Paris Agreement climate goals. From Figure 6, which demonstrates the annual "CO_2" emissions globally, we conclude that the building sector accounts for 27% of global "CO_2" emissions [35]. Figure 7 presents the heating degree days (*y*-axis) for the United States, European Union, and China. Figure 8 illustrates the cooling degree days in summer (*y*-axis) for the United States, European Union, and China.

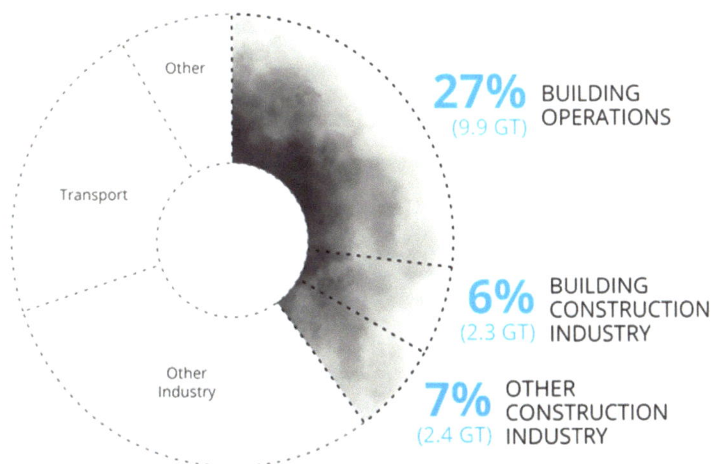

Figure 6. Global "CO_2" emissions, annually–globally [35].

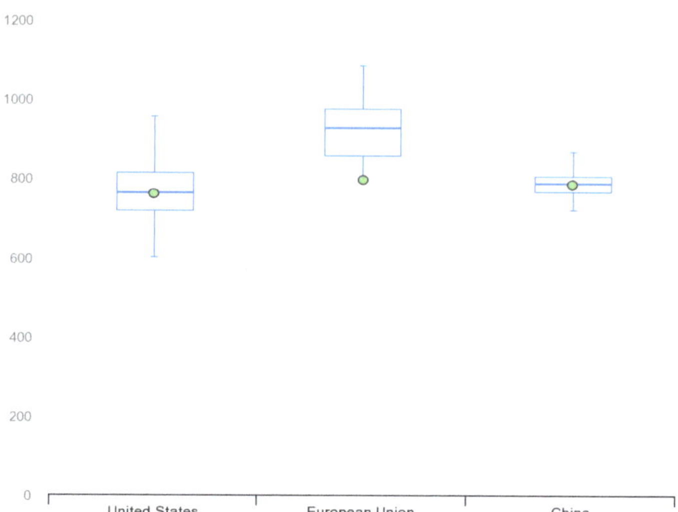

Figure 7. Heating degree days in winter months. The blue box plot indicates the period 2000–2021 on average. The green bullet indicates the year 2022 [36].

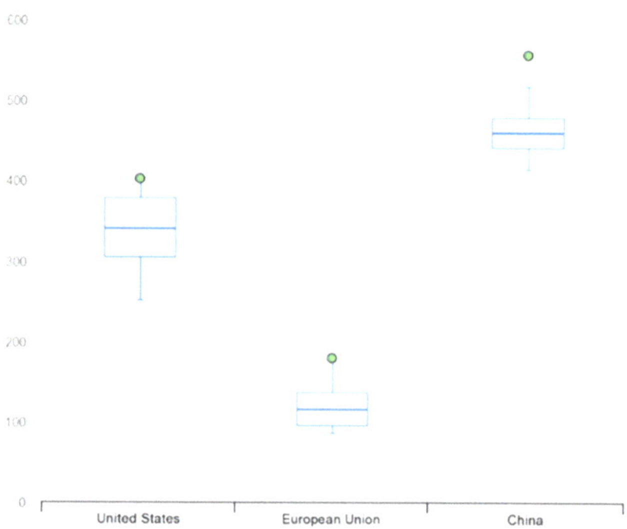

Figure 8. Cooling degree days in summer months. The blue box plot indicates the period 2000–2021 on average. The green bullet points for the year 2022 [36].

According to ref. [36], emissions were increased due to fossil fuel power plants covering consumption needs for excess cooling demand during extreme summer heat. Cooling degree days in 2022 exceeded typical levels or even the maximum level seen for 2000–2021. Furthermore, for the year 2021, cooling and heating consumption needs from extreme weather increased global emissions by around 60 Mt "CO_2". Two-thirds of this are due to additional cooling needs. The remaining one-third came from heating needs. This accounted for almost one-fifth of the total worldwide rise in "CO_2" releases. Improving energy efficiency in buildings seems a promising way to reach, or at least considerably approach, the carbon neutrality target by 2050. From this perspective, ref. [26] asserts that improvements in the building stock and advancements concerning commercial equipment and household appliances can positively impact energy use and building services. This will result in limiting "CO_2" emissions. Experts should specifically set minimum levels of energy performance requirements (standards), such as appliance and equipment standards or building energy codes [25]. Furthermore, ref. [37] argues that energy efficiency efforts should be categorized into the following sections: envelope design, form, orientation and height, ventilation, carbon emission, renewable energies, and occupant behavior. The review concerned 48 studies considering the energy and carbon performance of high-rise buildings (HRB) between 2005–2020. In the literature review in ref. [38], 134 studies were systematically reviewed. The focus was on multiple topics for improving energy efficiency by limiting devastating impacts on the environment with socio-economic concerns. This research interrelates Sustainability Development Goals, namely SDG11, which considers sustainability in cities and communities, and SDG13, which concerns climate action.

Literature on climate change impacts on building energy consumption is increasing, driven by the need to process adaptation measures since they can greatly safeguard the built environment's long-term integrity and effective operation [39]. According to ref. [40], studies related to the impacts of climate change on buildings can be grouped into five categories: (i) estimation of impacts concerning energy consumption; (ii) adaptation and mitigation measures for buildings toward combating adverse effects of climate change; (iii) models for building retrofitting and renovation to handle the climate change; (iv) creation of new methods and tools for making projections for future conditions; (v) handling and estimating uncertainty concerning climate projection models and relevant impacts on building simulation results. Ref. [23] highlights the role of uncertainties when making projections and relevant estimations for energy consumption patterns and "CO_2" emissions in the case of buildings. Notably, ref. [41] presents three methodological phases to evaluate climate change impacts on buildings. The first phase includes the study context identification, which concerns the geographical context and the building typology. The second phase refers to future weather prediction. This phase considers the selection of emission scenarios, global circulation models (GSMs), downscaling techniques, weather file types, and study periods. The third phase relates to energy consumption prediction, and concerns dynamical simulation models and regression models to compare future time slices with a reference period.

The energy transition concept is widely acknowledged in the literature as a shift in the so-called 'energy paradigm', namely replacing fossil fuels with renewable energy sources to decarbonize energy systems [42]. In this effort, authors stress the importance of the 'energy triangle' approach: (i) generate electricity directly from renewables; (ii) use electricity as the core energy vector; and (iii) electrification of end-use. This 'jump' from fossil fuels to renewables constitutes an answer provider, a fundamental response against 'quick fixes' or 'easy solutions' that treat only symptoms of problems. Tackling effectively (e.g., building planning) and efficiently (e.g., using wisely resource materials) the impacts of climate change requires deep knowledge of the current situation. Forecasts for future scenarios and proactive rather than reactive behavior from all stakeholders are essential. This series of events will provide spatial planners, policy and decision-makers, and officials an advantage to prevent worse situations. The role of buildings in this process is fundamental. In this regard, ref. [43] emphasizes the energy efficiency benchmarking of buildings. It is an accurate technique to measure, track, and limit end-use energy usage of buildings by adopting comparative scenarios. This approach discloses opportunities to order energy-saving processes, such as modifications to end-use appliances or building operations. The proposed approach employs machine-learning techniques to maximize accuracy and precision compared to other benchmarking methods [44–46]. Data gathering and availability of relevant information to process simulation models and use tools and techniques to evaluate building performance is crucial. For instance, ref. [47] asserts that the precise provision of data (e.g., daily, monthly) concerning a typical meteorological year (TMY) is a requirement and important task. With this procedure, we can evaluate building energy consumption, which impacts the good use of outdoor data for building energy conservation. Consequently, in the absence of adequate data provision, the predictive power of models and the dependability of results are in question. To overcome these difficulties, ref. [48] proposes a building information modeling (BIM) and building energy modeling (BEM) process grounded on a 3D laser scanning process. Geometric information on the existing building can be implemented in the case of inadequate information to run building energy models. Moreover, ref. [49] developed a new approach that combines machine learning and a domain knowledge-based expert system. This system is helpful to increase building energy flexibility supported by a rule-based expert system and a decision tree model. Authors conceptualize energy flexibility as indicators related to cost and energy-saving margins (potential), load, and peak shaving efficiency.

Officials in ref. [6] state that by 2030, GHG emissions of buildings within the European Union must be limited by 60%, final energy consumption by at least 14%, and energy consumption for heating and cooling purposes by 18%, compared to 2015 levels. Researchers who recognize such a need constantly develop tools to evaluate energy efficiency and take corrective actions for embedding sustainability into the building sector. Ref. [50] completed one of these works. The authors highlight buildings' spatial and functional dimensions and incorporate them into urban building energy modeling (UBEM). They apply such an approach to forecasting hourly heat load profiles of residential buildings using detailed building simulation tools. This effort is vital for high-resolution results concerning spatial and temporal dimensions. The literature stresses the significance of UBEM, especially in modeling large-scale buildings. For instance, ref. [51] systematically processed a literature review considering physics-based modeling techniques. The main purpose was to assess conservation energy-related measures.

Given the multiple outstanding studies concerning sufficiency, efficiency, and renewables for attaining goals for reducing GHGs and energy demand, ref. [52] identified a gap in achieving building energy sufficiency (BES) in the building operational phase. They considered not only energy or emissions requirements but also addressed occupant demand. The definition of BES varies in the relevant literature. In the building sector, occupant demands are categorized into four categories: time and space, quality and quantity, control and adjustment, and flexibility, matching human well-being with building energy sufficiency. Energy sufficiency is defined as "a state in which the population's basic needs for energy services are met equitably and ecological limits are respected" [53]. An issue that is more than challenging, contemporary, and important to achieve sustainability. Refs. [54,55] mention that lifestyle and occupant behavior can be recognized as crucial determinants impacting buildings' final energy use.

Technological advancements and innovations in the construction and use of buildings are important for experiencing sustainability goals. This is how smart technology enters the equation of building energy efficiency. Notably, ref. [56] states that data-driven models for occupancy prediction are appropriate (e.g., indoor environmental data-driven model) with machine learning techniques. In this context, Bluetooth Low Energy (BLE) technology promises to increase energy efficiency in buildings. Notably, such a smart technology approach identifies a set of occupancy profiles representing the varied occupancy patterns observed in the research area [57]. Interestingly, technology-oriented solutions help to reduce energy consumption with a positive impact on protecting the built environment. However, technological solutions and innovations concerning materials used are demanding and complex issues since buildings comprise dynamic systems, and the occupants demonstrate different behaviors in a complex mode [54].

Current and future researchers should motivate, inspire, and guide further innovative achievements, models, and applications to maximize space for energy efficiency and drastically limit energy use in buildings. Given the conditions in socio-economic systems worldwide, this is a multidisciplinary task with many variables in the 'equation' of sustainable development (GHGs, "CO2" emissions) and predictors of building energy consumption and efficiency.

5. Results

The review process discloses the results of selected articles. This study aimed to release contemporary issues from reviewed articles concerning climate change impacts derived from building energy use and related GHGs and "CO_2" emissions. These results can be further examined for inclusion in decision-making processes. They can also contribute to formulating energy management schemes and building planning for energy-efficient buildings. Ref. [58] concludes that using remotely sensed data when making predictions for energy efficiency levels of buildings brings opportunities for future work. This work can integrate additional data sources compared to on-site, in-field visits of certified energy auditors, which might make the whole process slow, costly, and geographically incomplete. The research concerned data from 40,000 buildings in the United Kingdom. Accordingly, technology plays a unique role in promptly getting things done efficiently, accurately, and cost-effectively. For instance, using the Internet of Things (IoT) smart ecosystems helps reach decisions that can benefit all stakeholders in the energy system [59]. Supportively, ref. [60] proposes a novel IoT-based occupancy-driven plug load management system. Energy use reduction is feasible with these systems' help, whereas their applicability promises a building-wide implementation. Review results show that these issues deeply interrelate with the concepts of 'smart' or 'intelligent' buildings. Interestingly, advanced technology helps make predictions, define occupancy profiles, and adjust heating, ventilating, and air conditioning (HVAC) operations. Then, we expand our ability in light of limiting building energy consumption [61]. This is especially the case in building lighting, a fundamental issue in the literature. Since artificial lighting accounts for 19% of energy consumption in building environments, advanced lighting control systems facilitate occupants to regulate or customize their luminance preferences (indicatively see ref. [62]).

Climate change mainly impacts building-energy demand by increasing or decreasing the demand needs for cooling and heating. Building technologies (e.g., building equipment and shell, renovations to the building stocks) contribute primarily to achieving energy-efficient buildings [25]. Ref. [22] concludes that climate change affects residential demand due to average temperature rise, weather conditions, and space heating and cooling needs. Future energy and electricity consumption demand considerations are associated with numerous factors: environmental (e.g., energy mix and renewables inclusion) and socioeconomic factors (e.g., severe market competition and energy use, production lines, and innovations). The main methods adopted to estimate the future residential demand use are parametric, energy balance, and degree day models [22]. Another method is the building energy simulation technique [21]. Various energy simulation tools are processed to elaborate on energy and "CO_2" building performance and energy efficiency gains. All are targeted to enrich strategies and plans for decreasing the environmental impact of buildings due to climate change. Ref. [63] stated that no validated tool could precisely and explicitly simulate buildings' power demand; for instance, at the city level. Thus, space for further improvements and deployments of new models is present and comparable to existing ones.

Optimization methods and settings always play a significant role in processing scenarios. They help draw safe conclusions about how buildings will behave and evaluate their resilience and mitigation capacity [40]. Energy efficiency issues are also critical [64]. Interestingly, machine learning and a domain knowledge-based expert system ease building demand-side management while they advance the building's energy design and control systems for greater demand flexibility [49]. Review results show that energy flexibility is vital for keeping a power grid sustainable and resilient. Furthermore, it is a significant measure to decrease utility costs for building owners [65]. Moreover, we receive information for building characteristics (e.g., energy consumption) based on machine learning methods from various authors, such as [66–68], as well as for energy efficiency inputs based on deep learning-based multi-source data fusion frameworks [69].

Energy sufficiency is highly important since it comprises one of the three energy sustainability strategies, following energy efficiency and renewables [70]. The authors elaborated on 230 sufficiency-related policy measures from a systematic document analysis. They searched the European national energy and climate plans (NECPs) and long-term strategies (LTSs). They concluded that relevant regulatory frameworks comprise a valuable instrument to achieve great sufficiency rates concerning national energy management plans in European Union countries.

Mitigation and adaptation alternatives challenge the potential to handle changing conditions of climate. Mitigation measures can be applied to building envelopes and internal loads [71,72]. Dropping the lighting load density is a great energy-saving option, mainly applied in cooling-dominated buildings in warmer climates [73]. Ref. [74] found that an improved artificial light source (e.g., LED lamp technology) will support constant solar lighting and energy efficiency in indoor illumination. The role of technology always remains crucial in using less energy without losing the desired output. Climate adaptation measures should be appropriately planned when designing buildings and at the operation stages to limit significantly negative impacts [75].

The preceding results stemming from the reviewed studies focused on minimizing the devastating impacts of buildings and energy needs regarding climate change conditions. Consequently, links exist among energy demand, the building and construction industry, and climate change impacts. These interrelations question the achievement of a nation's goals toward a sustainable future—an issue that needs continuous efforts, multifaceted approaches, and cooperation. These issues need partnerships in academia and business environment, within countries, across nations, always with a long-term perspective. A crucial issue for receiving benefits from all research efforts remains the proper and ethical circulation of gained knowledge among scientists. Review studies offer this opportunity in favor of advancing the flow of research results, conceptual frameworks, and any other scientific input.

In this context, Figure 9 illustrates the technical aspect at the interface of buildings' efficiency and climate change impacts. This figure showcases how the technical relationships interact with building energy performance and behavior to reduce relevant "CO_2" emissions. Table 3 presents the more popular methods and models processed by reviewed studies to accomplish robust research and make forecasts and projections based on simulations and scenarios, for instance, reduction of energy consumption and relevant "CO_2" emissions. Practically, we wish to increase the contribution of renewables in the energy mix for residential and non-residential buildings, and reviewed studies with technical aspects concerns, including the fundamental role of building design and building envelope and materials to experience building efficiency. A wide range of technical factors should be put together to achieve the outcome of using less energy without losing quality. For instance, the thermal performance of materials, buildings' thermal insulation (e.g., walls, ceilings, roofs), and buildings' systems (e.g., HVAC control systems and occupants' energy use profiles) were the subject of research to increase efficiency, avoid diminished comfort, identify energy use patterns, gather data, and prevent energy loss. The reviewed studies stress the importance of using eco-friendly materials and replacing traditional or conventional ones. Indicatively, the authors highlight the need to use insulated concrete forms instead of traditional poured concrete in building foundations. Furthermore, the authors propose to replace spray-foam insulation with structured insulated panels in buildings' structural framing.

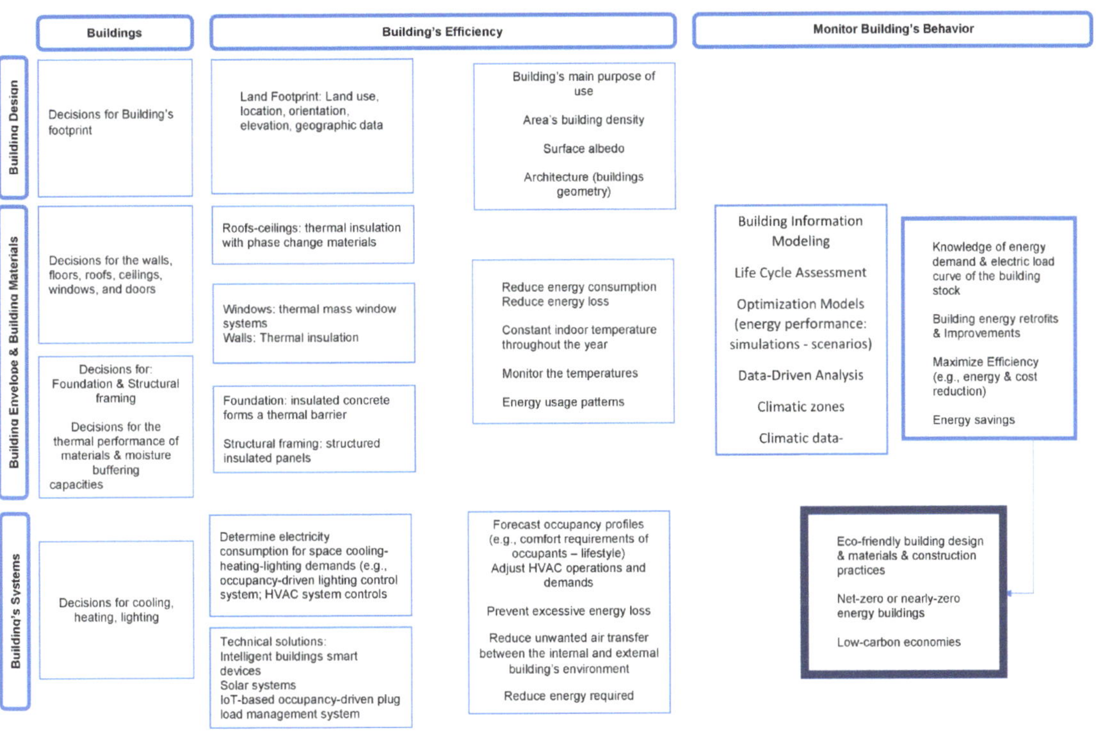

Figure 9. Technical aspects at the interface of buildings' energy efficiency and climate.

Table 3. Reviewed data-driven econometric methods and optimization models.

Data-Driven Analysis			Optimization Methods
Econometric models—specifications (e.g., regression models—statistical analysis)	Qualitative and quantitative analysis	*Simulations Scenarios (e.g., energy modeling, occupancy prediction models, building stock models)*	Classifications algorithms
	Panel data analysis		Computational methods and software
	Parametric and sensitivity analysis		GIS-driven statistical models
	Social network analysis Quadratic assignment procedure		Benchmarking models
	Data mining methods and analysis		Machine learning models

Energy optimization models (e.g., scenarios and simulations) will direct the distribution and transmission endeavors to reduce linkages and power grid problems (e.g., overconsumption, overloading). This sequence of events requires alignment of the energy sector with reduction targets of carbon emissions (e.g., replacement of fossil fuel to produce energy and generate electricity). For this reason, technological advancements and innovations (e.g., intelligent buildings, smart technology, Internet of Things (IoT)-run devices) might keep buildings' energy performance, consumer behavior, and energy use patterns in the desired equilibrium with positive environmental impacts.

6. Recommendations

This study's review covered a wide range of issues and topics related to energy efficiency and the energy footprint of buildings as a function of climate change impacts. All researchers' efforts concentrate on technologies, regulatory frameworks, models, and instruments that help reduce "CO_2" emissions and energy use of buildings. Many studies focus on energy savings and the energy performance of buildings to embed sustainability in the construction phase, operation, and lifecycle of buildings. Ref. [76] underlines the necessity for highly energy-efficient and decarbonized building stock toward a decrease of 19%, at least by 2050. Supportively, it is central to advance the energy efficiency of buildings to reach the targets of a carbon emission peak by 2030 and carbon neutrality by 2060 [77,78]. Indicatively, in the case of the European Union, energy performance certificates provide a pathway to determine the energy efficiency of buildings [79]. These figures are not simply numerical values or percentages, but mirror the current reality; they mark future objectives and call for immediate action to advance the built environment.

At a time of increasing interest in developing 'green' consumption patterns, the relationship between energy and high-leverage market sectors (e.g., the building sector) seems to be a motivating topic for research. All efforts should focus on managing natural and technical resources to meet all environmental, economic, and social needs; for instance, from construction to building operations and occupant behavior. Perceiving these responses' direction, magnitude, linkages, and causalities allow researchers to anticipate environmental changes better and adapt as necessary. Interestingly, switching focus from short-term management plans to long-term strategies based on comprehensive and sophisticated research efforts is a promising way to bring sustainability to the building sector.

Keeping the momentum active, methodologies and econometric models (e.g., panel data or time series analysis) are significant. These methodologies investigate linkages, causalities, and long-term relationships. They decode impacts between growth variables at a macro level and energy-related variables. In turn, researchers can use variables or proxies or indicators that reflect building performance, efficiency, sufficiency, flexibility, demand, end-use, resilience, and request of the energy grid operator. This approach needs further development since it is scarce, untested, or insufficiently mentioned.

For instance, a set of variables for further elaboration could be building energy consumption rates or British thermal units (BTUs) from cooling and heating devices (air-conditioning) in different climate zones and seasons. This approach could impact environmental degradation or growth rates in the context of the Environmental Kuznets Curve (EKC) hypothesis and energy growth nexus discussion. This approach can be adopted for a group of countries (e.g., eurozone member states, OECD countries, G7 countries, G20 countries, Asian countries, and USA states. Another interesting point would be the inclusion of high-leverage and profitable market sectors (e.g., the construction sector) under the same econometric modeling. In this approach, data received from various techniques mentioned in this study could widely benefit such an approach. This perspective might have needed to be more visible to the broader community within natural and socioeconomic systems for energy-related issues.

Indicatively, ref. [80] states that the energy-growth nexus concentrates on the contribution of energy as a factor of production in the economic sector. Consequently, this approach helps to reach results concerning the sensitivity of the growth process against energy conservation measures. In particular, concerns are visible regarding the optimum equilibrium between use—users and demand—growth [81] (Ekonomou and Halkos, 2023). Hence, we obtain feedback for regulating energy consumption. For instance, for limiting greenhouse emissions and fossil fuel resource depletion in the presence of climate change. This is an unexplored area in the case of buildings, and future opportunities for thorough research are present, particularly for highly energy-dependent economies.

Another interesting point is the EKC hypothesis test. Refs. [82,83] explored the linkage between environmental quality and the economy in the EKC hypothesis context. They determined a specific point after which the growth process does not impact environmental quality levels. In this strand of literature, variables that determine building energy-related variables are absent. For instance, "CO_2" emissions from buildings could enter the EKC equation for further research. Given the importance of the building sector in the economic system, researchers should grasp the opportunity and open a new debate based on this approach. Building materials should be eco-friendly with thermal insulation characteristics to 'arm' buildings capable of becoming efficient. Indicatively, we mention replacing poured concrete with insulated concrete and using structured insulated panels instead of spray foam insulation in buildings' structural framing (see [84,85]).

Many researchers utilize building model simulations, use databases, and establish scientific arguments based on forecasts and projections with environmental concerns. One additional research field that can be matched with these research findings would be their impacts on welfare status concerning the Index of Sustainable Economic Welfare (ISEW). These results can interrelate with relevant climate change impacts. Ref. [1] elaborates further on inputs and insights we gain when investigating the role of ISEW in the interaction of energy and economic growth.

Individual behavior regarding energy use and building appliances and devices is a crucial issue that deserves our scientific attention. This issue directly connects pro-environmental behavior, environmental awareness, and everyday life's eco-friendly attitude. Lifestyle trends and ways of thinking and acting (e.g., mindset, culture) affect building energy demand and use. In this perspective, one could process empirical research focusing on willingness to pay for energy quality improvements (e.g., renewables, solar panels, photovoltaics, smart technologies) in buildings (e.g., residential and non-residential buildings). For instance, this approach can be processed on a city scale or neighborhood. Moreover, the willingness to accept living and acting in conventional, traditional buildings that impact environmental quality levels can be explored. The received results can be matched with climate change impacts. These preference-stated methods can benefit climate change mitigation and adaptation plans. Furthermore, estimating the total economic value concerning the effects of climate change on building an environmental footprint is a valuable addition to scientific research. Consequently, they can guide the relevant absorption of economic resources and utilization of financial instruments. Indicatively, in the case of the European Union, an option could be the National Strategic Reference Framework (ESPA). This financial instrument can advance building environmental performance against climate change. These issues and topics remain less visible in the relevant literature.

Last but not least, we must act individually and collectively under interdisciplinary teams. The goal is to reach tangible and measurable results and yield prosperity in human life, in which a practical role is assigned to the built environment.

We should note that many authors have adopted the PRISMA flowchart (indicatively see [86]) to visualize and conduct their review process. This approach is highly referenced and recommended in the relevant literature and is dependable for conducting similar reviews.

7. Conclusions

The present integrative review study concerns the climate change impacts in the presence of energy-related issues attributed to buildings. Buildings play a fundamental role in preserving air quality (e.g., "CO_2" emissions), type of energy resource use (e.g., fossil fuels against renewables), and energy demand and end-use issues. Reviewed articles resulted from a comprehensive review process from well-acknowledged databases: SCOPUS, ScienceDirect, and MDPI. All reviewed articles contributed to relevant literature on a wide range of issues. Indicatively, studies presented in this work concern building simulation modeling, energy efficiency issues, technology and innovations, and energy sufficiency matters.

Results indicate that energy efficiency is an issue under continuous research and optimization methods to receive data and make projections and forecasts for future scenarios. This is a demanding and challenging issue. Sustainable energy use is not an issue of customization, but an integrated concept profoundly related to energy efficiency. Gathered knowledge suggests that building stocks and materials must limit devastating environmental effects in light of climate change conditions. Mitigation and adaptation strategies call for the integration of 'green' patterns in the building sector and consider maximizing the percentage of renewables in the energy mix related to building consumption. Environmental benefits from reducing energy consumption rely on improving machine learning and knowledge-based methods and techniques. Researchers constantly improve these issues by offering new understandings of building environmental performance.

Future challenges call for demonstrating a proactive character, individually and collectively, if we wish to experience a better future in the built environment. New areas for further research arise. Empirical studies can be implemented to investigate linkages of building environmental indicators with economic growth rates and environmental degradation regarding climate change impacts.

Considering all of the above, the role of buildings in preserving the natural and human environment is vital. We anticipate that the present review study will benefit current and future research to move closer, safer, and faster to sustainable building environments and combat climate change drastically.

Author Contributions: Conceptualization, A.N.M. and G.E.; literature review, G.E. and A.N.M.; investigation, G.E.; writing—original draft preparation, G.E.; writing—review and editing, A.N.M.; supervision, A.N.M. All authors have read and agreed to the published version of the manuscript.

Funding: This research received no external funding.

Data Availability Statement: The data generated during the current review are available upon reasonable request.

Conflicts of Interest: The authors declare no conflict of interest.

References

1. Menegaki, A.N.; Tugcu, C.A. Energy consumption and Sustainable Economic Welfare in G7 countries; A comparison with the conventional nexus. *Renew. Sustain. Energy Rev.* **2017**, *69*, 892–901. [CrossRef]
2. Halkos, G.E.; Paizanos, E.A. The effects of fiscal policy on CO_2 emissions: Evidence from the USA. *Energy Policy* **2016**, *88*, 317–328. [CrossRef]
3. Hailu, G. Energy systems in buildings. Energy Services and Management. *Energy Serv. Fundam. Financ.* **2021**, *8*, 181–209. [CrossRef]
4. Ropo, M.; Mustonen, H.; Knuutila, M.; Luoranen, M.; Kosonen, A. Considering embodied CO_2 emissions and carbon compensation cost in life cycle cost optimization of carbon-neutral building energy systems. *Environ. Impact Assess. Rev.* **2023**, *101*, 107100. [CrossRef]
5. Min, J.; Yan, G.; Abed, A.M.; Elattar, S.; Khadimallah, M.A.; Jan, A.; Ali, H.A. The effect of carbon dioxide emissions on the building energy efficiency. *Fuel* **2022**, *326*, 124842. [CrossRef]
6. European Commission. Communication from the commission to the EUROPEAN parliament. In *The EUROPEAN Economic and Social Committee and the Committee of the Regions: A Renovation Wave for Europe—Greening Our Buildings, Creating Jobs*; improving lives; The European Council: Brussels, Belgium, 2020.
7. Whittemore, R.; Knafl, K. The integrative review: Updated methodology. *J. Adv. Nurs.* **2005**, *52*, 546–553. [CrossRef]
8. Oermann, M.H.; Knafl, K.A. Strategies for completing a successful integrative review. *Nurse Author Ed.* **2021**, *31*, 65–68. [CrossRef]
9. Toronto, C.E.; Remington, R. *A Step-by-Step Guide to Conducting an Integrative Review*, 1st ed.; Springer International Publishing AG: Berlin/Heidelberg, Germany, 2020.
10. Lubbe, W.; Ham-Baloyib, E.; Smit, K. The integrative literature review as a research method: A demonstration review of research on neurodevelopmental supportive care in preterm infants. *J. Neonatal Nurs.* **2020**, *26*, 308–315. [CrossRef]
11. Soares, C.B.; Hoga, L.A.; Peduzzi, M.; Sangaleti, C.; Yonekura, T.; Silva, D.R. Revisão integrativa: Conceitos e métodos utilizados na enfermagem [Integrative review: Concepts and methods used in nursing]. *Rev. Esc. Enferm. USP* **2014**, *48*, 335–345. (In Portuguese) [CrossRef]
12. de Souza, M.T.; Silva, M.D.; Carvalho, R.D. Integrative review: What is it? How to do it? *Einstein* **2005**, *8*, 102–106. (In English) (In Portuguese) [CrossRef]
13. Grove, S.K.; Burns, N.; Gray, J.R. *The Practice of Nursing Research: Appraisal, Synthesis, and Generation of Evidence*, 7th ed.; Elsevier Saunders: St Louis, MI, USA, 2013.
14. Torraco, R.J. Writing integrative literature reviews: Using the past and present to explore the future. *Hum. Resour. Dev. Rev.* **2016**, *15*, 404–428. [CrossRef]

15. Cronin, M.A.; George, E. The Why and How of the Integrative Review. *Organ. Res. Methods* **2023**, *26*, 168–192. [CrossRef]
16. Broome, M.E. Integrative literature reviews for the development of concepts. In *Concept Development in Nursing*, 2nd ed.; Rodgers, B.L., Knafl, K.A., Eds.; W.B. Saunders Co.: Philadelphia, PA, USA, 1993.
17. Torraco, R.J. Writing Integrative Literature Reviews: Guidelines and Examples. *Hum. Resour. Dev. Rev.* **2005**, *4*, 356–367. [CrossRef]
18. Russell, C.L. An overview of the integrative research review. *Prog. Transplant.* **2005**, *15*, 8–13. [CrossRef] [PubMed]
19. Bowden, V.R.; Purper, C. Types of Reviews—Part 3: Literature Review, Integrative Review, Scoping Review. *Pediatr. Nurs.* **2022**, *48*, 97–100.
20. Kutcher, A.M.; LeBaron, V.T. A simple guide for completing an integrative review using an example article. *J. Prof. Nurs.* **2022**, *40*, 13–19. [CrossRef]
21. Li, D.H.W.; Wang, L.; Lam, J.C. Impact of climate change on energy use in the built environment in different climate zones: A review. *Energy* **2012**, *42*, 103–112. [CrossRef]
22. Figueiredo, R.; Nunes, P.; Panoa, M.; Brito, M.C. Country residential building stock electricity demand in future climate—Portuguese case study. *Energy Build.* **2020**, *209*, 109694. [CrossRef]
23. Gou, Y.; Uhde, H.; Wen, W. Uncertainty of energy consumption and CO_2 emissions in the building sector in China. *Sustain. Cities Soc.* **2023**, *97*, 104728. [CrossRef]
24. Harvey, L.D.D. Global climate-oriented building energy use scenarios. *Energy Policy* **2014**, *67*, 473–487. [CrossRef]
25. Scott, M.J.; Daly, D.S.; Hathaway, J.E.; Lansing, C.S.; Liu, Y.; McJeon, H.C.; Moss, R.H.; Patel, P.L.; Peterson, M.J.; Rice, J.S.; et al. Calculating impacts of energy standards on energy demand in US buildings with uncertainty in an integrated assessment model. *Energy* **2015**, *90*, 1682–1694. [CrossRef]
26. Scott, M.J.; Daly, D.S.; Zhou, Y.; Rice, J.S.; Patel, P.L.; McJeon, H.C.; Page Kyle, G.; Kim, S.H.; Eom, J.; Clarke, L.E. Evaluating sub-national building-energy efficiency policy options under uncertainty: Efficient sensitivity testing of alternative climate, technological, and socioeconomic futures in a regional integrated-assessment model. *Energy Econ.* **2014**, *43*, 22–33. [CrossRef]
27. Zhou, Y.; Clarke, L.; Eom, J.; Kyle, P.; Patel, P.; Kim, S.H.; Dirks, J.; Jensen, E.; Liu, Y.; Rice, J.; et al. Modeling the effect of climate change on U.S. state-level buildings energy demands in an integrated assessment framework. *Appl. Energy* **2014**, *113*, 1077–1088. [CrossRef]
28. Clift, R. Climate change and energy policy: The importance of sustainability arguments. *Energy* **2007**, *32*, 262–268. [CrossRef]
29. Levermore, G.J. A review of the IPCC assessment report four, part 1: The IPCC process and greenhouse gas emission trends from buildings worldwide. *Build. Serv. Eng. Res. Technol.* **2008**, *29*, 349–361. [CrossRef]
30. Climate Action Network (CAN). How to Roll out the Energy Transition in Buildings. 2021. Available online: https://caneurope.org/energy_transition_buildings_factsheet/ (accessed on 15 July 2023).
31. European Council. Fit for 55': Council Agrees on Stricter Rules for Energy Performance of Buildings. 2022. Available online: https://www.consilium.europa.eu/en/press/press-releases/2022/10/25/fit-for-55-council-agrees-on-stricter-rules-for-energy-performance-of-buildings/ (accessed on 30 June 2023).
32. European Commission. *A Green Deal Industrial Plan for the Net-Zero Age*; COM (2023) 62 final; European Commission: Brussels, Belgium, 2023.
33. Sesana, M.M.; Salvalai, G. A review on Building Renovation Passport: Potentialities and barriers on current initiatives. *Energy Build.* **2018**, *173*, 195–205. [CrossRef]
34. United Nations Environment Programme. *Global Status Report Buildings and Construction (Global ABC)*; UNEP: Washington, DC, USA, 2017.
35. Architecture 2030. The Built Environment. Available online: https://architecture2030.org/why-the-built-environment/ (accessed on 17 July 2023).
36. IEA. *CO2 Emissions in 2022*; International Energy Agency: Paris, France, 2022.
37. Mostafavi, F.; Tahsildoost, M.; Zomorodian, Z.S. Energy efficiency and carbon emission in high-rise buildings: A review (2005–2020). *Build. Environ.* **2021**, *206*, 108329. [CrossRef]
38. Hafez, F.S.; Sadi, B.; Safa-Gamal, M.; Taufiq-Yap, Y.H.; Alrifaey, M.; Seyedmahmoudian, M.; Stojcevski, A.; Horan, B.; Mekhilef, F. Energy Efficiency in Sustainable Buildings: A Systematic Review with Taxonomy, Challenges, Motivations, Methodological Aspects, Recommendations, and Pathways for Future Research. *Energy Strategy Rev.* **2023**, *45*, 101013. [CrossRef]
39. Stagrum, A.E.; Andenæs, E.; Kvande, T.; Lohne, J. Climate change adaptation measures for buildings—A scoping review. *Sustainability* **2020**, *12*, 1721. [CrossRef]
40. Nguyen, A.T.; Rockwood, D.; Doan, M.K.; Dung Le, T.K. Performance assessment of contemporary energy-optimized office buildings under the impact of climate change. *J. Build. Eng.* **2021**, *35*, 102089. [CrossRef]
41. Campagna, L.M.; Fiorito, F. On the Impact of Climate Change on Building Energy Consumptions: A Meta-Analysis. *Energies* **2022**, *15*, 354. [CrossRef]
42. Bompard, E.; Botterud, A.; Corgnati, S.; Huang, T.; Jafari, M.; Leone, P.; Mauro, S.; Montesano, G.; Papa, C.; Profumo, F. An electricity triangle for energy transition: Application to Italy. *Appl. Energy* **2020**, *277*, 115525. [CrossRef]
43. Gupta, G.; Mathur, S.; Mathur, J.; Nayak, B.K. Comparison of energy-efficiency benchmarking methodologies for residential buildings. *Energy Build.* **2023**, *285*, 112920. [CrossRef]

44. Gupta, G.; Mathur, S.; Mathur, J.; Nayak, B.K. Blending of energy benchmarks models for residential buildings. *Energy Build.* **2023**, *292*, 113195. [CrossRef]
45. Konhauser, K.; Wenninger, S.; Werner, T.; Wiethe, C. Leveraging advanced ensemble models to increase building energy performance prediction accuracy in the residential building sector. *Energy Build.* **2022**, *269*, 112242. [CrossRef]
46. Gupta, G.; Mathur, J.; Mathur, S. A new approach for benchmarking of residential buildings: A Case study of Jaipur city, BuildSys 2022. In Proceedings of the 2022 9th ACM International Conference on Systems for Energy-Efficient Buildings, Cities, and Transportation, Boston, MA, USA, 9–10 November 2022; pp. 443–449. [CrossRef]
47. Li, H.; Huo, Y.; Fu, Y.; Yang, Y.; Yang, L. Improvement of methods of obtaining urban TMY and application for building energy consumption simulation. *Energy Build.* **2023**, *295*, 113300. [CrossRef]
48. Jung, D.E.; Kim, S.; Han, S.; Yoo, S.; Jeong, S.; Ho Lee, K.; Kim, J. Appropriate level of development of in-situ building information modeling for existing building energy modeling implementation. *J. Build. Eng.* **2023**, *69*, 106233. [CrossRef]
49. Zhou, H.; Du, H.; Sun, Y.; Ren, H.; Cui, P.; Ma, Z. A new framework integrating reinforcement learning, a rule-based expert system, and decision tree analysis to improve building energy flexibility. *J. Build. Eng.* **2023**, *71*, 106536. [CrossRef]
50. Heidenthaler, D.; Deng, Y.; Leeb, M.; Grobbauer, M.; Kranzl, L.; Seiwald, L.; Mascherbauer, P.; Reindl, P.; Bednar, T. Automated energy performance certificate based urban building energy modelling approach for predicting heat load profiles of districts. *Energy* **2023**, *278*, 128024. [CrossRef]
51. Kamel, E. A Systematic Literature Review of Physics-Based Urban Building Energy Modeling (UBEM) Tools, Data Sources, and Challenges for Energy Conservation. *Energies* **2022**, *15*, 8649. [CrossRef]
52. Hu, S.; Zhou, X.; Yan, D.; Guo, F.; Hong, T.; Jiang, Y. A systematic review of building energy sufficiency towards energy and climate targets. *Renew. Sustain. Energy Rev.* **2023**, *181*, 113316. [CrossRef]
53. Darby, S.; Fawcett, T. Energy sufficiency: An introduction. 2018. Available online: https://www.energysufficiency.org/static/media/uploads/site-8/library/papers/sufficiency-introduction-final-oct2018.pdf (accessed on 17 July 2023).
54. Hong, T.; Yan, D.; D'Oca, S.; Fei Chen, C. Ten questions concerning occupant behavior in buildings: The big picture. *Build. Environ.* **2017**, *114*, 518–530. [CrossRef]
55. Hu, S.; Yan, D.; Azar, E.; Guo, F. A systematic review of occupant behavior in building energy policy. *Build. Environ.* **2020**, *175*, 106807. [CrossRef]
56. Ryu, S.H.; Jun Moon, H. Development of an occupancy prediction model using indoor environmental data based on machine learning techniques. *Build. Environ.* **2016**, *107*, 1–9. [CrossRef]
57. Tekler, Z.D.; Low, R.; Gunay, B.; Korsholm Andersen, R.; Blessing, L. A scalable Bluetooth Low Energy approach to identify occupancy patterns and profiles in office spaces. *Build. Environ.* **2020**, *171*, 106681. [CrossRef]
58. Mayer, K.; Haas, L.; Huang, T.; Bernabı-Moreno, J.; Rajagopal, R. Estimating building energy efficiency from street view imagery, aerial imagery, and land surface temperature data. *Appl. Energy* **2023**, *333*, 120542. [CrossRef]
59. García-Monge, M.; Zalba, B.; Casas, S.; Cano, E.; Guillén-Lambea, S.; López-Mesa, B.; Martinez, I. Is IoT monitoring key to improve building energy efficiency? Case study of a smart campus in Spain. *Energy Build.* **2023**, *285*, 112882. [CrossRef]
60. Tekler, Z.D.; Low, R.; Yuen, C.; Blessing, L. Plug-Mate: An IoT-based occupancy-driven plug load management system in smart buildings. *Build. Environ.* **2022**, *223*, 109472. [CrossRef]
61. Esrafilian-Najafabadi, M.; Haghighat, F. Occupancy-based HVAC control systems in buildings: A state-of-the-art review. *Build. Environ.* **2021**, *197*, 107810. [CrossRef]
62. Zou, H.; Zhou, Y.; Jiang, H.; Chien, S.-C.; Xie, L.; Spanos, C.J. WinLight: A WiFi-based occupancy-driven lighting control system for smart building. *Energy Build.* **2018**, *158*, 924–938. [CrossRef]
63. Frayssinet, L.; Merlier, L.; Kuznik, F.; Hubert, J.-L.; Milliez, M.; Roux, J.-J. Modeling the heating and cooling energy demand of urban buildings at city scale. *Renew. Sustain. Energy Rev.* **2018**, *81*, 2318–2327. [CrossRef]
64. Xiao, Y.; Zhang, T.; Liu, Z.; Fei, F.; Fukuda, H. Optimizing energy efficiency in HSCW buildings in China through temperature-controlled PCM Trombe wall system. *Energy* **2023**, *278*, 128015. [CrossRef]
65. Li, H.; Hong, T. On data-driven energy flexibility quantification: A framework and case study. *Energy Build.* **2023**, *296*, 113381. [CrossRef]
66. Pham, A.D.; Ngo, N.T.; Truong, T.T.H.; Huynh, N.; Truong, N.S. Predicting energy consumption in multiple buildings using machine learning for improving energy efficiency and sustainability. *J. Clean. Prod.* **2020**, *260*, 121082. [CrossRef]
67. Streltsov, A.; Malof, J.M.; Huang, B.; Bradbury, K. Estimating residential building energy consumption using overhead imagery. *Appl. Energy* **2020**, *280*, 116018. [CrossRef]
68. Rosenfelder, M.; Wussow, M.; Gust, G.; Cremades, R.; Neumann, D. Predicting residential electricity consumption using aerial and street view images. *Appl. Energy* **2021**, *301*, 117407. [CrossRef]
69. Sun, M.; Han, C.; Nie, Q.; Xu, J.; Zhang, F.; Zhao, Q. Understanding building energy efficiency with administrative and emerging urban big data by deep learning in Glasgow. *Energy Build.* **2022**, *273*, 112331. [CrossRef]
70. Zell-Ziegler, C.; Thema, J.; Best, B.; Wiese, F.; Lage, J.; Schmidt, A.; Toulouse, E.; Stagl, S. Enough? The role of sufficiency in European energy and climate plans. *Energy Policy* **2021**, *157*, 112483. [CrossRef]
71. Bojic, M.; Yik, F.; Leung, W. Thermal insulation of cooled spaces in high rise residential buildings in Hong Kong. *Energy Convers. Manag.* **2002**, *43*, 165–183. [CrossRef]

72. Lam, J.C.; Wan, K.K.W.; Liu, D.; Tsang, C.L. Multiple regression models for energy use in air-conditioned office buildings in different climates. *Energy Convers. Manag.* **2010**, *51*, 2692–2697. [CrossRef]
73. Li, D.H.W.; Lam, J.C.; Wong, S.L. Daylighting and its implications to overall thermal transfer value determination. *Energy* **2002**, *27*, 991–1008. [CrossRef]
74. Han, H.J.; Mehmood, M.U.; Ahmed, R.; Kim, Y.; Dutton, S.; Lim, S.H.; Chun, W. An advanced lighting system combining solar and an artificial light source for constant illumination and energy saving in buildings. *Energy Build.* **2010**, *203*, 109204. [CrossRef]
75. Ren, Z.; Chen, Z.; Wang, X. Climate change adaptation pathways for Australian residential buildings. *Build. Environ.* **2011**, *46*, 2398–2412. [CrossRef]
76. Lopez, L.M.; Las-Heras-Casas, J.; Gonzalez-Caballín, J.M.; Carpio, M. Towards nearly zero-energy residential buildings in Mediterranean countries: The implementation of the Energy Performance of Buildings Directive 2018 in Spain. *Energy* **2023**, *276*, 127539. [CrossRef]
77. Huo, T.; Cao, R.; Xia, N.; Hu, X.; Cai, W.; Liu, B. Spatial correlation network structure of China's building carbon emissions and its driving factors: A social network analysis method. *J. Environ. Manag.* **2022**, *320*, 115808. [CrossRef]
78. Goubran, S.; Walker, T.; Cucuzzella, C.; Schwartz, T. Green building standards and the united nations' sustainable development goals. *J. Environ. Manag.* **2023**, *326*, 116552. [CrossRef]
79. Spudys, P.; Afxentiou, N.; Georgali, P.-Z.; Klumbyte, E.; Jurelionis, A.; Fokaides, P. Classifying the operational energy performance of buildings with the use of digital twins. *Energy Build.* **2023**, *290*, 113106. [CrossRef]
80. Menegaki, A.N.; Tugcu, C.A. Two versions of the Index of Sustainable Economic Welfare (ISEW) in the energy-growth nexus for selected Asian countries. *Sustain. Prod. Consum.* **2018**, *14*, 21–35. [CrossRef]
81. Ekonomou, G.; Halkos, G. Exploring the Impact of Economic Growth on the Environment: An Overview of Trends and Developments. *Energies* **2023**, *16*, 4497. [CrossRef]
82. Halkos, G.E. Environmental Kuznets Curve for sulfur: Evidence using GMM estimation and random coefficient panel data models. *Environ. Dev. Econ.* **2003**, *8*, 581–601. [CrossRef]
83. Halkos, G. Exploring the economy–environment relationship in the case of sulphur emissions. *J. Environ. Plan. Manag.* **2013**, *56*, 159–177. [CrossRef]
84. Renneboog, R.M. Energy-Efficient Building. In *Salem Press Encyclopedia of Science*; Salem Press: Hackensack, NJ, USA, 2016.
85. Rizzi, A. Technological aspects of energy efficient building: Design, architecture, and building envelopes. *Tech. Rep.* **2019**.
86. Immonen, J.A.; Richardson, S.J.; Sproul Bassett, A.M.; Garg, H.; Lau, J.D.; Nguyen, L.M. Remediation practices for health profession students and clinicians: An integrative review. *Nurse Educ. Today* **2023**, *127*, 105841. [CrossRef] [PubMed]

Review

Anomaly Detection in Power System State Estimation: Review and New Directions

Austin Cooper [1], Arturo Bretas [2,3,*] and Sean Meyn [1]

[1] Electrical and Computer Engineering Department, University of Florida, Gainesville, FL 32603, USA; austin.cooper@ufl.edu (A.C.); meyn@ece.ufl.edu (S.M.)
[2] Distributed Systems Group, Pacific Northwest National Laboratory, Richland, WA 99354, USA
[3] G2Elab, Grenoble INP, CNRS, Université Grenoble Alpes, 38000 Grenoble, France
[*] Correspondence: arturo.bretas@pnnl.gov

Abstract: Foundational and state-of-the-art anomaly-detection methods through power system state estimation are reviewed. Traditional components for bad data detection, such as chi-square testing, residual-based methods, and hypothesis testing, are discussed to explain the motivations for recent anomaly-detection methods given the increasing complexity of power grids, energy management systems, and cyber-threats. In particular, state estimation anomaly detection based on data-driven quickest-change detection and artificial intelligence are discussed, and directions for research are suggested with particular emphasis on considerations of the future smart grid.

Keywords: anomaly detection; cyber-security; false data injection; hypothesis testing; machine learning; power system monitoring; quickest-change detection; state estimation

Citation: Cooper, A.; Bretas, A.; Meyn, S. Anomaly Detection in Power System State Estimation: Review and New Directions. *Energies* **2023**, *16*, 6678. https://doi.org/10.3390/en16186678

Academic Editors: Boštjan Polajžer, Davood Khodadad, Younes Mohammadi and Aleksey Paltsev

Received: 24 August 2023
Revised: 13 September 2023
Accepted: 14 September 2023
Published: 18 September 2023

1. Introduction

Since its introduction by Schweppe in the late 1960s [1,2], power system state estimation has proved an integral component of Energy Management Systems (EMSs). Schweppe's proposed nonlinear static state estimation (SSE) provides estimates of the actual network status, which could then be leveraged for subsequent analysis, including contingency evaluation and power flow studies [3]. Soon after, strategies for mitigating erroneous measurement data [4,5] were developed to ensure the fidelity of the power system state estimates. SSE and dynamic state estimation (DSE) both share a rich history of research [6–8]; however, SSE has seen more real-world implementation. Nevertheless, DSE shows great promise in having an enhancing role in legacy SSE-based EMS [9], especially with the increased adoption of synchrophasor measurements [10], and thus, anomaly-detection methods using both approaches are surveyed.

Numerous sources of state estimation error have been identified and formulated in the literature, including measurement, parameter, and topology discrepancies with respect to the system model. More recently, with the integration of EMS into sophisticated computer networks, the potential for cyber-security vulnerabilities became apparent. What new considerations must be made when bad data are malicious? Stealthy false data injection attacks [11], for example, were formulated as an exercise in fooling legacy bad-data-detection schemes. That said, attacks on cyber-physical systems have yielded very real consequences, including equipment damage and rolling blackouts [12]. Anomaly-detection techniques that can properly handle these manufactured instances of bad data, and thus improve bad data processing in state estimation generally, are surveyed in this review. This review also hopes to highlight some considerations for future approaches to anomaly detection in state estimation, including implementation-based research in the face of increasingly dynamic load and generation profiles, the complexity of distributed cyber-physical infrastructure, and pushes for combined SSE and DSE approaches for higher-fidelity EMS information to improve control, efficiency, and stability in the future smart grid. Because the field of anomaly detection covers a wide range of approaches, this survey limits its scope to power system state estimation, which is a central component of EMS and is expected to remain as such well into the future [9].

Articles selected for this review were chosen based on their impact on the power state estimation anomaly detection field. For earlier foundational works, the authors sought to include papers with lasting influence and citation impact for bad data detection generally. Particular emphasis was placed on real-world implementation in modern EMSs. More recent works required consideration of cyber-attacks and/or error types designed to circumvent the approaches of older works. Because many of these approaches have yet to be implemented in EMSs, selected papers required notable metrics of improvement compared to legacy detection methods.

The contributions of this work include:

- Providing a history of legacy bad data detection and error types in power system state estimation and the connection to newer detection approaches and cyber-attack types.
- Surveying various sources of state estimation cyber-threats and the challenges they pose to anomaly detection schemes.
- An overview of newer approaches for anomaly detection based on quickest-change detection and AI.
- Considerations for future research, including the incorporation of dynamic load profiles, autocorrelated data, and asynchronous measurements.

This review is organized as follows. Section 2 provides a brief theory of static and dynamic state estimation generally and the components used for bad data detection. Section 3 describes the theory and physical meaning behind three main types of error in state estimation: measurement, parameter, and topology. Section 4 outlines the traditional methodologies developed for bad data detection and identification, which often serve as a basis for many modern approaches. Section 5 discusses malicious data attacks designed specifically to circumvent traditional bad data detection. Section 6 describes more modern approaches that aim to overcome these pitfalls. Section 7 provides a summary and considerations for future work.

2. Power System State Estimation

2.1. Static State Estimation

One of the most used models to perform power system SE is the Weighted Least Squares (WLS) estimator [7]. A power system with n buses and d measurements can be modeled through a set of nonlinear algebraic equations in the measurement model:

$$\mathbf{z} = h(\mathbf{x}) + \mathbf{e} \tag{1}$$

where $\mathbf{z} \in \mathbb{R}^{1 \times d}$ is the measurement vector, $\mathbf{x} \in \mathbb{R}^{1 \times N}$ the state variables vector, $h : \mathbb{R}^{1 \times N} \to \mathbb{R}^{1 \times d}$ is a continuous nonlinear differentiable function, and $\mathbf{e} \in \mathbb{R}^{1 \times d}$ is the measurement error vector. Each measurement error e_i is assumed to follow a zero mean Gaussian distribution. $N = 2n - 1$ is the number of unknown state variables, i.e., the complex voltages at each bus.

In the traditional WLS approach, the state vector estimate in (1) is determined by minimizing the weighted norm of the residual [13], represented with the cost function $J(\mathbf{x})$:

$$J(\mathbf{x}) = \|\mathbf{z} - h(\mathbf{x})\|_{\mathbf{W}}^2 = [\mathbf{z} - h(\mathbf{x})]^T \mathbf{W}[\mathbf{z} - h(\mathbf{x})] \tag{2}$$

where $\mathbf{W} = \mathbf{R}^{-1}$ is the inverse covariance matrix of the measurements, otherwise known as the weight matrix.

Linearizing the measurement model (1) yields

$$\Delta \mathbf{z} = \mathbf{H} \Delta \mathbf{x} + \mathbf{e} \tag{3}$$

where $\mathbf{H} = \frac{\partial h}{\partial \mathbf{x}}$ is the Jacobian matrix of h at the current state estimate. The estimate of the linearized state vector is then given by

$$\Delta\hat{\mathbf{x}} = (\mathbf{H}^T\mathbf{W}\mathbf{H})^{-1}\mathbf{H}^T\mathbf{W}\Delta\mathbf{z}. \tag{4}$$

The estimated value of the measurement vector mismatch $\Delta\mathbf{z}$ is given by

$$\Delta\hat{\mathbf{z}} = \mathbf{H}\Delta\hat{\mathbf{x}} = \mathbf{P}\Delta z. \tag{5}$$

where $\mathbf{P} = \mathbf{H}(\mathbf{H}^T\mathbf{W}\mathbf{H})^{-1}\mathbf{H}^T\mathbf{W}$ denotes the linear projection or "hat" matrix. The idempotent matrix $\mathbf{P}$ also has the following properties [7]:

$$\mathbf{PH} = \mathbf{H} \tag{6a}$$
$$(\mathbf{I} - \mathbf{P})\mathbf{H} = 0. \tag{6b}$$

These properties facilitate an expression for the measurement residuals [8]:

$$\mathbf{r} = \Delta\mathbf{z} - \Delta\hat{\mathbf{z}} \tag{7a}$$
$$= (\mathbf{I} - \mathbf{P})\Delta\mathbf{z} \tag{7b}$$
$$= (\mathbf{I} - \mathbf{P})(\mathbf{H}\Delta\mathbf{x} + \mathbf{e}) \tag{7c}$$
$$= (\mathbf{I} - \mathbf{P})\mathbf{e} \quad \text{[Using Equation (6b)]} \tag{7d}$$
$$= \mathbf{S}\mathbf{e} \tag{7e}$$

where $\mathbf{S}$ is known as the residual sensitivity matrix, which was first recognized in [5] for representing the sensitivity of the measurement residual to the measurement error during bad data processing. Also useful is the residual covariance matrix Ω [7]:

$$[\mathbf{r}] = [\mathbf{S}\mathbf{e}] = 0 \tag{8a}$$
$$Cov[\mathbf{r}] = \left[\mathbf{r}\mathbf{r}^T\right] = \mathbf{S}\left[\mathbf{e}\mathbf{e}^T\right]\mathbf{S}^T \tag{8b}$$
$$= \mathbf{S}\mathbf{R} = \Omega. \tag{8c}$$

The residual covariance matrix is used for the detection and identification of bad data, as well as providing insight into the degree of interaction; these concepts will be elaborated upon further in Section 3.

2.2. Dynamic State Estimation

SSE does not consider any history of the measurement vector $\mathbf{z}$, but instead provides a snapshot of the system. This "memoryless" assumption of SSE proved sufficient for real-time monitoring in early EMS. For one, power networks were not as regimented at the distribution level, with far fewer microgrids, distributed energy resources, and net load dynamics compared to today's systems. Secondly, the measurement data fed to the state estimator almost always came from measurement devices with slow sampling rates, such as the 2–4 s range of SCADA. One might argue, then, that the true bottleneck for capturing dynamic behavior in state estimation was slow metering rates. That said, Schweppe's formulation arrived just shortly after the introduction of the Kalman filter in 1961 [14], which inspired power researchers to seek formulations beyond the still-developing SSE. The practical hangup of slow meter sampling rates would be relieved somewhat with the introduction of synchronized phasor measurements in the 1980s [10]. Phasor Measurement Units (PMUs) provide higher sampling rates compared to SCADA but also GPS coordination to avoid the uncertainty associated with asynchronicity.

Like SSE, dynamic state estimation (DSE) encompasses a wide range of methods. Early DSE formulations considered the same set of measurements and state variables as those used in SSE: active and/or reactive power flow and injections and complex bus voltages. Other approaches seek to better capture load dynamics by considering generator rotor angle and speed as differential-algebraic state variables [9,15,16]; however, this review will primarily consider DSE-based anomaly-detection implementations that use algebraic state variables.

DSE can be accomplished by modeling the power system as a discrete-time dynamic system. The Kalman filter is used [17] to estimate the state variables at time k through prediction and measurement update steps upon each iteration:

Predict:

$$\hat{\mathbf{x}}_{k|k-1} = \mathbf{A}_k \hat{\mathbf{x}}_{k-1|k-1} \tag{9}$$

$$\mathbf{F}_{k|k-1} = \mathbf{A}_k \mathbf{F}_{k-1|k-1} \mathbf{A}_k^T + \mathbf{Q}_k. \tag{10}$$

Update:

$$\mathbf{K}_k = \mathbf{F}_{k|k-1} \mathbf{H}_k^T \left(\mathbf{H}_k \mathbf{F}_{k|k-1} \mathbf{H}_k^T + \mathbf{R}_k \right)^{-1} \tag{11}$$

$$\hat{\mathbf{x}}_{k|k} = \hat{\mathbf{x}}_{k|k-1} + \mathbf{K}_k \left(\mathbf{z}_k - \mathbf{H}_k \hat{\mathbf{x}}_{k|k-1} \right) \tag{12}$$

$$\mathbf{F}_{k|k} = \mathbf{F}_{k|k-1} - \mathbf{K}_k \mathbf{H}_k \mathbf{F}_{k|k-1} \tag{13}$$

where, at time k, $\mathbf{A}_k$ is the state transition matrix, $\mathbf{K}_k$ is the Kalman gain matrix, and $\mathbf{H}_k$ is the measurement matrix. $\mathbf{F}_{k|k}$ and $\mathbf{F}_{k|k-1}$ denote the state covariance matrix estimates based on measurements up to times k and $k-1$. $\mathbf{Q}_k$ and $\mathbf{R}_k$ are the process and observation noise covariance matrices, respectively.

The authors of the first Kalman filter power system DSE approach [18] hinted at its compatibility with anomaly-detection methods, which, at the time, were being studied for SSE. Early work soon after [19,20] formulated bad data detection by analyzing the innovation process:

$$\mathbf{v}_k = \mathbf{y}_k - \mathbf{h}(\hat{\mathbf{x}}_{k|k-1}). \tag{14}$$

Additional approaches for bad data processing in DSE include asymmetry analysis based on the skewness of the normalized estimation error [17,21]. DSE anomaly detection research remains an active field [16,22], especially since dynamic load and generation profiles are commonplace in microgrid systems with distributed energy resources (DERs).

3. Bad Data Types and Considerations

Bad data can be classified as either single or multiple. For single bad data, one measurement in the system is corrupted with a large error. Multiple bad data describe more than one measurement being in error and can be further classified by the degree of interaction and conformity [7]. Multiple bad data are said to interact when the residuals are highly correlated, whereas conformity describes the degree to which gross errors are "masked" in the residual (i.e., nonconforming errors present as highly normalized residuals) [8]. Another illustration of how error is not always fully reflected in the residual is the concept of leverage points [23–26], which can hinder the effectiveness of the largest residual methods. Leverage points arise as a consequence of system topology, parameter values, and measurement placement and are usually caused by the following: (i) injection and flow measurements near branches with a small X/R ratio; (ii) injection measurements near buses with a large number of incident branches; and (iii) a measurement with a large weight [6,27]. Even a single leverage point can compromise bad data detectability.

Gross errors that exist beyond the acceptable noise limit of the state estimation model can be categorized into three types: measurement, parameter, and topology. Each of these errors suggests a discrepancy between the measurement data and model and are described further in the following.

3.1. Measurement Error

Measurement error is inevitable given the limitations of metering equipment accuracy. Meters can fail or degrade, introducing bias and compromising both accuracy and Gaussian error assumption: empirical studies of synchrophasor errors have yielded heavy-tailed error distributions such as Cauchy, Student's t, logistic, and Laplace [28,29]. Further, the communications infrastructure itself may contribute to measurement error in the case of failure or interference [7]. Particularly egregious measurement errors that suggest physically impossible grid conditions, such as negative bus voltage magnitudes or magnitudes several times larger or shorter than nominal values, are filtered through pre-processing [8], but more "agreeable" measurement errors can nevertheless affect the accuracy of state estimates.

3.2. Parameter Error

Parameter errors suggest discrepancies between measurement data and the system model. While Schweppe in his original formulation [1] did recognize the impact of erroneous model parameters, such errors were not considered in the network model. For example, a parameter error might arise when the variability in a line-impedance value due to extreme weather conditions is not taken into account. The mismatch between the measurement data and the line impedance database value, which is used in the Y-admittance matrix for power flow calculations, would be reflected in the state estimation result.

A simple alteration of (1) yields an augmented model [30] and linearization:

$$z_i = h_i(x, p_0) + e_i \approx h_i(x, p) + \frac{\partial h_i}{\partial p} \Delta p + e_i \tag{15}$$

where p is the true parameter value, p_0 is the erroneous parameter value, and $\Delta p = p_0 - p$ is the parameter error.

Stuart and Herget [31] investigated the impact of parameter errors on SSE by simulating erroneous values for line impedance, measurement error variance, and transformer tap settings. Of particular note was an observed relationship between the severity of error and lightly loaded lines.

Parameter errors can be thought of as a special case of multiple bad data in which only the measurements pertaining to the erroneous model parameter are in error. As such, studies have been performed with the goal of differentiating between the two. In [32], it was shown through analysis of the state estimation error distribution that parameter errors are reflected only in the measurement functions with erroneous parameter values. Parameter estimation itself has been treated as a process separate from state estimation. A practical implementation of this was first developed in [33], in which a sensitivity-based WLS estimation approach is used to both identify and estimate parameter error.

3.3. Topology Error

Like parameter errors, topology errors suggest discrepancies in the measurement model. System topology describes the bus-branch network configuration at the time of state estimation. Topology processing, which precedes state estimation, normally determines the correct status of manual switching and the circuit-breaking apparatus. A topological discrepancy, such as a branch outage unaccounted for by the topology processor, would be reflected in the Jacobian measurement matrix **H**, which requires accurate bus-branch connection logic for the calculation of power flow. Topology errors can significantly compromise state estimation accuracy through multiple conforming bad data [7]. Early work showed that such topology errors can be reflected in the state estimation error [34,35] and that normalized residual methods could be used for detection. Other approaches suggest incorporating the statuses of switching devices themselves as additional state variables [36], aiding in the *identification* of topology errors as such.

4. Bad Data Detection

To preserve the accuracy of state variable estimates, bad data must be detected, identified, and either eliminated or corrected. Whether the source of the bad data is measurement-, parameter-, or topology-based, detection is the first step. The classical components of bad data detection can be broadly categorized into three main branches and are often used in conjunction with one another: chi-square χ^2 testing, residual-based methods, and hypothesis testing.

4.1. Chi-Squared χ^2 Test

For a set of d random variables $\{X_i,\ i = 1, 2, \ldots, d\}$ with unit Gaussian distribution $X_i \sim \mathcal{N}(0,1)$, a new random variable with χ^2 distribution is defined as $Y = \sum_{i=1}^{d} X_i^2$ [6]. This follows the form of the cost function defined in (2) and can be written as the performance index [8]

$$J(\hat{\mathbf{x}}) = \sum_{i=1}^{d} \left(\frac{z_i - h_i(\hat{x})}{\sigma_i} \right)^2 \tag{16}$$

assuming that the measurement errors are independent and distributed $e_i \sim \mathcal{N}(0,\sigma^2)$. $J(\hat{\mathbf{x}})$ then follows a χ^2 distribution with $d - N$ degrees of freedom, where d is the number of measurements and N is the number of unknown state variables.

A critical value $C = \chi^2_{(d-N),p}$ can then be obtained based on the degrees of freedom $d - N$ and the desired detection confidence with probability $p = 1 - \alpha$, where α is a constraint on false probability. If $J(\hat{\mathbf{x}}) \geq C$, then bad data are suspected; otherwise, the measurements are assumed to be free of bad data. χ^2 testing has proved valuable for the detection of bad data even in the early history of SSE [5], where it was quickly realized that χ^2 and normalized residual methods can outperform one another generally, but that χ^2 often proved better for multiple bad data.

4.2. Residual-Based Methods

The χ^2 test soon became commonplace for the *detection* of bad data detection in WLS SSE for a specified constraint on false probability α, after which residual analysis could be performed for the *identification* of the measurement(s) in error [37]. However, in the case of single bad data in larger networks, the analysis of both the weighted and normalized residuals also proved viable for detection due to a more pronounced response in the presence of gross errors when compared to χ^2 testing. The use of normalized residuals for bad data detection was introduced in [5]. Using the residual covariance matrix $\Omega_{ii} = diag(\Omega)$, the normalized residuals can be defined

$$r_i^N = \frac{|r_i|}{\sqrt{\Omega_{ii}}} \tag{17}$$

It was shown in [5] that, after bad data had been detected through means such as the χ^2 test, a list of the normalized residuals in descending order could be obtained. The *largest* normalized residual could be used to identify the measurement in error, after which the measurement was removed and the state estimation re-run. If bad data were still detected, the procedure would repeat until all erroneous measurements were eliminated. Further techniques were developed to correct measurements contaminated with bad data, rather than eliminating them [8]. Correction keeps the measurement structure intact, which is especially important in cases of limited redundancy.

Both the detection and identification of bad data can be achieved without χ^2 testing by comparing the largest normalized residual to a statistical threshold depending on the desired sensitivity [7]. The case studies in [5] demonstrated that, in the case of multiple bad data, either interacting or noninteracting, no consensus could be developed as to whether χ^2 testing or the largest normalized residual test proved superior for bad data detection. A geometric interpretation of the normalized residuals was developed in [38], significantly improving the generalizability of multiple interacting bad data detection. The residual difference between estimated and actual measurements continues to be a vital component in state estimation anomaly detection, including in newer formulations to be expanded upon in Section 6.

4.3. Hypothesis Testing

Hypothesis testing is a statistical method for deciding between accepting a null hypothesis H_0 or an alternative hypothesis H_1 based on available observations. In power system state estimation, the hypotheses are formulated as follows:

H_0:　z_i is a valid measurement.
H_1:　z_i is a measurement in error.

The first work to use hypothesis testing identification (HTI) for bad data in power system state estimation [39] developed regions of acceptance between H_0 and H_1 by comparing the estimation error to a threshold dependent on the measurement standard deviation and a pre-selected constraint on false probability α. New results of this HTI method were presented in [40], where the optimality of the linear estimator is established along with a decision strategy based on a constraint for missed detection β. In [41], the authors bridge the gaps between theory and practice by implementing the HTI on eight test systems, showcasing its strengths in detecting multiple interacting bad data. For bad data identification, HTI methods show significant advantages over methods based on normalized residuals, which may be strongly correlated [7]. HTI techniques also demonstrated potential for discerning error type, such as in topology error identification [42,43].

5. When Bad Data Become Malicious

The introduction of the concept of false data injection attacks (FDIAs) [11] helped to highlight the limitations of classical bad-data-detection methods. What if bad data are malicious and/or statistically derived to avoid conventional detection? The basic idea of FDIAs is that an attacker can design an injection of multiple interacting bad data, which is then applied to the measurement vector $\mathbf{z}$. Consider the representation $\mathbf{z_a} = \mathbf{z} + \mathbf{a}$, where $\mathbf{a} = (a_1, a_2, \ldots, a_m)^T$ is a vector of malicious data. The attacker's goal is to design $\mathbf{a}$ to alter the state estimates, which EMSs use to make operating decisions, but without triggering bad data detection. Ramifications of undetected attacks include compromised system stability [12] and negative economic impact [44]. The success of such attacks is largely dependent on the information available to the attacker, such as the number of meters compromised, state estimates, system topology, and Jacobian structure, to name a few.

Denial-of-service (DoS) attacks are another source of mismatch between the measurement data fed to the state estimator and the true power system state. Causes for DoS attacks are numerous [45], including communication channel jamming, packet flooding, and compromising of metering devices such as SCADA and PMUs so that data are not updated for that region of the power grid. For state estimation, DoS attacks are typically modeled as a set of measurements that are no longer available, which can negatively impact state variable accuracy. If stealthiness is desired, care would need to be taken on the attacker's part so as not to render the system unobservable. FDIAs can also be designed to create a topology error attack [46–48], in which a conventionally nondetectable mismatch between measurement data and topology processing can lead to compromised system stability and cost-effective operation.

The authors of [49] present FDIA strategies from the attacker and defender perspectives. For the attacker, it is typically assumed that there is a cost associated with the information obtained. With this in mind, an algorithm is presented to find the minimal set of measuring devices required to manufacture an unobservable attack. In [50], a comparative analysis of the FDIA impact between so-called DC and AC SSE is conducted. DC SSE considers active power measurements only, with bus voltage angles as the state variables. In contrast, the complete AC SSE considers both active and reactive power measurements, with bus voltage magnitudes and angles as the state variables. Such a study was important due to the DC model warranting far more attention in the FDIA research space at the time, despite the full nonlinear AC model finding use in real-world EMS applications [51,52].

Impacts of FDIAs on Kalman filter DSE approaches were studied in [53], where it was found that the unscented Kalman filter (UKF) [54] yielded better performance compared to the extended Kalman filter (EKF) [55] and the enhanced EKF [56]. Further, an online nonparametric cumulative sum (CUSUM) approach was proposed to detect anomalies based on distribution changes in the state estimation error. This is related to quickest-change detection approaches, which will be elaborated upon further in Section 6.1. A Kalman filter state estimation approach was proposed in [57], where a Euclidean detector was used to overcome the shortcomings of the χ^2 test for detecting statistically derived FDIAs as well as DoS attacks.

The FDIA formulation highlighted a need for improved bad data detection. The classification of bad data as such would also need improvement. Common confusion matrix metrics like false negatives and false positives become harder to minimize when stealth FDIAs can closely resemble power system events like transients, switching, and sudden load changes. Further, with the increasing push towards the cyber-physical operation of the smart grid [58], many new points of entry for cyber-attack became apparent, such as Internet of Things (IoT) infrastructure [59], communication channels [60], and distributed computing [61]. The intersection of model-based and data-driven solutions should grow to better handle the bad data detection limitations posed by FDIAs. With state estimation anticipated to remain a vital component of EMSs, new formulations based on quickest-change detection and AI should be developed for improved anomaly detection.

6. Recent Approaches

6.1. Quickest-Change Detection

Quickest-change detection (QCD) is concerned with detecting a possible change in the distribution of a monitored observation sequence [62], which is indicative of an anomaly in a stochastic environment. The general goal of QCD theory is to design algorithms to detect these changes with the smallest detection delay possible, subject to a constraint on false alarms.

Three main ingredients are needed in the QCD problem [63]: an observed stochastic process $\{X_n,\ n = 1, 2, \ldots\}$, a change time τ^a at which the statistical properties of the process undergo change, and a decision maker that declares a change time τ^s based on observations of the stochastic process. A false alarm is defined as an instance of the decision maker declaring a change before the change occurs: $I\{\tau^s < \tau^a\}$. The constraint on false alarm follows from the Neyman–Pearson hypothesis testing formulation [64], which is foundational to the QCD problem.

The Neyman–Pearson Lemma [65] establishes the optimal test for binary hypothesis testing, involving the null (H_0) and alternate (H_1) hypotheses. For a single observation X:

H_0: X has pdf p.
H_1: X has pdf q.

Then, comparing the *likelihood ratio* $q(X)/p(X)$ to a threshold value is the most powerful test in terms of deciding which hypothesis is true while minimizing missed detection subject to a constraint on false alarms [66]. The likelihood ratio plays a fundamental role in recursive sequential-change-detection algorithms such as Page's CUSUM [67] and the Shiryaev–Roberts procedure [68], each of which enjoys optimality properties in terms of minimizing false alarm and detection delay $(\tau^s - \tau^a)_+ \max(0, \tau^s - \tau^a)$. These properties are given proper discussion in [62].

QCD approaches have shown great promise for power system anomaly detection applications, such as line outage detection and identification [69–71]. QCD has further application in detecting changes in the state estimation error, which has been proposed for fault and FDIA detection. The first QCD approach for state estimation FDIA detection implemented an adaptive approach using the CUSUM statistic:

$$S_n = \max\{0, S_{n-1} + L(Z_n)\}, \qquad n \geq 1, \ \text{with } S_0 = 0. \tag{18}$$

where $\{Z_n, \ n = 1, 2, \dots\}$ is the observed stochastic process and L is the log-likelihood ratio. Sample plots of a subtle change in a Gaussian observation process, along with the corresponding CUSUM statistic, are included in Figure 1.

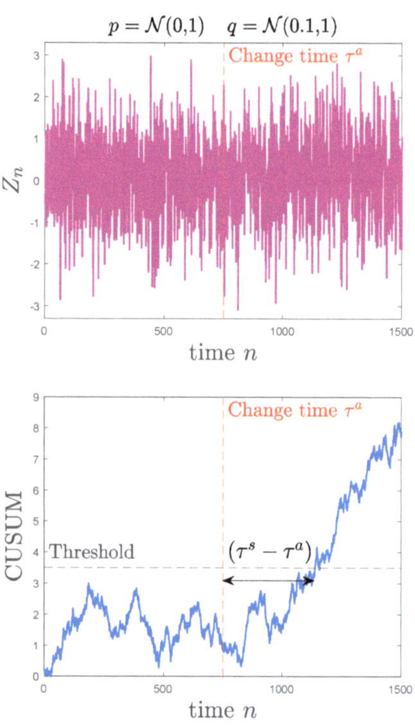

Figure 1. Example of a small mean shift observation sequence with the corresponding CUSUM evolution.

Because the exact form of the post-change distribution q is not known, the authors in [72,73] used a Rao test-based approximation [74] of the generalized likelihood ratio test for CUSUM-based FDIA detection. A low-complexity Orthogonal Matching Pursuit CUSUM (OMP-CUSUM) approach in [75] accounts for the unknown change distribution by maximizing the cumulative log-likelihood ratio to detect FDIAs that are sparse (i.e., only a small number of meters are assumed accessible to the attacker).

Both centralized and distributed CUSUM-based approaches for FDIA detection are proposed in [76], replacing the unknown parameters of the post-change distribution with their maximum likelihood estimates (MLEs). For the centralized case, the observed stochastic process of interest is the projection of the measurement vector on the orthogonal Jacobian space component $\mathcal{R}^{\perp}(\mathbf{H})$. This is expressed as $\tilde{\mathbf{y}}_n \triangleq \mathbf{P}_n\mathbf{y}_n$, where $\mathbf{P}$ is the previously defined linear projection matrix. The distributed case partitions the power system into areas and estimates the state variables through the alternating direction method of multipliers (ADMM) [77], where each area i has its own observed process $\{\tilde{\mathbf{y}}_n^i, \ n = 1, 2, \dots\}$. These approaches outperformed the adaptive-CUSUM approach in [72,73], due in part to the improved detection of FDIAs with negative and larger elements of the attack vector $\mathbf{a}$.

The work in [78] incorporates a Kalman filter approach and separately evaluates DoS attacks and FDIAs. Better detection performance was observed for stealth FDIAs in particular, in which perfect system topology knowledge allows an attacker to inject false data along the column space of **H**. Four Kalman filtering techniques in [53] were evaluated using nonparametric CUSUM, in which both pre- and post-change distributions p and q are unknown. Hybrid FDIA/jamming attacks are assessed for the Kalman filter CUSUM-based detector in [79]. The distinction between persistent and non-persistent attacks was made as well. Most CUSUM-based detectors assume persistence in the change in the observed stochastic process, and so an intermittent attack series could be designed to increase the detection delay. Thus, the Generalized Shewhart Test, which can detect significant increases in L, is presented as a countermeasure against stealthy, non-persistent FDIAs. A relaxed generalized CUSUM (RGCUSUM) algorithm is presented in [80] for FDIA detection. A relaxation on maximizing the post-change likelihood over the unknown parameters yielded a more computationally efficient algorithm than its generalized CUSUM counterpart. A normalized Rao CUSUM-based detector with a time-varying dynamic model was employed in [81] to better distinguish between FDIA and sudden load changes.

The work in [82] also assesses the Shiryaev–Roberts (SR) procedure, along with CUSUM for change detection. In contrast to CUSUM, the optimality of the SR procedure pertains to detecting τ at a distant time horizon [83,84]. The SR procedure is defined recursively as

$$T_n = \exp\big(L(Z_n)\big)\big[T_{n-1} + 1\big], \qquad n \geq 1, \ \ with \ T_0 = 0. \tag{19}$$

Further, the modified CUSUM and SR procedure algorithms [85] are employed in the same work as evaluation benchmarks for a so-called DeepQCD algorithm for online cyber-attack detection, which uses deep recurrent neural networks to detect changes in transient cases and with autocorrelated observations.

6.2. AI Approaches

FDIA detection can be framed as a binary classification problem in which the measurement vector $\mathbf{z}$ is determined to be either normal (negative class) or anomalous (positive class). One of the first to use semi-supervised and supervised learning for FDIA detection [86] explored perceptron, support vector machine (SVM), k-nearest neighbors (k NN), and sparse logistic regression algorithms for supervised learning. Semi-supervised learning, in which unlabelled test data are incorporated in training, was explored with semi-supervised SVMs. Many valuable takeaways were garnered from this work, including considerations of power system size and and computational complexity; however, stealthy FDIAs were not considered. An Extended Nearest Neighbors (ENN) algorithm was proposed in [87] to better handle the imbalanced data problem (i.e., cases in which the number of negative class samples greatly exceeds or is significantly less than the number of positive class samples). Classification performance was then compared to SVM and k-NN algorithms. The work in [88] used a method based on the margin-setting algorithm, typically used in image processing applications, in which hypersphere decision boundaries were formed through labeled PMU time-series data. The MSA approach yielded superior classification performance compared to standard artificial neural networks (ANNs) and SVM.

Unsupervised principal component analysis (PCA) showed utility in the construction of stealthy and blind FDIAs, as well as in developing robust detection methods [89,90]. PCA is again employed in [91] as a preprocessing step to project higher-dimensional correlated measurement data to a lower dimension, removing the correlation between data and magnifying the distance between normal and anomalous measurements. For performance comparison, the authors implemented a supervised distributed ADMM-based SVM, which could only outperform the PCA-based anomaly detection when the training set was large. Mahalanobis distance-based ensemble detection methods demonstrated success for FDIA detection in [92–95], including in high-fidelity real-time simulation.

Reinforcement learning (RL)-based QCD approaches are explored in [82,96]. The QCD problem can be formulated as a case of optimal stopping, in which a decision to exercise must be made to minimize cost [97,98]. In QCD, this is understood as declaring a stop time τ^s at a cost relative to the actual stop time τ^a. For the Markov Decision Process (MDP) component of RL, one can either seek to maximize reward or minimize cost [99]. Two components for the cost are constructed [97]: one for continuing (associated with missed detection) and one for stopping (associated with false alarm). The authors in [96] use a model-free state–action–reward–state–action (SARSA) approach to learn the expected future cost for each state–action pair in a Q-table. The authors opt for a quantization scheme for learning when faced with the continuous observation space. Because the actual change time τ^a is a hidden state, a partially observable Markov decision process (POMDP) formulation is used. This RL approach significantly outperformed the Euclidean [57] and cosine-similarity metric [100]-based detectors in terms of minimizing the mean probability of false alarm and detection delay for various cyber-attack types, including hybridFDI/jamming, DoS, and network topology attacks.

Neural network and deep learning approaches also show promise for malicious and standard bad data detection. A Deep-Belief-Network-based classifier is proposed in [101] using Conditional Gaussian–Bernoulli Restricted Boltzmann Machines in the hopes of revealing higher-dimensional temporal features of stealthy FDIAs. The temporal correlation between measurements with the state estimator is analyzed through Recurrent Neural Networks (RNNs) for FDIA detection in [102]. A nonlinear autoregressive exogenous (NARX) model configuration for ANNs is explored in [103] for stealthy optimized FDIA detection. The authors in [104] consider a limited set of target labels for attacked measurement data, an example of semi-supervised learning. Autoencoders, used for dimensionality reduction and feature extraction, are integrated into a generative adversarial network. The framework compensates for the limited labeled data set by using two neural networks: one generative, responsible for creating fake samples, and the other discriminative, responsible for distinguishing between real and generated samples.

7. Conclusions and Suggestions for Future Work

A survey of legacy bad-data-detection procedures has been presented along with limitations with respect to malicious bad data. Cyber-attack formulations such as FDIA highlight the need for better data detection by pointing out the theoretical manipulation of grid-operating procedures by bad actors. Even if one argues that the FDIA formulation is more of a theoretical exercise than a practical concern, it still points to shortcomings in legacy bad data detection. Standard bad data and physical line faults under the leverage point conditions discussed earlier are difficult to detect for similar reasons as statistically derived stealth FDIAs. Newer methods such as QCD and AI seek to overcome legacy bad-data-detection techniques by leveraging features such as measurement data temporal patterns and probability density changes in the state estimation error.

Increased access to real state estimation measurement data would aid greatly in accessing the practicality of QCD and AI anomaly-detection formulations. For example, a QCD formulation assuming independent and identically distributed (i.i.d.) observations may be compromised under dynamic load and generation profiles, in which case the measurement data exhibit complicating factors like autocorrelation, as investigated in [82]. The robustness of newer anomaly detection strategies to asynchronous measurement data should also be investigated. Until synchronized measurement data for state estimation become standard, uncertainty quantification of this type should considered so as not to be considered a false-positive source of anomalous behavior. The availability of time-series data such as SCADA and/or PMU measurements for multi-bus systems would aid state estimation researchers in quantifying uncertainty and measurement correlation. It is also recommended that future work incorporate dynamic load and generation profiles to better reflect the future directions of the modern smart grid. This was a motivation in the work [81], which highlighted the importance of discerning anomalies from dynamic behavior such as large load shifts. Such conditions are expected to increase with more DER penetration in the future smart grid and should be included when evaluating detection and identification metrics.

Author Contributions: Conceptualization, A.C. and A.B.; methodology, A.C.; software, A.C.; validation, A.C., A.B. and S.M.; formal analysis, A.C.; investigation, A.C.; resources, A.C.; data curation, A.C.; writing—original draft preparation, A.C.; writing—review and editing, A.B.; visualization, A.C.; supervision, A.B.; project administration, A.C. All authors have read and agreed to the published version of the manuscript.

Funding: This research was partially funded by the U.S. Department of Energy under Contract DE-AC05-76RL01830.

Data Availability Statement: No new data were created or analyzed in this study. Data sharing is not applicable to this article.

Conflicts of Interest: The authors declare no conflict of interest.

References

1. Schweppe, F.C.; Wildes, J. Power System Static-State Estimation, Part I: Exact Model. *IEEE Trans. Power Appar. Syst.* **1970**, *PAS-89*, 120–125. [CrossRef]
2. Filho, M.; da Silva, A.; Falcao, D. Bibliography on power system state estimation (1968–1989). *IEEE Trans. Power Syst.* **1990**, *5*, 950–961. [CrossRef]
3. Schellstede, G.; Beissler, G. A Software Package for Security Assessment Functions. In *Power Systems and Power Plant Control*; Pingyang, W., Ed.; IFAC Symposia Series; Pergamon: Oxford, UK, 1987; pp. 277–284. [CrossRef]
4. Merrill, H.M.; Schweppe, F.C. Bad Data Suppression in Power System Static State Estimation. *IEEE Trans. Power Appar. Syst.* **1971**, *PAS-90*, 2718–2725. [CrossRef]
5. Handschin, E.; Schweppe, F.; Kohlas, J.; Fiechter, A. Bad data analysis for power system state estimation. *IEEE Trans. Power Appar. Syst.* **1975**, *94*, 329–337. [CrossRef]
6. Monticelli, A. *State Estimation in Electric Power Systems: A Generalized Approach*; Springer: Berlin, Germany, 2012.
7. Abur, A.; Expósito, A.G. *Power System State Estimation: Theory and Implementation*, 1st ed.; CRC Press: Boca Raton, FL, USA, 2004. [CrossRef]
8. Bretas, A.; Bretas, N.; London, J.B., Jr.; Carvalho, B. *Cyber-Physical Power Systems State Estimation*; Elsevier: Amsterdam, The Netherlands, 2021.
9. Zhao, J.; Gómez-Expósito, A.; Netto, M.; Mili, L.; Abur, A.; Terzija, V.; Kamwa, I.; Pal, B.; Singh, A.K.; Qi, J.; et al. Power System Dynamic State Estimation: Motivations, Definitions, Methodologies, and Future Work. *IEEE Trans. Power Syst.* **2019**, *34*, 3188–3198. [CrossRef]
10. Phadke, A. Synchronized phasor measurements-a historical overview. In Proceedings of the IEEE/PES Transmission and Distribution Conference and Exhibition, Yokohama, Japan, 6–10 October 2002; Volume 1, pp. 476–479. [CrossRef]
11. Liu, Y.; Ning, P.; Reiter, M.K. False Data Injection Attacks against State Estimation in Electric Power Grids. *ACM Trans. Inf. Syst. Secur.* **2011**, *14*, 1–33. [CrossRef]
12. Musleh, A.S.; Chen, G.; Dong, Z.Y. A Survey on the Detection Algorithms for False Data Injection Attacks in Smart Grids. *IEEE Trans. Smart Grid* **2020**, *11*, 2218–2234. [CrossRef]
13. Bretas, N.G.; Bretas, A.S. The Extension of the Gauss Approach for the Solution of an Overdetermined Set of Algebraic Non Linear Equations. *IEEE Trans. Circuits Syst. Ii Express Briefs* **2018**, *65*, 1269–1273. [CrossRef]
14. Kalman, R.E.; Bucy, R.S. New Results in Linear Filtering and Prediction Theory. *J. Basic Eng.* **1961**, *83*, 95–108. [CrossRef]
15. Zhao, J.; Singh, A.K.; Mir, A.S.; Taha, A.; Rouhani, A.; Gomez-Exposito, A.; Meliopoulos, A.; Pal, B.; Kamwa, I.; Qi, J.; et al. *Power System Dynamic State and Parameter Estimation-Transition to Power Electronics-Dominated Clean Energy Systems: IEEE Task Force on Power System Dynamic State and Parameter Estimation*; IEEE: Piscataway, NJ, USA, 2021.
16. Liu, Y.; Singh, A.K.; Zhao, J.; Meliopoulos, A.P.S.; Pal, B.; Ariff, M.A.b.M.; Van Cutsem, T.; Glavic, M.; Huang, Z.; Kamwa, I.; et al. Dynamic State Estimation for Power System Control and Protection. *IEEE Trans. Power Syst.* **2021**, *36*, 5909–5921. [CrossRef]
17. Bretas, N. An iterative dynamic state estimation and bad data processing. *Int. J. Electr. Power Energy Syst.* **1989**, *11*, 70–74. [CrossRef]
18. Debs, A.S.; Larson, R.E. A Dynamic Estimator for Tracking the State of a Power System. *IEEE Trans. Power Appar. Syst.* **1970**, *PAS-89*, 1670–1678. [CrossRef]
19. Nishiya, K.I.; Takagi, H.; Hasegawa, J.; Koike, T. Dynamic state estimation for electric power systems—introduction of a trend factor and detection of innovation processes. *Electr. Eng. Jpn.* **1976**, *96*, 79–87. [CrossRef]
20. Nishiya, K.; Hasegawa, J.; Koike, T. Dynamic state estimation including anomaly detection and identification for power systems. *IEE Proc. Gener. Transm. Distrib.* **1982**, *129*, 192–198. [CrossRef]
21. Bretas, A.S.; Bretas, N.G.; Massignan, J.A.D.; London Junior, J.B.A. Hybrid Physics-Based Adaptive Kalman Filter State Estimation Framework. *Energies* **2021**, *14*, 6787. [CrossRef]

22. Jin, Z.; Zhao, J.; Ding, L.; Chakrabarti, S.; Gryazina, E.; Terzija, V. Power system anomaly detection using innovation reduction properties of iterated extended kalman filter. *Int. J. Electr. Power Energy Syst.* **2022**, *136*, 107613. [CrossRef]
23. Mili, L.; Phaniraj, V.; Rousseeuw, P. Least median of squares estimation in power systems. *IEEE Trans. Power Syst.* **1991**, *6*, 511–523. [CrossRef] [PubMed]
24. Celik, M.; Abur, A. A robust WLAV state estimator using transformations. *IEEE Trans. Power Syst.* **1992**, *7*, 106–113. [CrossRef]
25. Majumdar, A.; Pal, B.C. Bad Data Detection in the Context of Leverage Point Attacks in Modern Power Networks. *IEEE Trans. Smart Grid* **2018**, *9*, 2042–2054. [CrossRef]
26. Mili, L.; Cheniae, M.; Vichare, N.; Rousseeuw, P. Robust state estimation based on projection statistics [of power systems]. *IEEE Trans. Power Syst.* **1996**, *11*, 1118–1127. [CrossRef]
27. Zhao, J.; Mili, L. Vulnerability of the Largest Normalized Residual Statistical Test to Leverage Points. *IEEE Trans. Power Syst.* **2018**, *33*, 4643–4646. [CrossRef]
28. Wang, S.; Zhao, J.; Huang, Z.; Diao, R. Assessing Gaussian Assumption of PMU Measurement Error Using Field Data. *IEEE Trans. Power Deliv.* **2018**, *33*, 3233–3236. [CrossRef]
29. Huang, C.; Thimmisetty, C.; Chen, X.; Stewart, E.; Top, P.; Korkali, M.; Donde, V.; Tong, C.; Min, L. Power Distribution System Synchrophasor Measurements with Non-Gaussian Noises: Real-World Data Testing and Analysis. *IEEE Open Access J. Power Energy* **2021**, *8*, 223–228. [CrossRef]
30. Zarco, P.; Exposito, A. Power system parameter estimation: A survey. *IEEE Trans. Power Syst.* **2000**, *15*, 216–222. [CrossRef]
31. Stuart, T.A.; Herczet, C.J. A Sensitivity Analysis of Weighted Least Squares State Estimation for Power Systems. *IEEE Trans. Power Appar. Syst.* **1973**, *PAS-92*, 1696–1701. [CrossRef]
32. Bretas, A.S.; Bretas, N.G.; Carvalho, B.E. Further contributions to smart grids cyber-physical security as a malicious data attack: Proof and properties of the parameter error spreading out to the measurements and a relaxed correction model. *Int. J. Electr. Power Energy Syst.* **2019**, *104*, 43–51. [CrossRef]
33. Liu, W.H.E.; Lim, S.L. Parameter error identification and estimation in power system state estimation. *IEEE Trans. Power Syst.* **1995**, *10*, 200–209. [CrossRef]
34. Costa, I.; Leao, J. Identification of topology errors in power system state estimation. *IEEE Trans. Power Syst.* **1993**, *8*, 1531–1538. [CrossRef]
35. Wu, F.; Liu, W.H. Detection of topology errors by state estimation (power systems). *IEEE Trans. Power Syst.* **1989**, *4*, 176–183. [CrossRef]
36. Korres, G.N.; Manousakis, N.M. A state estimation algorithm for monitoring topology changes in distribution systems. In Proceedings of the 2012 IEEE Power and Energy Society General Meeting, San Diego, CA, USA, 22–26 July 2012; pp. 1–8. [CrossRef]
37. Koglin, H.J.; Neisius, T.; Beißler, G.; Schmitt, K. Bad data detection and identification. *Int. J. Electr. Power Energy Syst.* **1990**, *12*, 94–103. [CrossRef]
38. Clements, K.A.; Davis, P.W. Multiple Bad Data Detectability and Identifiability: A Geometric Approach. *IEEE Trans. Power Deliv.* **1986**, *1*, 355–360. [CrossRef]
39. Cutsem, T.V.; Ribbens-Pavella, M.; Mili, L. Hypothesis Testing Identification: A New Method For Bad Data Analysis In Power System State Estimation. *IEEE Trans. Power Appar. Syst.* **1984**, *PAS-103*, 3239–3252. [CrossRef]
40. Mili, L.; Van Cutsem, T.; Ribbens-Pavella, M. Decision Theory Applied to Bad Data Identification in Power System State Estimation. In Proceedings of the 7th IFAC/IFORS Symposium on Identification and System Parameter Estimation, York, UK, 3–7 July 1985; Volume 18, pp. 945–950. [CrossRef]
41. Mili, L.; Van Cutsem, T. Implementation of the hypothesis testing identification in power system state estimation. *IEEE Trans. Power Syst.* **1988**, *3*, 887–893. [CrossRef]
42. Lourenco, E.; Costa, A.; Clements, K. Bayesian-based hypothesis testing for topology error identification in generalized state estimation. *IEEE Trans. Power Syst.* **2004**, *19*, 1206–1215. [CrossRef]
43. Wu, W.B.; Cheng, M.X.; Gou, B. A Hypothesis Testing Approach for Topology Error Detection in Power Grids. *IEEE Int. Things J.* **2016**, *3*, 979–985. [CrossRef]
44. Xie, L.; Mo, Y.; Sinopoli, B. Integrity Data Attacks in Power Market Operations. *IEEE Trans. Smart Grid* **2011**, *2*, 659–666. [CrossRef]
45. Amin, S.; Cárdenas, A.A.; Sastry, S.S. Safe and Secure Networked Control Systems under Denial-of-Service Attacks. In Proceedings of the Hybrid Systems: Computation and Control, San Francisco, CA, USA, 13–15 April 2009; Majumdar, R., Tabuada, P., Eds.; Springer: Berlin/Heidelberg, Germany, 2009; pp. 31–45.
46. Kim, J.; Tong, L. On Topology Attack of a Smart Grid: Undetectable Attacks and Countermeasures. *IEEE J. Sel. Areas Commun.* **2013**, *31*, 1294–1305. [CrossRef]
47. Liu, X.; Li, Z. Local Topology Attacks in Smart Grids. *IEEE Trans. Smart Grid* **2017**, *8*, 2617–2626. [CrossRef]
48. Liang, G.; Weller, S.R.; Luo, F.; Zhao, J.; Dong, Z.Y. Generalized FDIA-Based Cyber Topology Attack with Application to the Australian Electricity Market Trading Mechanism. *IEEE Trans. Smart Grid* **2018**, *9*, 3820–3829. [CrossRef]
49. Kosut, O.; Jia, L.; Thomas, R.J.; Tong, L. Malicious Data Attacks on Smart Grid State Estimation: Attack Strategies and Countermeasures. In Proceedings of the 2010 First IEEE International Conference on Smart Grid Communications, Gaithersburg, MD, USA, 4–6 October 2010; pp. 220–225. [CrossRef]

50. Hug, G.; Giampapa, J.A. Vulnerability Assessment of AC State Estimation with Respect to False Data Injection Cyber-Attacks. *IEEE Trans. Smart Grid* **2012**, *3*, 1362–1370. [CrossRef]

51. Nuthalapati, S. State Estimation Performance Monitoring. 2015. Available online: https://www.nerc.com/pa/rrm/Resources/Monitoring_and_Situational_Awareness_Conference1/10 (accessed on 3 June 2023).

52. ETAP State Estimation Software. 2015. Available online: https://etap.com/product/state-estimation-software (accessed on 3 June 2023).

53. Yang, Q.; Chang, L.; Yu, W. On false data injection attacks against Kalman filtering in power system dynamic state estimation. *Secur. Commun. Netw.* **2016**, *9*, 833–849. [CrossRef]

54. Valverde, G.; Terzija, V. Unscented Kalman filter for power system dynamic state estimation. *Iet Gener. Transm. Distrib.* **2011**, *5*, 29–37. [CrossRef]

55. Ghahremani, E.; Kamwa, I. Dynamic State Estimation in Power System by Applying the Extended Kalman Filter with Unknown Inputs to Phasor Measurements. *IEEE Trans. Power Syst.* **2011**, *26*, 2556–2566. [CrossRef]

56. Shih, K.R.; Huang, S.J. Application of a robust algorithm for dynamic state estimation of a power system. *IEEE Trans. Power Syst.* **2002**, *17*, 141–147. [CrossRef]

57. Manandhar, K.; Cao, X.; Hu, F.; Liu, Y. Detection of Faults and Attacks Including False Data Injection Attack in Smart Grid Using Kalman Filter. *IEEE Trans. Control. Netw. Syst.* **2014**, *1*, 370–379. [CrossRef]

58. Faheem, M.; Shah, S.; Butt, R.; Raza, B.; Anwar, M.; Ashraf, M.; Ngadi, M.; Gungor, V. Smart grid communication and information technologies in the perspective of Industry 4.0: Opportunities and challenges. *Comput. Sci. Rev.* **2018**, *30*, 1–30. [CrossRef]

59. Yilmaz, Y.; Uludag, S. Mitigating IoT-based Cyberattacks on the Smart Grid. In Proceedings of the 2017 16th IEEE International Conference on Machine Learning and Applications (ICMLA), Cancun, Mexico, 18–21 December 2017; pp. 517–522. [CrossRef]

60. Yan, Y.; Qian, Y.; Sharif, H.; Tipper, D. A Survey on Cyber Security for Smart Grid Communications. *IEEE Commun. Surv. Tutor.* **2012**, *14*, 998–1010. [CrossRef]

61. Kurt, M.N.; Yılmaz, Y.; Wang, X. Secure Distributed Dynamic State Estimation in Wide-Area Smart Grids. *IEEE Trans. Inf. Forensics Secur.* **2020**, *15*, 800–815. [CrossRef]

62. Xie, L.; Zou, S.; Xie, Y.; Veeravalli, V.V. Sequential (Quickest) Change Detection: Classical Results and New Directions. *IEEE J. Sel. Areas Inf. Theory* **2021**, *2*, 494–514. [CrossRef]

63. Veeravalli, V.V.; Banerjee, T. Quickest-change detection. In *Academic Press Library in Signal Processing*; Elsevier: Amsterdam, The Netherlands, 2014; Volume 3, pp. 209–255. [CrossRef]

64. Poor, H.V. *An Introduction to Signal Detection and Estimation*, 2nd ed.; Springer Texts in Electrical Engineering; Springer: New York, NY, USA, 1994.

65. Neyman, J.; Pearson, E.S. IX. On the problem of the most efficient tests of statistical hypotheses. *Philos. Trans. R. Soc. London Ser. Contain. Pap. Math. Phys. Character* **1933**, *231*, 289–337. [CrossRef]

66. Moulin, P.; Veeravalli, V.V. *Statistical Inference for Engineers and Data Scientists*; Cambridge University Press: Cambridge, UK, 2018. [CrossRef]

67. Page, E.S. Continuous Inspection Schemes. *Biometrika* **1954**, *41*, 100–115. [CrossRef]

68. Polunchenko, A.S.; Tartakovsky, A.G. On Optimality of the Shiryaev-Roberrts Procedure for Detecting a Change in Distribution. *Ann. Stat.* **2010**, *38*, 3445–3457. [CrossRef]

69. Banerjee, T.; Chen, Y.C.; Dominguez-Garcia, A.D.; Veeravalli, V.V. Power system line outage detection and identification—A quickest change detection approach. In Proceedings of the 2014 IEEE International Conference on Acoustics, Speech and Signal Processing (ICASSP), Florence, Italy, 4–9 May 2014; pp. 3450–3454. [CrossRef]

70. Rovatsos, G.; Jiang, X.; Domínguez-García, A.D.; Veeravalli, V.V. Comparison of statistical algorithms for power system line outage detection. In Proceedings of the 2016 IEEE International Conference on Acoustics, Speech and Signal Processing (ICASSP), Shanghai, China, 20–25 March 2016; pp. 2946–2950. [CrossRef]

71. Yang, X.; Chen, N.; Zhai, C. A Control Chart Approach to Power System Line Outage Detection Under Transient Dynamics. *IEEE Trans. Power Syst.* **2021**, *36*, 127–135. [CrossRef]

72. Huang, Y.; Li, H.; Campbell, K.A.; Han, Z. Defending false data injection attack on smart grid network using adaptive CUSUM test. In Proceedings of the 2011 45th Annual Conference on Information Sciences and Systems, Baltimore, MD, USA, 23–25 March 2011; pp. 1–6. [CrossRef]

73. Huang, Y.; Tang, J.; Cheng, Y.; Li, H.; Campbell, K.A.; Han, Z. Real-Time Detection of False Data Injection in Smart Grid Networks: An Adaptive CUSUM Method and Analysis. *IEEE Syst. J.* **2016**, *10*, 532–543. [CrossRef]

74. De Maio, A. Rao Test for Adaptive Detection in Gaussian Interference with Unknown Covariance Matrix. *IEEE Trans. Signal Process.* **2007**, *55*, 3577–3584. [CrossRef]

75. Akingeneye, I.; Wu, J. Low Latency Detection of Sparse False Data Injections in Smart Grids. *IEEE Access* **2018**, *6*, 58564–58573. [CrossRef]

76. Li, S.; Yılmaz, Y.; Wang, X. Quickest Detection of False Data Injection Attack in Wide-Area Smart Grids. *IEEE Trans. Smart Grid* **2015**, *6*, 2725–2735. [CrossRef]

77. Kekatos, V.; Giannakis, G.B. Distributed Robust Power System State Estimation. *IEEE Trans. Power Syst.* **2013**, *28*, 1617–1626. [CrossRef]

78. Kurt, M.N.; Yılmaz, Y.; Wang, X. Distributed Quickest Detection of Cyber-Attacks in Smart Grid. *IEEE Trans. Inf. Forensics Secur.* **2018**, *13*, 2015–2030. [CrossRef]

79. Kurt, M.N.; Yılmaz, Y.; Wang, X. Real-Time Detection of Hybrid and Stealthy Cyber-Attacks in Smart Grid. *IEEE Trans. Inf. Forensics Secur.* **2019**, *14*, 498–513. [CrossRef]

80. Zhang, J.; Wang, X. Low-Complexity quickest-change detection in Linear Systems with Unknown Time-Varying Pre- and Post-Change Distributions. *IEEE Trans. Inf. Theory* **2021**, *67*, 1804–1824. [CrossRef]

81. Nath, S.; Akingeneye, I.; Wu, J.; Han, Z. Quickest Detection of False Data Injection Attacks in Smart Grid with Dynamic Models. *IEEE J. Emerg. Sel. Top. Power Electron.* **2022**, *10*, 1292–1302. [CrossRef]

82. Kurt, M.N. Data-Driven Quickest-Change Detection. Ph.D. Thesis, Columbia University, New York, NY, USA, 2020. [CrossRef]

83. Moustakides, G.V.; Polunchenko, A.S.; Tartakovsky, A.G. Numerical Comparison of CUSUM and Shiryaev–Roberts Procedures for Detecting Changes in Distributions. *Commun. Stat. Theory Methods* **2009**, *38*, 3225–3239. [CrossRef]

84. Pollak, M.; Tartakovsky, A.G. Exact optimality of the Shiryaev-Roberts procedure for detecting changes in distributions. In Proceedings of the 2008 International Symposium on Information Theory and Its Applications, Auckland, New Zealand, 7–10 December 2008; pp. 1–6. [CrossRef]

85. Polunchenko, A.S.; Raghavan, V. Comparative performance analysis of the Cumulative Sum chart and the Shiryaev-Roberts procedure for detecting changes in autocorrelated data. *Appl. Stoch. Model. Bus. Ind.* **2018**, *34*, 922–948. [CrossRef]

86. Ozay, M.; Esnaola, I.; Yarman Vural, F.T.; Kulkarni, S.R.; Poor, H.V. Machine Learning Methods for Attack Detection in the Smart Grid. *IEEE Trans. Neural Netw. Learn. Syst.* **2016**, *27*, 1773–1786. [CrossRef]

87. Yan, J.; Tang, B.; He, H. Detection of false data attacks in smart grid with supervised learning. In Proceedings of the 2016 International Joint Conference on Neural Networks (IJCNN), Vancouver, BC, Canada, 24–29 July 2016; pp. 1395–1402. [CrossRef]

88. Wang, Y.; Amin, M.M.; Fu, J.; Moussa, H.B. A Novel Data Analytical Approach for False Data Injection Cyber-Physical Attack Mitigation in Smart Grids. *IEEE Access* **2017**, *5*, 26022–26033. [CrossRef]

89. Yu, Z.H.; Chin, W.L. Blind False Data Injection Attack Using PCA Approximation Method in Smart Grid. *IEEE Trans. Smart Grid* **2015**, *6*, 1219–1226. [CrossRef]

90. Hao, J.; Piechocki, R.J.; Kaleshi, D.; Chin, W.H.; Fan, Z. Sparse Malicious False Data Injection Attacks and Defense Mechanisms in Smart Grids. *IEEE Trans. Ind. Inform.* **2015**, *11*, 1–12. [CrossRef]

91. Esmalifalak, M.; Liu, L.; Nguyen, N.; Zheng, R.; Han, Z. Detecting Stealthy False Data Injection Using Machine Learning in Smart Grid. *IEEE Syst. J.* **2017**, *11*, 1644–1652. [CrossRef]

92. Trevizan, R.D.; Ruben, C.; Nagaraj, K.; Ibukun, L.L.; Starke, A.C.; Bretas, A.S.; McNair, J.; Zare, A. Data-driven Physics-based Solution for False Data Injection Diagnosis in Smart Grids. In Proceedings of the 2019 IEEE Power & Energy Society General Meeting (PESGM), Atlanta, GA, USA, 4–8 August 2019; pp. 1–5. [CrossRef]

93. Ruben, C.; Dhulipala, S.; Nagaraj, K.; Zou, S.; Starke, A.; Bretas, A.; Zare, A.; McNair, J. Hybrid data-driven physics model-based framework for enhanced cyber-physical smart grid security. *IET Smart Grid* **2020**, *3*, 445–453. [CrossRef]

94. Nagaraj, K.; Zou, S.; Ruben, C.; Dhulipala, S.; Starke, A.; Bretas, A.; Zare, A.; McNair, J. Ensemble CorrDet with adaptive statistics for bad data detection. *IET Smart Grid* **2020**, *3*, 572–580. [CrossRef]

95. Vega-Martinez, V.; Cooper, A.; Vera, B.; Aljohani, N.; Bretas, A. Hybrid Data-Driven Physics-Based Model Framework Implementation: Towards a Secure Cyber-Physical Operation of the Smart Grid. In Proceedings of the 2022 IEEE International Conference on Environment and Electrical Engineering and 2022 IEEE Industrial and Commercial Power Systems Europe (EEEIC / I&CPS Europe), Prague, Czech Republic, 28 June–1 July 2022; pp. 1–5. [CrossRef]

96. Kurt, M.N.; Ogundijo, O.; Li, C.; Wang, X. Online Cyber-Attack Detection in Smart Grid: A Reinforcement Learning Approach. *IEEE Trans. Smart Grid* **2019**, *10*, 5174–5185. [CrossRef]

97. Tsitsiklis, J.; van Roy, B. Optimal stopping of Markov processes: Hilbert space theory, approximation algorithms, and an application to pricing high-dimensional financial derivatives. *IEEE Trans. Autom. Control.* **1999**, *44*, 1840–1851. [CrossRef]

98. Chen, S.; Devraj, A.M.; Bušić, A.; Meyn, S. Zap Q-Learning for Optimal Stopping. In Proceedings of the 2020 American Control Conference (ACC), Denver, CO, USA, 1–3 July 2020; pp. 3920–3925. [CrossRef]

99. Meyn, S. *Control Systems and Reinforcement Learning*; Cambridge University Press: Cambridge, UK, 2022. [CrossRef]

100. Chen, P.Y.; Yang, S.; McCann, J.A.; Lin, J.; Yang, X. Detection of false data injection attacks in smart-grid systems. *IEEE Commun. Mag.* **2015**, *53*, 206–213. [CrossRef]

101. He, Y.; Mendis, G.J.; Wei, J. Real-Time Detection of False Data Injection Attacks in Smart Grid: A Deep Learning-Based Intelligent Mechanism. *IEEE Trans. Smart Grid* **2017**, *8*, 2505–2516. [CrossRef]

102. Ayad, A.; Farag, H.E.Z.; Youssef, A.; El-Saadany, E.F. Detection of false data injection attacks in smart grids using Recurrent Neural Networks. In Proceedings of the 2018 IEEE Power & Energy Society Innovative Smart Grid Technologies Conference (ISGT), Washington, DC, USA, 19–22 February 2018; pp. 1–5. [CrossRef]

103. Ganjkhani, M.; Fallah, S.N.; Badakhshan, S.; Shamshirband, S.; Chau, K.w. A Novel Detection Algorithm to Identify False Data Injection Attacks on Power System State Estimation. *Energies* **2019**, *12*, 2209. [CrossRef]

104. Zhang, Y.; Wang, J.; Chen, B. Detecting False Data Injection Attacks in Smart Grids: A Semi-Supervised Deep Learning Approach. *IEEE Trans. Smart Grid* **2021**, *12*, 623–634. [CrossRef]

Article

How to Decarbonize Greece by Comparing Wind and PV Energy: A Land Eligibility Analysis

Qilin Wang [1], Evangelia Gontikaki [2], Peter Stenzel [3], Vasilis Louca [1], Frithjof C. Küpper [1,4,5,*] and Martin Spiller [6,*]

[1] School of Biological Sciences, University of Aberdeen, Cruickshank Building, St. Machar Drive, Aberdeen AB24 3UU, Scotland, UK; q.wang2.22@abdn.ac.uk (Q.W.); v.louca@abdn.ac.uk (V.L.)

[2] Institute of Geoenergy, Foundation for Research and Technology Hellas, 73100 Chania, Greece; egontikaki@ig.forth.gr

[3] Cologne Institute for Renewable Energy (CIRE), Faculty of Process Engineering, Energy and Mechanical Systems, TH Köln, Betzdorfer Str. 2, 50679 Köln, Germany; peter.stenzel@th-koeln.de

[4] Marine Biodiscovery Centre, Department of Chemistry, University of Aberdeen, Aberdeen AB24 3UE, Scotland, UK

[5] Department of Biology, San Diego State University, San Diego, CA 92182-4614, USA

[6] ISATEC GmbH, Rathausstraße 10, 52072 Aachen, Germany

* Correspondence: fkuepper@abdn.ac.uk (F.C.K.); m.spiller@isatec-aachen.de (M.S.)

Abstract: To achieve sustainable development, the energy transition from lignite burning to renewable energy resources for electric power generation is essential for Greece. Wind and solar energy have emerged as significant sources in this transition. Surprisingly, numerous studies have examined the potential for onshore wind based on land eligibility, while few studies on open-field photovoltaic (PV) installations have been conducted. Therefore, based on the Specific Framework for Spatial Planning and Sustainable Development for Renewable Energy Sources (SFSPSD-RES), along with insights from previous relevant studies, this work conducts a land eligibility analysis of onshore wind and open-field PV installations in Greece using the software Geospatial Land Availability for Energy Systems (GLAES 1.2.1) and ArcGIS 10.2. Additionally, through an in-depth exploration of wind and solar PV energy potential in decommissioned lignite mines integrated with wind power density (WPD) and global horizontal irradiation (GHI) maps, this study compares the suitability of wind versus solar as energy sources for the decarbonization of Greece. Overall, despite the greater spatial eligibility for onshore wind turbines compared to open-field PV power plants, the relatively lower wind energy potential and operational limitations of wind turbines lead to the study's conclusion that solar energy (PV) is more suitable for the decarbonization of Greece.

Keywords: land eligibility; renewable energy resources; onshore wind; open-field PV; GIS

Citation: Wang, Q.; Gontikaki, E.; Stenzel, P.; Louca, V.; Küpper, F.C.; Spiller, M. How to Decarbonize Greece by Comparing Wind and PV Energy: A Land Eligibility Analysis. *Energies* **2024**, *17*, 567. https://doi.org/10.3390/en17030567

Academic Editors: Ahmed Abu-Siada and Adam Cenian

Received: 26 September 2023
Revised: 30 December 2023
Accepted: 3 January 2024
Published: 24 January 2024

1. Introduction

In order to keep the global average temperature rise well below 2 K above pre-industrial levels and to pursue efforts to limit the temperature rise to 1.5 K, the Paris Climate Agreement was signed by 196 countries in 2015 [1]. Greece, as one of the signatory countries, has set national energy policies in different periods to achieve a phased sustainable development. Since the early 1960s, Greece has met most of its electric power demand from thermal power stations burning either lignite on the mainland or heavy fuel oil on the islands. Based on the European Energy Policy, the goal of renewable energy penetration in 2020 (Law 3851/2020) formulated at the national level in Greece in 2010 marked the start of the Greek energy transition [2]. As Figure 1 shows, Greek electric power generation shifted from coal to natural gas by 2020 and wind and solar PV gradually became important sources of electric power generation recently [3]. In terms of power generation resources for Greece, solar accounted for 10.7% of the total installed capacity, while wind accounted for 23% of the total installed capacity in 2023 [4]. For the country's sustainable development, further electric power generation from renewable energy sources (RESs) is needed. The Greek National Energy and Climate Plan (NECP) aims to deploy renewable energy generation systems on the islands, interconnect them with the electricity grid on the mainland, and phase out most of the heavy fuel oil-based power generation on the islands by 2030. Another policy, the National Climate Law, enacted in May 2022, set targets to achieve a 55% reduction in the total greenhouse gas emissions by 2030, an 80% reduction by 2040, and ultimately reach net zero emissions by 2050 [3]. It also outlined the essential emissions reduction measures and set a binding target to end lignite power generation by 2028.

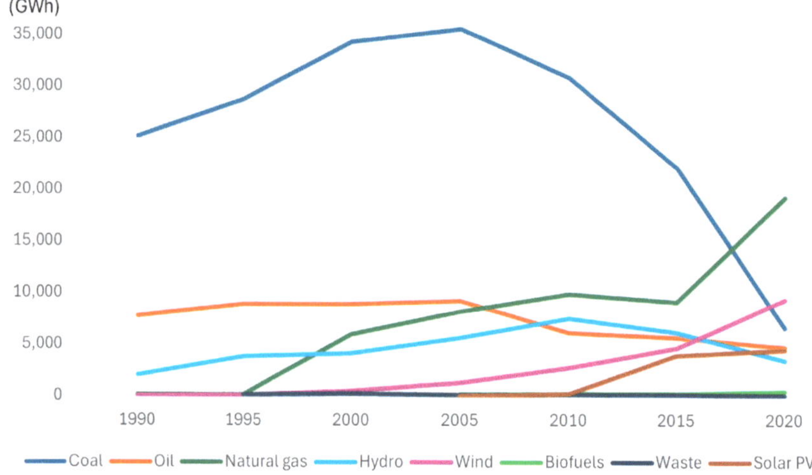

Figure 1. Electric power generation by variable sources in Greece from 1990 to 2020 [5].

It is evident that the need for RESs to achieve energy transition in Greece is growing rapidly. According to the National Climate Law [3], the energy transition plan of Greece focuses on wind and solar PV. Therefore, it is very timely and topical to study Greece's wind and PV potential. While there are a plethora of studies about wind potential using the Geographic Information System (GIS) for Greece [6–10], relatively few studies, such as the one by Vagiona [11], have been conducted about the country's PV potential.

This work aims to explore the prospective wind and PV energy prospects in Greece by identifying the optimal locations for wind turbines and PV installations using geospatial data. The approach contrasts with those of previous studies [6,8–11] in Greece, which did not incorporate economic criteria into the land eligibility analysis, thereby providing a more comprehensive evaluation of the viability of RESs. It also aims to compare the electric-generating capacity of the existing wind turbines and PV power plants at their present state with the predicted wind and PV potential to assess the most suitable RESs for Greece, either wind or solar. Recognizing their untapped value, while mining areas are normally excluded from such studies, special attention is dedicated to the potential of wind and PV in former lignite mining areas that could markedly increase the efficiency of RESs in Greece. This will not only broaden the scope of land use for energy production but also support the transition to sustainable development by turning neglected sites into productive assets.

2. Materials and Methods

2.1. Study Areas

Greece, a country located in southeastern Europe, consists of 13 regions with a total area of about 132,000 km^2. By 2023, according to statistics [12], the total population of Greece was 10,497,595 inhabitants and, among the more than 2000 Greek islands, only about 170 islands were inhabited. Although many uninhabited islands host established nature reserves for flora and fauna, other areas that meet the designated standards are worthy of becoming sites of investment in renewable energy due to the abundance of wind and solar resources [10]. Based on the Archaeological Cadastre of Greece [13], the current inventory reveals the presence of 220 museums, 420 historical sites, 844 protected areas, 3100 archaeological sites, and 17,000 monuments in the entire territory of the country. In the established network of nature protection areas in the European Union, Greece has a total of 446 protected areas, including 239 Special Areas of Conservation (SAC) and 181 Special Protection Areas for birds (SPAs), with the remaining 26 areas falling into both categories [14].

In this work, the operations were based on a shapefile of Greece [15]. Before the land eligibility analysis, Natura 2000 areas were excluded first according to SFSPSD-RES [16]. In order to determine the land position that can support the construction of wind turbines and PV power plants in Greece, the open-source tool GLAES [17] and the Prior datasets containing typical criteria for variable RESs were utilized. A geographic analysis was conducted to assess the wind and solar potential in Greece. This analysis involved the utilization of the Digital Elevation Model of Greece obtained from the European Environment Agency Digital Elevation Model (EU-DEM) [18]. Additionally, WPD data at a height of 100 m were obtained from the Global Wind Atlas [19], while GHI data were obtained from the Global Solar Atlas [20]. These datasets were employed to compare the approximate wind and solar potential in Greece using a Geographic Information System digital platform, specifically ArcGIS 10.8 (Esri, Aylesbury, UK).

2.2. Current State Analysis

At present, Greece has established many wind and PV farms based on previous or recently designated standards. Data affecting the construction of wind turbines and PV power plants under the current conditions were analyzed to develop a series of land constraints for the suitable placing of wind turbines and PV power plants in Greece. Meanwhile, the data of 2023 from the Geoinformation Map of the Regulatory Authority for Energy [21] were used to present the existing construction areas of wind and solar energy. Subsequently, the areas with installed wind turbines and PV power plants were combined with the DEM map and mean WPD and GHI maps of Greece to understand the topography of the current construction locations and assess the current electric-generating capacity of Greece.

2.3. Geospatial Land Availability for Energy Systems (GLAES)

Land eligibility is a process that evaluates the suitability of a land parcel for implementing a specific technology based on a predetermined set of exclusion constraints and serves as a fundamental and widely utilized procedure through which geospatial criteria shape the distribution patterns across a given region [22]. Since not all open fields are eligible for the installation of wind turbines and PV power plants, land eligibility analysis based on geospatial data is an essential step before analyzing RES potential.

GLAES is based on Python 3 language, providing a simple and efficient way to analyze land eligibility using the Prior datasets [17]. Ryberg et al. [23] examined more than 50 literature sources that independently conducted a land eligibility analysis for prevalent variable RES technologies, documenting the approaches and frequencies used in defining the criteria. In this study, 28 typical criteria were identified, which included the distance from settlements and the distance from airports. Meanwhile, depending on the underlying motivations driving their exclusion, the identified criteria were divided into 4 distinct groups, namely, socio-political, physical, conservation, and economic. The typical criteria were further subdivided into multiple sub-criteria, for example, the exclusion distance from settlements is different for urban and rural areas. Finally, a collection of standardized datasets called Prior was developed [23], which defined common criteria related to variable RESs in the European context. There were 46 Priors in total, each of which represented the values of a criterion or sub-criterion across Europe and was georeferenced using the EPSG:3035 spatial reference system at a spatial resolution of 100 m by 100 m. The comprehensive documentation of each Prior dataset can be found in the work of Ryberg et al. [23]. The GLAES model and the Prior datasets can be obtained via GitHub [17]. The threshold of each criterion in the Prior datasets was determined considering the social, technical, environmental, and economic factors specific to Greece. These are discussed in Sections 2.4 and 2.5.

2.4. Onshore Wind Land Eligibility Analysis

For onshore wind energy, there are already many studies on land eligibility considering different series of land constraints. For instance, the European Environment Agency conducted a land eligibility analysis of onshore wind only excluding protected areas, such as Natura 2000, and found that the available area for onshore wind is 85.3% in Europe [24]. In contrast, other studies opted to apply multiple land constraints to analyze land eligibility. McKenna et al. [25] and Eurek et al. [26] both selected agricultural areas, settlement areas, protected areas, forests, waterbodies, slope, and elevation as land constraints, although the set thresholds for each exclusion constraint were different. Eurek et al. [26] found that 40% of the area in Europe is eligible for wind energy. In comparison, in addition to the land constraints mentioned above, McKenna et al. [25] also excluded buffer areas around airports, harbors, roads, and railways and used higher resolution maps, revealing that 23% of land surface in Europe is eligible. Moreover, Ryberg et al. [27] reviewed 53 land eligibility studies to develop a set of common land constraints for Europe considering social, technical, environmental, and economic factors and found that the total eligible area amounts to 1,352,260 km^2 for onshore wind turbines in Europe overall, which includes an eligible area of 28,326 km^2 in Greece. However, according to the specific circumstances of different countries, there will be certain differences in the formulated exclusion criteria and thresholds. At the national level in Greece, a Special Framework for the Spatial Planning and the Sustainable Development of Renewable Energy Sources (SFSPSD-RES) was formulated, which in addition to considering common land constraints, such as settlements and protected areas, also excluded archaeological reserves and considered visual factors, i.e., the esthetic impact of wind turbines on the landscape [16]. Several subsequent studies on wind farms' site selection in Greece were based on this framework. Tsoutsos et al. [7] conducted a study in Crete and found that 2517 km^2 on the island are available. Latinopoulos and Kechagia [8] also conducted a study in the region of Kozani based on SFSPSD-RES, but they excluded the areas where the average wind speed is below 4.5 m/s and the slope above 25%, concluding that there are 550 km^2 available for wind farms.

In this study, the series of land constraints that were used to analyze the land eligibility of onshore wind in Greece are summarized in Table 1. These were based on SFSPSD-RES, a previous study conducted in Greece [16], and a generalized land constraints list for onshore wind developed by Ryberg [27]. In order to estimate the eligible area and distributions for onshore wind turbines, a reference wind turbine was used (Table 2). The parameters of this reference wind turbine correspond to the Vestas V136 wind turbines available at present and to technology changes for future wind turbines by 2050 [28].

Table 1. Land constraints for onshore wind turbines.

Constraint	Threshold	Data Source
Socio-political:		
Distance from Rural Settlements	>500 m	CLC [29]
Distance from Urban Settlements	>1000 m	EuroStat [30]
Distance from Roadways	>120 m	Open Street Map [31]
Distance from Railways	>120 m	Open Street Map [31]
Distance from Power Lines	>120 m	Open Street Map [31]
Distance from Touristic Areas	>1500 m	Open Street Map [31]
Distance from Airports	>3000 m	CLC [29], EuroStat [30]
Distance from Agricultural Areas	Excluded	CLC [29]
Physical:		
Terrain Slope	<17°	EuroDEM [18]
Distance from Coastlines	>1500 m	CLC [29]
Distance from Sandy Areas	Excluded	CLC [29]
Distance from Woodlands	Excluded	CLC [29]
Distance from Waterbodies	Excluded	CLC [29]
Distance from Rivers	Excluded	EuroStat [30]
Distance from Wetlands	Excluded	CLC [29]
Conservation:		
Distance from Natural Monuments	>800 m	WDPA [32]
Distance from Parks	>800 m	WDPA [32]
Distance from Landscapes	>1000 m	WDPA [32]
Natura 2000 Areas	Excluded	Natura 2000 [14]
Economic:		
Wind Speed	<4 m/s	Global Wind Atlas [19]
Access Distance	<5 km	Open Street Map [31]
Connection Distance	<20 km	Open Street Map [31]

Table 2. Summary of the parameters of the reference wind turbine [28].

Parameter	Value
Hub Height	120 m
Rotor Diameter	136 m
Capacity	4200 kW
Specific Power	289 W/m^2

In the socio-political group of criteria, SFSPSD-RES set a safe distance to urban (population > 2000 inhabitants) and rural (population < 2000 inhabitants) settlements considering noise and safety factors that could cause a negative impact on society [33]. At the same time, it clearly stipulated that wind turbines must be installed at least 1500 m away from touristic areas. For safety reasons, wind turbines should be installed at a distance from airports due to the possible interference with aviation radar signals [33]. Also, they should be installed at a certain distance from roads, railways, and power lines, but to reduce transportation and transmission cost, the distance should not be too large [34]. Land use covers, such as agricultural, industrial, and mining areas, were excluded in related studies of Greece [6–8]. However, this work sought to explore the RES potential of deserted and decommissioned lignite mines in Greece without excluding mining sites.

In the physical group, slope was assessed as the third most significant factor affecting the construction of wind turbines in the study of Karamountzou and Vagiona [10]. Less steep land provides a better access to construct and maintain wind turbines [33,35]; therefore, the present study excluded areas with slopes greater than 17°. According to SFSPSD-RES [16], wind turbines are not allowed to be constructed in the areas of sand, wetland, and woodland and should be constructed 1500 m away from the coastline. Waterbodies and rivers are often covered by protected area status and are important for the functioning of biodiversity and ecosystems [36]. Therefore, waterbodies and rivers were excluded in this work.

Natura 2000 areas are explicitly excluded by SFSPSD-RES. However, Natura 2000 status only includes protected bird and habitat areas. Considering the cultural environment and heritage protection of Greece, it is worth setting a certain exclusion threshold around protected natural monuments, parks, and landscapes [8].

In order not to reduce the performance and increase the cost of construction and maintenance of wind turbines, the study by Karamountzou and Vagiona [10] combined technology and economics to evaluate the criteria, and found four important economic criteria, one of which was slope and the remainder were wind velocity, access distance (the distance from accessible roadways), and connection distance (the distance from a power line). Wind speed is an extremely important factor affecting the operation of wind turbines. The wind turbine starts to work when the wind speed reaches a certain value, at which the wind speed is called cut-in speed [28]. According to the power curve of the reference turbine, the cut-in speed is 4 m/s [28]. Therefore, areas where the average wind speed is below 4 m/s were excluded. Under the condition of ensuring that the wind turbines are at a certain safe distance from roadways and power lines, the distance should not be too large, because it will lead to increased construction, maintenance, and electricity production and transmission costs [9,35]. Finally, the thresholds for access and connection distance were determined based on the study by Ryberg et al. [27].

2.5. Open-Field PV Land Eligibility Analysis

While there are many studies on the land eligibility analysis of onshore wind, studies on the suitable construction of open-field PV power plants are relatively few. However, the analysis of land eligibility for PV in open areas adopts the analysis method of multi-criteria exclusion, considering the factors of society, technology, environment, and economy as well as that of onshore wind. A detailed study of European wind and solar energy potential by Ryberg [28] presented a consilient list that included 26 criteria that could be applied as exclusion constraints to select eligible areas for PV at the country level of Europe and found that the area eligible for open-field PV is 294,851 km^2 in Europe and the eligible area in Greece is 11,740 km^2. Although these established exclusion criteria cannot be fully generalized for open-field PV studies in a specific country, it lays the foundation for related studies in Europe. Based on exclusion criteria listed by Ryberg [28], Tlili et al. [37] conducted a literature review [38] about the areas that need to be excluded for the construction of PV power plants in France and found that the potential area for PV in France is 40,694 km^2. Likewise, on the basis of the general exclusion criteria list, Maestre et al. [39] reduced the threshold for some criteria, such as adjusting the distance from settlements areas from 200 m to 100 m, because the work was based on a hypothetical framework favorable to Spanish decarbonization goals between 2030 and 2050, and added some criteria in line with Spanish national conditions, such as historical sites. Finally, Maestre et al. [39] found that there was 143,820 km^2 eligible for open-field PV panels in Spain. Few studies have been conducted on the suitable sites of open-field PV in Greece. Vagiona [11] conducted a study on Rhodes Island (Greece), which considered 6 exclusion criteria, such as land cover, distance from protected areas, and altitude, based on SFSPSD-RES and showed that nine sites were eligible for open-field PV on Rhodes Island without mentioning the total eligible area. Unlike other studies [37,39], the criteria chosen for excluding land constraints in the study by Vagiona [11] did not include the slope and northward slope factors, which could lead to a poor performance of the PV panels due to shading [28].

Although the SFSPSD-RES of Greece provides exclusion criteria for wind turbines' construction and many related studies provide detailed land constraints and thresholds based on this framework, the exclusion criteria for PV power plants have not been elaborated. Therefore, in this section, the analysis of open-field PV land eligibility for Greece was mainly based on some criteria for the site selection of PV parks mentioned in the SFSPSD-RES and the detailed study by Ryberg [28] on analyzing PV potential in Europe

(Table 3). Meanwhile, as Greece is very close to Turkey and the climate of the two countries is similar, this section combined relevant studies conducted in Turkey to provide a more informative analysis. Additionally, a specific PV panel (Table 4) was used as a reference to determine the eligible areas for open-field PV power plants and to model their distribution in decommissioned lignite mines. The selected PV panel offered the optimal representation of distribution, while ensuring the highest number of full load hours [28].

Table 3. Land constraints for open-field PV.

Constraint	Threshold	Data Source
Socio-political:		
Distance from Settlements	>200 m	CLC [29]
Distance from Airports	>2000 m	CLC [29], EuroStat [30]
Distance from Industrial Areas	Excluded	CLC [29]
Distance from Agricultural Areas	Excluded	CLC [29]
Physical:		
Terrain Slope: Total	<10°	EuroDEM [18]
Terrain Slope: Northward	<3°	EuroDEM [18]
Distance from Elevation	<1500 m	EuroDEM [18]
Distance from Waterbodies	>500 m	CLC [29]
Distance from Woodlands	Excluded	CLC [29]
Distance from Wetlands	Excluded	CLC [29]
Distance from Coastlines	Excluded	CLC [29]
Distance from Sandy Areas	Excluded	CLC [29]
Conservation:		
Distance from Natural Monuments	>200 m	WDPA [32]
Distance from Landscapes	Excluded	WDPA [32]
Distance from Parks	Excluded	WDPA [32]
Natura 2000 Areas	Excluded	Natura 2000 [14]
Economic:		
Connection Distance	<20 km	Open Street Map [31]
Access Distance	<10 km	Open Street Map [31]

Table 4. Summary of the parameters of the reference open-field PV panels, Winaico WSx-240P6 [28].

Parameter	Value
Max Power	240.4 W
Area	1.663 m^2
Technology	Polycrystalline

Initially, according to SFSPSD-RES [16], the construction of PV parks is prohibited in areas of agriculture, wetlands, forests, natural monuments, protected landscapes, national parks, and Natura 2000 areas.

In the group of socio-political criteria, PV power plants should be built close enough to residential areas to provide a better energy demand and lower costs of electricity transmission without affecting the lives of residents [40]. In order to avoid accidents caused by the reflection of PV panels, based on the study by Vagiona [10], airports and the surrounding area within 2000 m were excluded. Most studies [11,28,41,42] considered some land covers, such as operational industrial and mining areas, as exclusion criteria since activities on them can stain PV panels leading to an inefficient performance. The same applies for land eligibility for onshore wind; however, this study did not exclude mining sites.

In physical criteria, in addition to considering the overall terrain slope (steep terrain can increase the construction cost), the northward slope should be also considered, because the self-shading losses of PV panels can be significantly high even with only slightly north-facing slopes [28]. Constructing PV power plants in high-altitude areas will increase the cost of the installation and transportation of materials; therefore, it is suggested to set the exclusion threshold of elevation to 1500 m [11]. At the same time, in order not to pollute environmental water resources when constructing PV panels, Ryberg's study [28] suggested that waterbodies and surrounding areas up to 500 m should be excluded. According to several studies [41–43], since the topography and geological structure of sands and beaches are not suitable for constructing PV power plants, sandy and beach areas were excluded.

Two significant economic indicators were considered in this section, namely, connection distance (the distance to power lines) and access distance (the distance to accessible roads). It is suggested that PV power plants should be constructed as close as possible to power lines, since the farther away from the power lines, the higher the cost and loss of electricity power transmission [44]. Moreover, it is also necessary to ensure that the distance between PV stations and the road is not too far, because a longer distance will increase the transportation cost of construction and the cost of operation and maintenance [45]. In summary, based on Ryberg's study [28], this section excluded the areas that were more than 20 km away from power lines and more than 10 km away from roads.

2.6. Deserted Lignite Mine Potential

According to the NECP, the Greek government has set a goal to completely eliminate coal-fired power generation in the country by 2028 [46]. Based on this plan, some lignite power stations, such as Kaida I and II as well as Amyndeo I and II, have been shut down since 2019 [47]. However, for Greece, a country long dependent on lignite for power generation, such an ideal and complete energy transition is difficult. In order to stabilize the electricity power supply system for a future entirely powered by natural gas and RESs, five lignite-burning power stations in Agios Dimitrios, Meliti, and Megalopolis are scheduled to have their operations extended to 2025 by the Public Power Corporation (PPC) in Greece [48]. It is obvious that the phasing out of lignite power generation is inevitable based on the energy transition plan of the Greek government. In the context of the shutdown of lignite-fired stations, lignite mines will be decommissioned, with potentially large areas without conservation value becoming available for other uses, such as renewable power generation. Meanwhile, there is an excellent connection to the power grid at these locations, which can be used by the newly installed PV parks. Therefore, it is important to analyze the RES potential of former lignite mines in Greece.

In this section, a lignite mining site in Megalopoli (Figure 2) and two lignite mining sites in Ptolemaida (Figure 3) were chosen as the study areas. The Ptolemaida Mine of Western Macedonia located in the northern part of Greece is the largest lignite mine in the whole of Greece, followed by what is the largest lignite mine in the Peloponnese Peninsula of Southern Greece, the Megalopoli Mine [49]. Due to the extension of operation at the Megalopoli lignite power station until 2025 and the expected closure of the Ptolemaida lignite power station in 2028, lignite mines in both regions are still in use. However, with the mandatory and inevitable shutdown of lignite power station in Megalopoli, the adjacent mine will be decommissioned in parallel, since lignite is not suitable for long-distance transport.

Without excluding mining sites using GLAES, this section aimed to explore the wind and solar energy potential in the mining areas of Megalopoli and Ptolemaida combined with WPD and GHI. Meanwhile, a reference wind turbine (Table 2) and solar panels (Table 4) were used to assess the renewable energy potential of these open-pit mines after their decommissioning. The separation distance for wind turbines was based on 8D × 4D, where D represents the rotor diameter of the turbine. The separation distance for PV parks was 1000 m.

Figure 2. Location and region of the Megalopoli lignite mine.

Figure 3. Location and region of the Ptolemaida lignite mine.

3. Results

3.1. Current State of Wind Turbines and PV Power Plants

Figure 4 shows the current distribution of installed wind turbines in Greece, with the areas featuring wind turbines highlighted in red. The total area occupied by wind turbines is 2095 km², accounting for 1.5% of Greece's total land area.

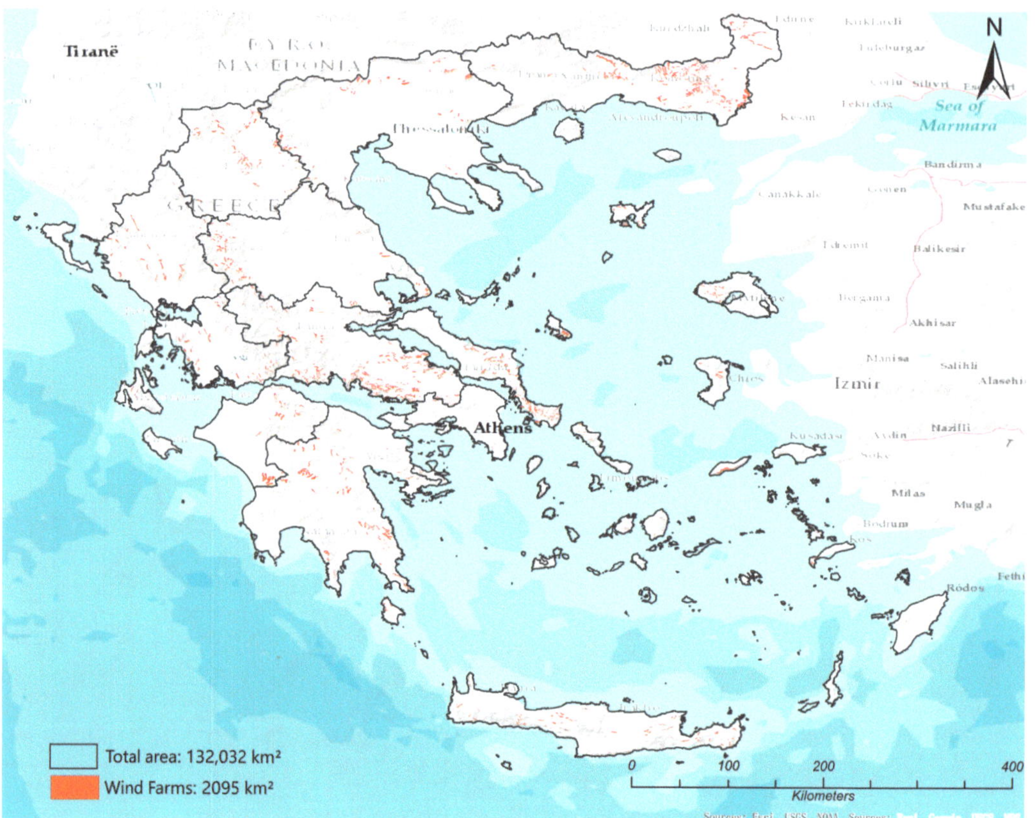

Figure 4. Currently installed wind farms' area in Greece.

By the end of 2022, the cumulative installed wind power capacity in Greece reached 4681.4 MW [50]. Wind turbines are predominantly distributed in Thrace, Western Macedonia, Central Greece, and the Peloponnese. Meanwhile, the altitude at which the wind farms were constructed is depicted in Figure 5a. The color of an area corresponds to its altitude, with high altitudes being represented by lighter shades and lower altitude by darker shades. The altitude values adhere to this color scheme in the subsequent diagrams related to DEM. Additionally, Figure 5b presents a WPD map combined with the locations of the existing wind farms. In this map, areas with a redder hue indicate higher mean WPD values, whereas areas with lighter shades of red correspond to lower mean WPD values. It is worth noting that the majority of the wind turbines were installed at relatively high altitudes, as shown in Figure 5a, specifically on mountain ridges. These areas exhibit higher mean WPD values in Greece, as indicated in Figure 5b, and higher average WPD values signify more abundant and favorable wind resources [19].

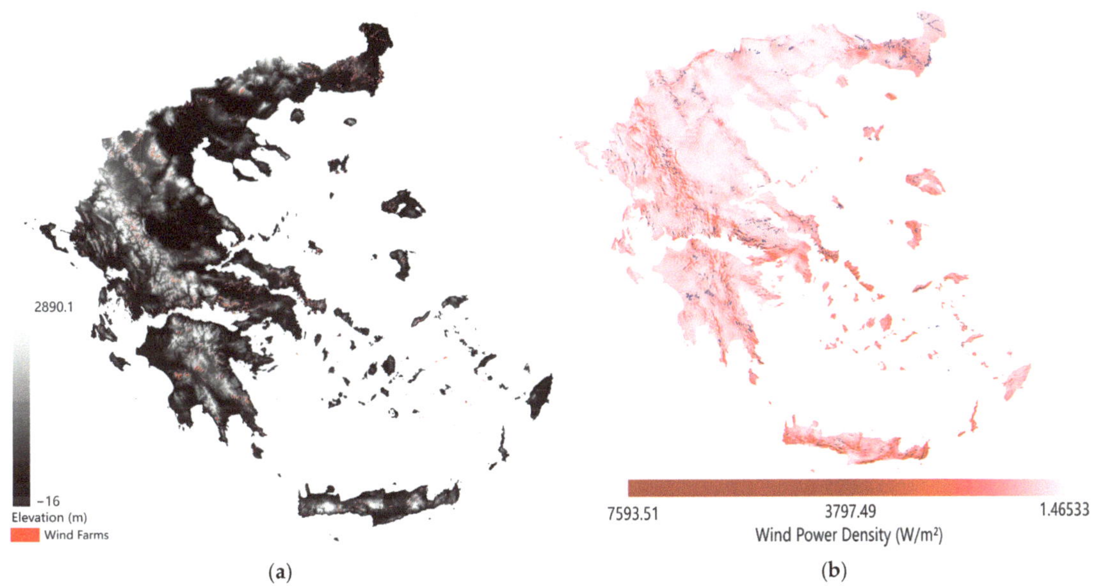

Figure 5. (**a**) The current wind farms combined with the DEM map of Greece. (**b**) Mean WPD map of Greece with the installed wind farms, where the blue areas indicate the areas with wind turbines.

Figure 6 displays the current areas in Greece where PV parks have been constructed. The total land area occupied by PV parks is 1984 km^2, accounting for 1.5% of the total land area of Greece. According to Petrova [51], the cumulative capacity of PV parks in Greece reached 5488 MWp in 2023. Meanwhile, it can be noted that the current PV parks are predominantly situated in Western Macedonia and Thessaly, where they are constructed on flat terrain with a relatively low elevation, as shown in Figure 7a. Since the GHI values can accurately quantify the solar energy potential for PV [52], higher GHI values correspond to a greater solar energy potential. Therefore, PV parks are strategically constructed in areas with relatively high GHI values, as shown in Figure 7b.

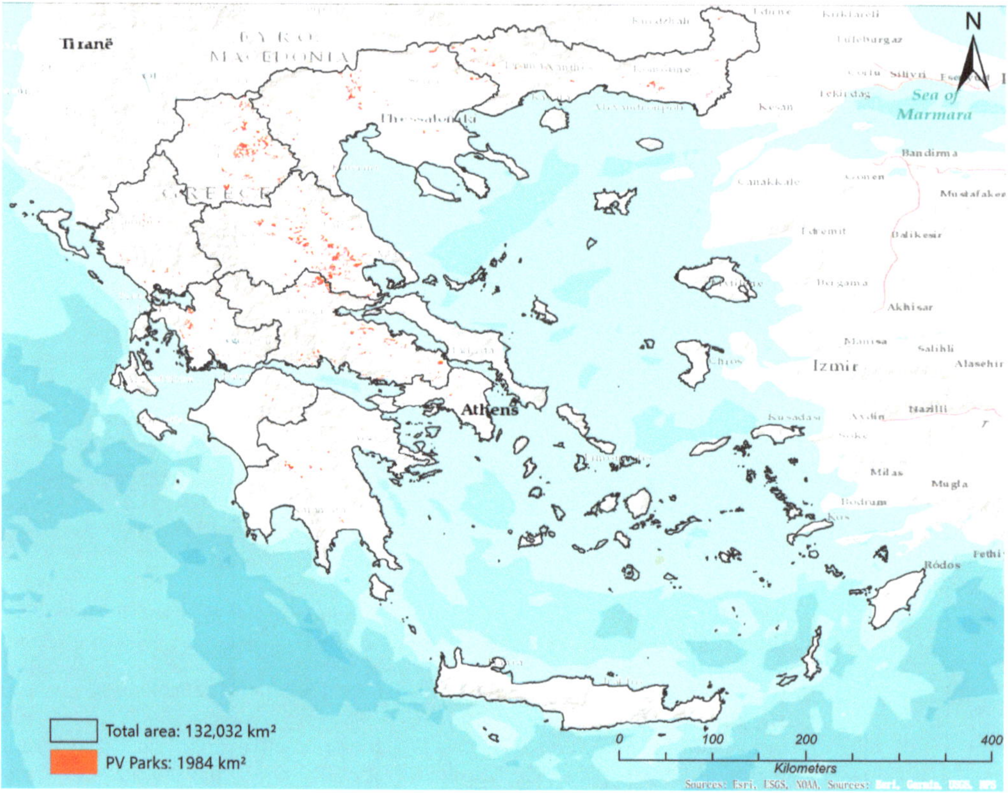

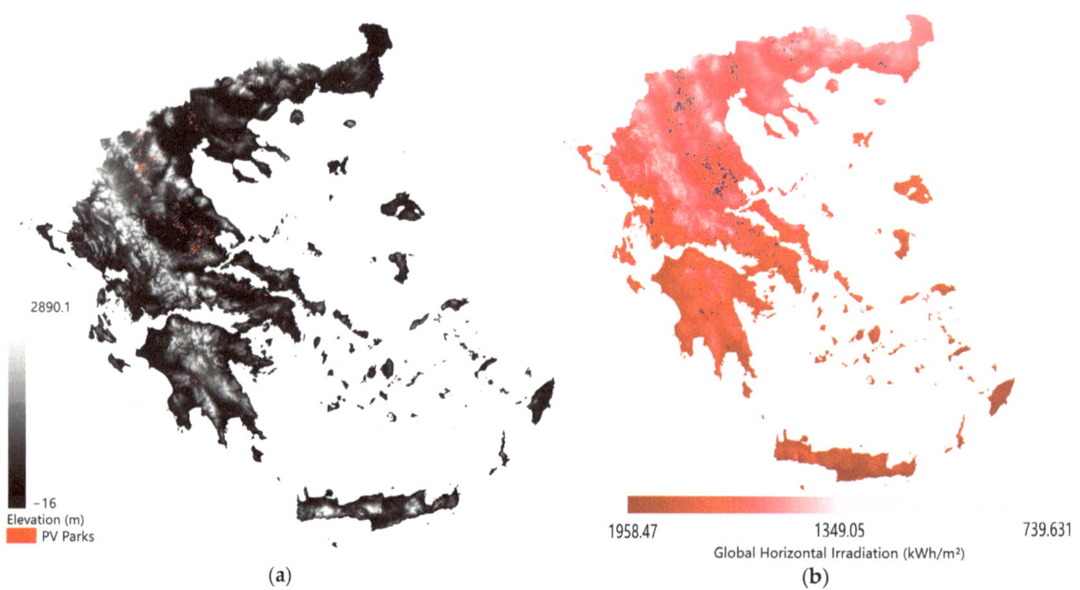

Figure 6. Areas of the currently installed PV parks in Greece.

Figure 7. (**a**) The current PV parks combined with the DEM map of Greece. (**b**) Mean GHI of Greece with the installed PV parks, where the blue areas indicate the areas with PV parks.

3.2. Onshore Wind Potential

As stated in Section 3.4, the Natura 2000 areas were excluded from the start of analysis. The remaining socio-political, physical, conservation, and economic constraints were excluded sequentially using GLAES (Figure 8). Upon the exclusion of the socio-political criteria, the land eligibility for onshore wind turbines in Greece was determined to be 44.58%. However, after excluding the physical criteria, the land eligibility ratio dramatically decreased to 16.21% and further dropped to 15.45% after excluding the conservation criteria. Finally, after excluding the economic criteria, the land eligibility for onshore wind turbines in Greece was determined to be 12.16%. Figure 8 clearly indicates that the physical criteria significantly impact the feasibility of constructing onshore wind turbines in Greece. Furthermore, among the economic criteria, the slope is the factor that most affects land eligibility; if the slope criterion is not excluded, land eligibility for onshore wind in Greece increases to 19.9%.

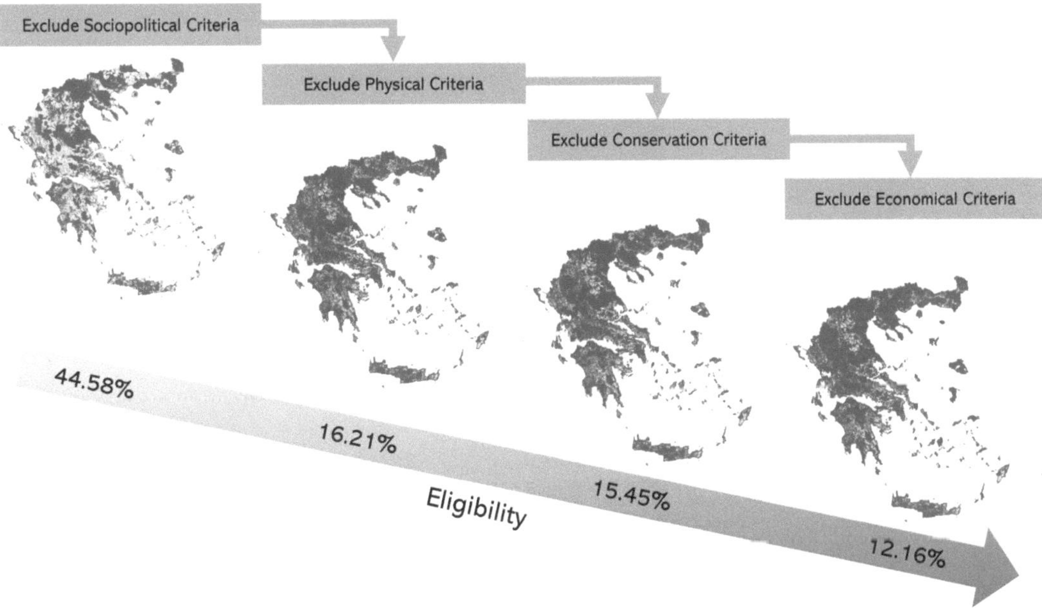

Figure 8. Land eligibility prediction process for onshore wind in Greece.

To provide a visual representation of the land eligibility for onshore wind turbines, Figure 9 presents a map where the eligible area constitutes 16,055 km^2, accounting for 12.16% of the total Greek area. Additionally, Figure 10 combines the predicted wind turbine locations with the mean WPD map of Greece. These figures demonstrate that the projected wind turbine sites are concentrated in Western Macedonia and Northern Thessaly, the eastern part of Central Greece, the Peloponnese, and the islands of Crete and Rhodes. As mentioned in Section 4.1, areas with a higher mean WPD are predominantly situated in mountainous regions with high altitudes. However, most eligible areas predicted by GLAES are distributed across flat terrain with relatively low mean WPD values. Only a few areas with a high mean WPD, such as the island of Crete and the junction area of the southern part of Central Greece and Attica, can support wind turbine construction.

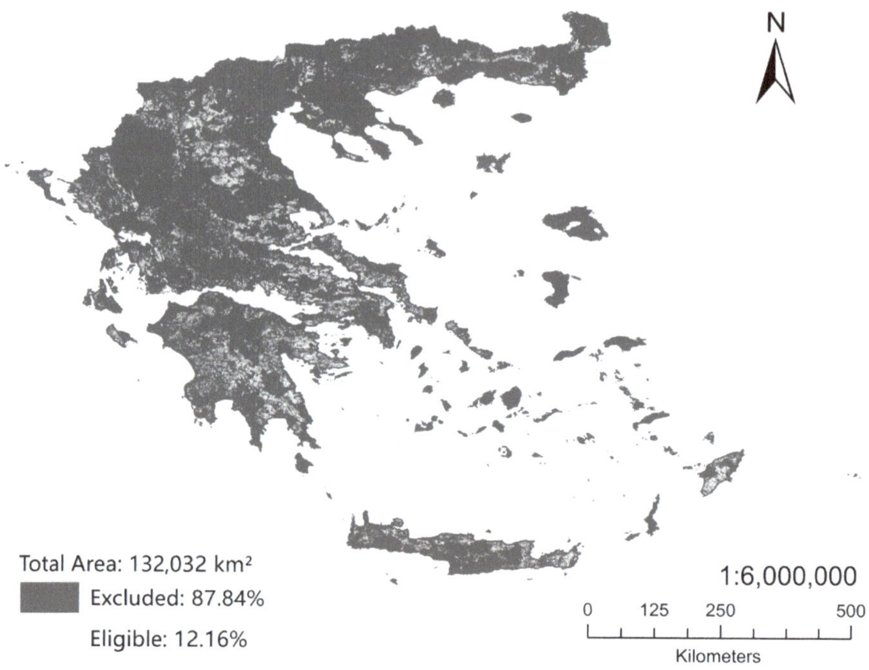

Figure 9. Onshore wind land eligibility excluding the land constraints from Table 1.

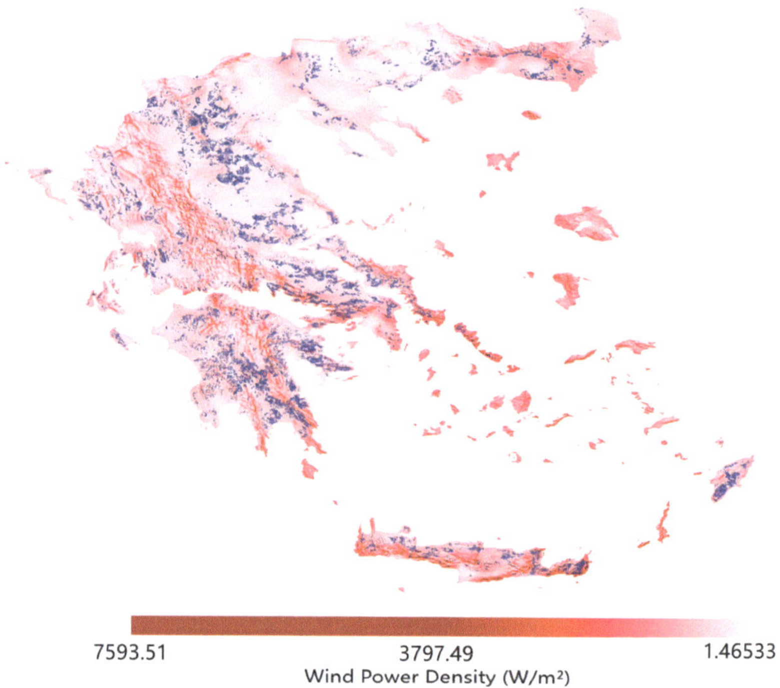

Figure 10. Predicted onshore wind turbine areas with the mean WPD map, in which blue areas indicate the areas with predicted wind turbines.

3.3. Open-Field PV Potential

The four categories of the exclusion criteria were successively excluded using GLAES, and the outcomes of each exclusion process are presented in Figure 11. Initially, the exclusion of the socio-political criteria resulted in 50.19% of the land being deemed eligible for open-field PV construction. However, this proportion drastically decreased to 7.71%, marking a reduction of 42.48%, when the physical criteria were excluded. The exclusion of the conservation criteria did not significantly impact the proportion of eligible land, with the value remaining at 7.44%. Finally, after excluding the economic criteria, the area suitable for PV power plants in open areas of Greece amounted to 4.67% of the country's total land area. Generally, physical criteria significantly influence the land eligibility for open-field PV power plants, with slope being the most important factor. Without excluding the slope criterion, the land eligibility for open-field PV increased to 18.17%.

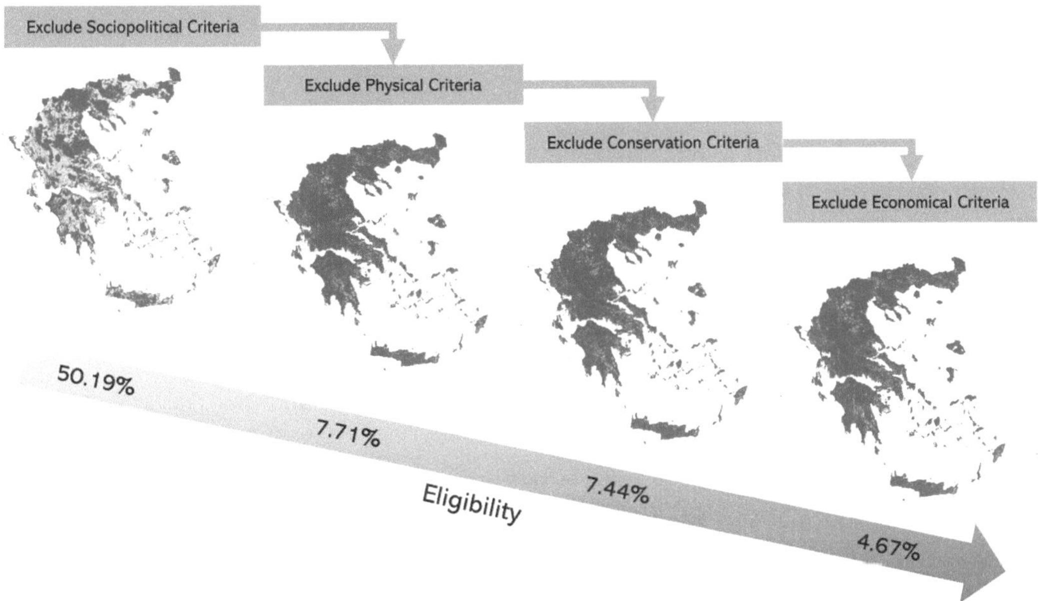

Figure 11. Land eligibility prediction process for open-field PV in Greece.

Figure 12 illustrates a map displaying the eligible areas for open-field PV power plants as predicted by GLAES. The total eligible area in Greece encompasses 6166 km^2, accounting for 4.67% of the country's total land surface. The predicted open-field PV power plant locations are predominantly on flat terrain. Moreover, Figure 13 clearly shows the relatively high mean GHI values in the eligible areas, particularly in the southern regions of Greece, such as the Peloponnese and the islands of Crete and Rhodes. These regions present peak GHI values, indicating the substantial solar energy potential of open-field PV installations in the predicted eligible land areas.

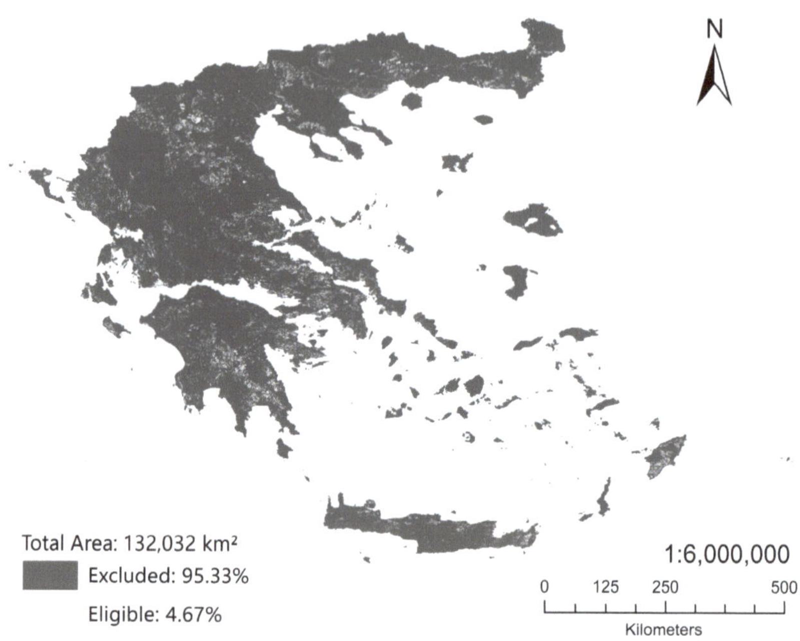

Figure 12. Open-field PV land eligibility excluding the land constraints from Table 3.

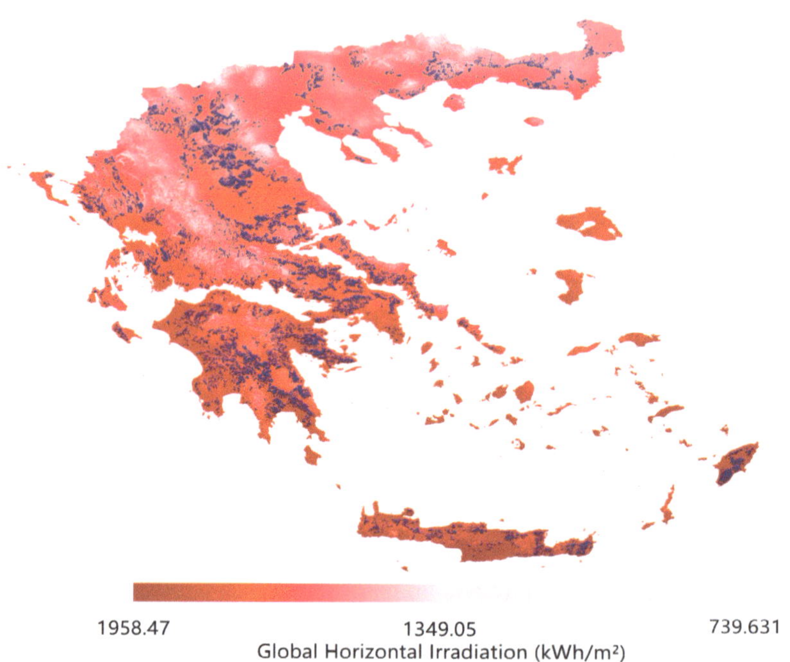

Figure 13. Predicted open-field PV area with the mean GHI map, in which blue areas indicate the areas with PV power plants.

3.4. Renewables' Potential of Decommissioned Lignite Mines

The potential installation of wind turbines and PV panels in the lignite mining areas of Megalopoli (25 km²), Ptolemaida I (134 km²), and Ptolemaida II (57 km²) was simulated using GLAES as the basis. The investigation employed a baseline wind turbine (Table 2) and a reference PV panel (Table 4) for the analysis. Table 5 presents the findings for wind turbines' and PV parks' placements, along with the respective annual energy potential.

Table 5. The predicted placements of wind turbines and PV parks along with the energy potential of both in the lignite mines of Megalopoli and Ptolemaida I and II.

Lignite Mine	Megalopoli	Ptolemaida I	Ptolemaida II	Total
Wind:				
Wind Turbines	34	155	61	250
Wind Turbine Power (kW)	4200	4200	4200	
Energy Yield (kWh/kW)	3000	3000	3000	
Wind Energy Potential (GWh)	428	1953	769	3150
Solar:				
PV Parks	21	102	37	160
Area for PV (km^2)	11.5	68.9	22.1	102.5
PV Peak Power (GWp)	1.7	9.9	3.2	
Energy Yield (kWh/kWp)	1400	1400	1400	
PV Energy Potential (GWh)	2324	13,901	4465	20,690

In the context of wind turbines, the Megalopoli Mine has a capacity of 34 turbines, possessing a total energy potential of 428 GWh annually. As for the Ptolemaida I Mine, it can accommodate 155 turbines, producing 1953 GWh of energy potential annually. Meanwhile, the Ptolemaida II Mine can host 61 turbines and is capable of producing 769 GWh of energy potential. Furthermore, the Megalopoli Mine can accommodate 21 PV parks, covering an area of 21 km^2 and generating an energy potential of 2324 GWh. The Ptolemaida I Mine can be covered by 102 PV parks, accounting for an area of 68.9 km^2 and yielding an impressive energy potential of 13,901 GWh annually. The Ptolemaida II Mine, on the other hand, has the capacity for 37 PV parks, covering an area of 22.1 km^2 with an energy potential of 4465 GWh per annum.

Overall, significantly, based on the results of the three lignite mines studied, the PV parks have an almost 7-fold greater potential for electric energy generation compared to wind turbines. A detailed breakdown of the distribution of the wind turbines and PV parks is provided in Section 4.2, followed by a comprehensive discussion of the comparative electric power generation potential of both technologies.

4. Discussion

4.1. Comparison between the GLAES-Defined Eligible Areas with the Current State

Compared to previous studies [6,8–11,16], this work incorporated wind speed, access, and connection distance directly into the exclusion criteria. Based on the land eligibility analysis conducted for onshore wind turbines and open-field PV power plants, it is evident that the area identified as eligible by the GLAES model was considerably larger than the areas presently developed for such use in Greece. Specifically, the land area deemed eligible for onshore wind turbines surpassed the extent of wind farms by 13,960 km^2, while the eligible area for PV power plants exceeded the current coverage of PV parks by 4182 km^2. These comparative findings emphasize the significant land capacity available for the deployment of wind and solar projects in Greece.

Figure 14 illustrates a spatial comparison between the areas suitable for wind turbines' construction according to GLAES and areas where wind turbines have already been installed. It is apparent that a significant number of predicted eligible areas are situated on low-elevation and flat terrain, which differs from the current state of wind turbines' placement. Notably, many wind turbines have been installed at high elevations, a practice not accounted for in the GLAES model. Nevertheless, the identification of unsuitable areas for wind turbine installation by GLAES does not imply that these areas are actually unsuitable for constructing wind turbines. While considering land constraints, the actual construction also needs to evaluate the wind energy potential of specific locations. Given that Greece exhibits a greater wind energy potential at higher elevations, the installation of wind turbines in such areas is justifiable.

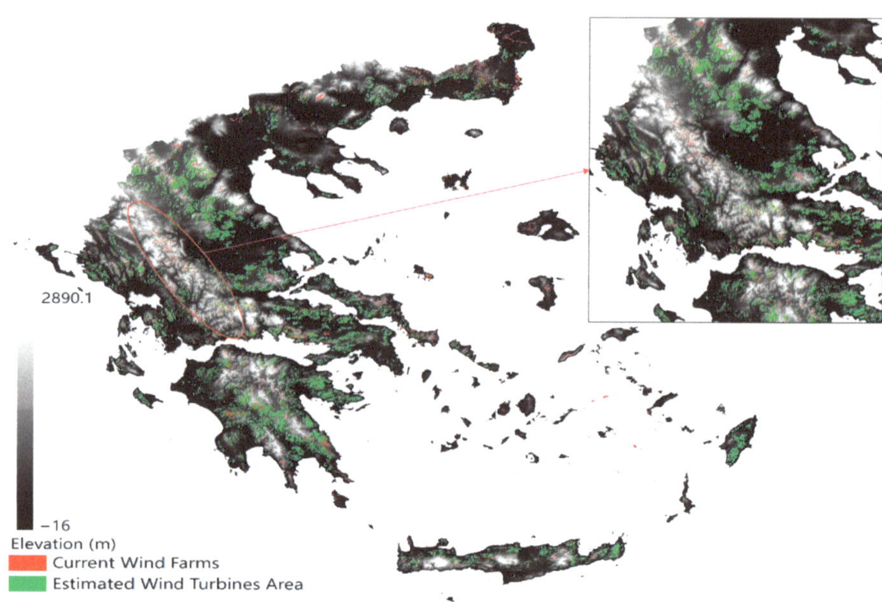

Figure 14. Current wind farms and predicted wind turbine areas shown in the DEM map of Greece.

Furthermore, the land eligibility maps, which indicate the predicted suitability of land for onshore wind and open-field PV, were merged with the existing wind farms and PV parks, as shown in Figures 15 and 16, respectively. Figure 15 reveals that a significant proportion, specifically 76.5% of the established wind farms, are situated in regions that were excluded from the onshore wind land eligibility analysis because of the terrain slope. Moreover, with potential implications for nature conservation, it was observed that 10.1% of wind farms have been constructed within Natura 2000 areas. Similarly, Figure 16 illustrates that 49.3% of the existing PV parks have been constructed in regions that were excluded by the land eligibility analysis for open-field PV installations. Additionally, 10.1% of these PV parks are situated in Natura 2000 areas. It should be noted that the reason for this can be attributed Greece's early initiative in the construction of wind farms, which date back to the early 1980s [53], before the implementation of Natura 2000. Additionally, the utilization of solar PV technology commenced in 2006 with the introduction of feed-in tariffs [54]. However, it was not until 2012 that a specific framework for the development of RESs in Greece, known as SFSPSD-RES, was proposed and implemented. Moreover, due to the later start in constructing PV power plants compared to wind turbines, the area covered by PV parks in the excluded regions is significantly smaller.

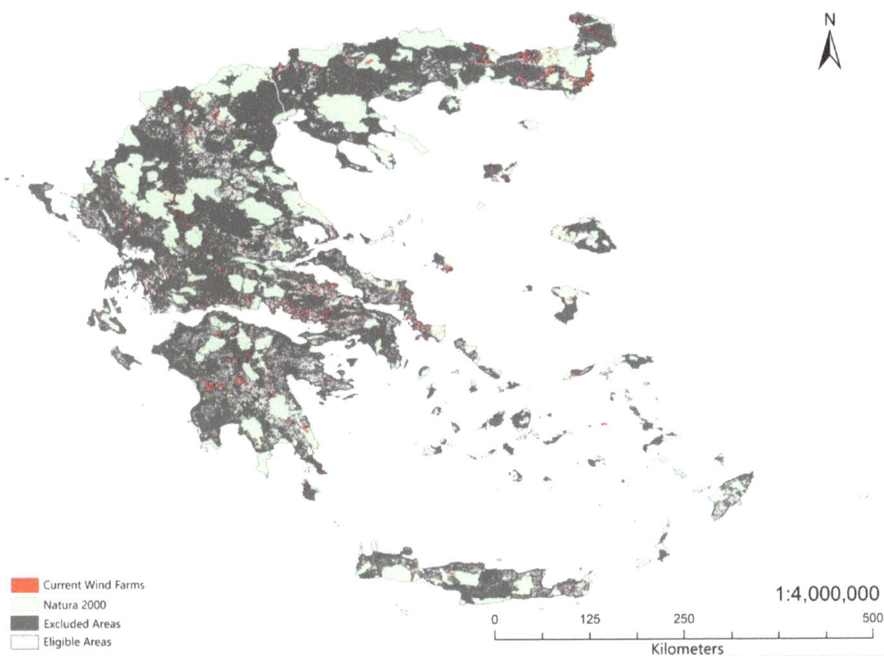

Figure 15. Unsuitable areas for wind turbines' construction and the existing wind farms in Greece. White areas present the eligible land, while gray areas present the unsuitable land. Green areas and red areas, respectively, stand for Natura 2000 sites and the current wind farms.

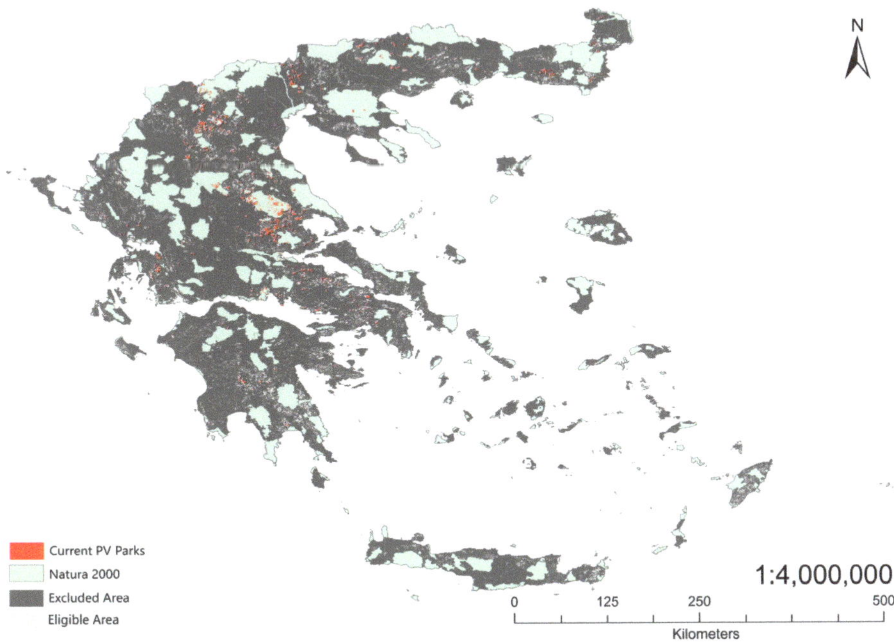

Figure 16. Unsuitable areas for the construction of PV power plants and the existing PV parks in Greece. White areas present the eligible land, while gray areas present the unsuitable land. Green areas and red areas, respectively, stand for Natura 2000 sites and the current PV parks.

4.2. Comparison of the Wind and Solar PV Energy Potential in the Mining Areas under Study

Figure 17 depicts the distribution of wind turbines in conjunction with the mean WPD for each mining area. Additionally, Figure 18 illustrates the distribution of PV parks combined with the mean GHI in each lignite mine.

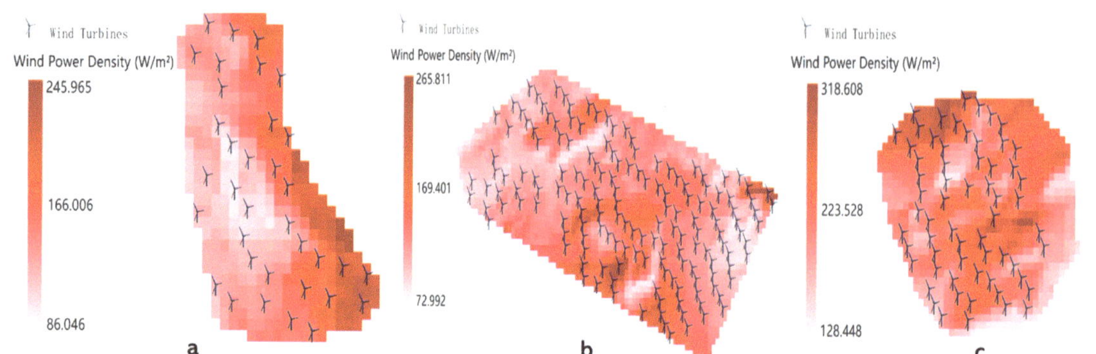

Figure 17. (**a**) Simulated placements of wind turbines with the WPD map of the Megalopoli Mine. (**b**) Simulated placements of wind turbines with the WPD map of the Ptolemaida I Mine. (**c**) Simulated placements of wind turbines with the WPD map of the Ptolemaida II Mine.

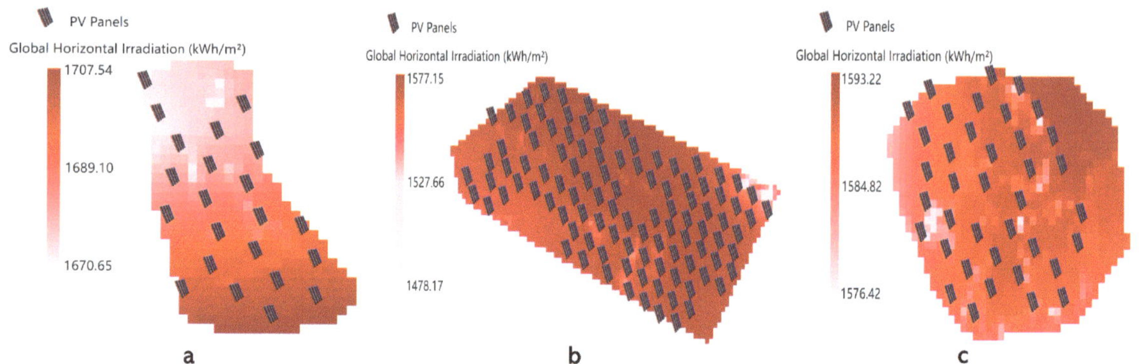

Figure 18. (**a**) Simulated placements of PV parks with the GHI map of the Megalopoli Mine. (**b**) Simulated placements of PV parks with the GHI map of the Ptolemaida I Mine. (**c**) Simulated placements of PV parks with the GHI map of the Ptolemaida II Mine.

Although there is potential for constructing additional wind turbines in decommissioned lignite mines, it should be noted that the wind energy potential of the three mining areas may not surpass the solar energy potential when comparing Figures 17 and 18. This observation is supported by the WPD and GHI maps for the entire country. The data revealed that the average WPD for Greece is 797 W/m^2 [19], whereas the average GHI is 1550 kWh/m^2 [20]. Interestingly, Figure 17 clearly illustrates that the WPD peaks of the three lignite mines correspond to low-value areas within the whole area of Greece, with values of 245.965 W/m^2, 265.881 W/m^2, and 318.608 W/m^2, which are significantly lower than the average WPD. Furthermore, it is evident that there are only a few wind turbines situated in the regions with the peak WPD values of each mine, with a mere total of 15 wind turbines having been observed. On the contrary, the extreme GHI values in the three lignite mines all exceed the average value of GHI for the entire region of Greece, amounting to 1707.5 kWh/m^2, 1577.2 kWh/m^2, and 1593.2 kWh/m^2, as shown in Figure 18. Notably, the peak GHI areas of the Megalopoli Mine and the Ptolemaida II Mine contain four PV parks each, while almost all PV parks are concentrated in the extreme GHI area of the Ptolemaida

I Mine. For robust verification, this study conducted a comprehensive examination by consolidating the electric energy potential WPD values achievable from wind turbines and the electric energy potential GHI values achievable from PV panels, revealing that the potential of electric generation by PV panels in the three lignite mines significantly surpasses that of wind turbines. The results from Table 5 further confirm that the potential of wind energy in the three lignite mines is lower than that of solar PV. Therefore, it is concluded that the total potential of solar PV in the lignite mines of Megalopoli, Ptolemaida I, and Ptolemaida II is 20,690 GWh, which is almost 7-fold greater than that (3150 GWh) of wind.

However, PV fields with such a high peak power output as the ones that can be built on the areas of the three lignite mines from this study could be a critical component in the electricity grid of Greece. Daily and seasonal variations in electricity production need to be smoothed by employing means for the storage of electricity, e.g., hydrogen or thermal energy storage. The seasonal storage of renewable energy using hydrogen involves capturing excess energy during the times of high renewable generation, such as sunny or windy days, and converting it into hydrogen through electrolysis. This hydrogen can then be stored for extended periods, i.e., several months, serving as a clean and efficient energy carrier. During low renewable energy periods, like calm or cloudy days, the stored hydrogen can be utilized in fuel cells or natural-gas-fired power stations to generate electricity, providing a reliable energy source. This approach addresses the intermittency of renewable sources, enabling a more consistent and sustainable power supply throughout the year. Additionally, the deployment of hydrogen in seasonal storage contributes to the decarbonization of the energy sector by offering a versatile solution for large-scale energy storage and distribution. What is more, the study of Schmidt et al. [55] showed that, for a few load cycles per year, hydrogen storage is more cost effective than using batteries. Another promising option for storing surplus renewable electrical energy is the application of the so-called Carnot batteries, which can store large amounts of energy in the form of high-temperature heat in inexpensive materials, such as water, stone, or molten salt, and are much cheaper than batteries [56]. In conjunction with seasonal heat storage systems, they can store heat energy for months. The size, capacity, and energy management of Carnot batteries can be adapted to the specific demand, i.e., the capacities of several megawatts are expected to be available from 2030 onwards.

In general, at the regional level in Greece, the potential for open-field PV is greater than that of onshore wind. It is safe to assume that the potential is even greater when roof areas, parking sites, and industrial areas are included (as they should be). At the same time, although the adoption of PV power generation in Greece came after wind power generation, a recent report indicated that the total installed PV capacity has exceeded the total installed wind capacity for Greece in 2022 [57]. Additionally, the abundance of sunshine in Greece allows for less restricted operations of PV power plants. In contrast, wind turbine operations face more complex limitations. For instance, low wind speeds generally make the installation of wind turbines unviable, while wind speeds that are too high necessitate the shutdown of wind turbines for their protection. Moreover, the construction of wind farms generally entails higher material and financial costs compared to that of PV power plants [57]—the cost of photovoltaic energy has declined by about 90% over the last decade, resulting in a remarkable 30% growth per year [58]. Furthermore, the focus on PV power plants will potentially allow for the further enhancement of biodiversity conservation in Greece. This is because the current focal areas for wind power production (Thrace, Western Macedonia, Central Greece, and the Peloponnese as well as, overall, the higher elevation areas) are regions of potentially high biodiversity value [59]. A focus on PV power production provides an opportunity for the Natura network of protected areas to expand further in those regions. In conclusion, the combination of the present analysis and considering the wider efficiency, economic, and biodiversity protection factors, this study concluded that solar energy (PV) is better suited than wind power for the decarbonization efforts of Greece.

Last but not least, a very important advantage of the proposed use of decommissioned lignite mining areas is that no valuable agricultural land is consumed for the generation of renewable electric energy.

4.3. Limitations

Based on the SFSPSD-RES of Greece, this study incorporated land eligibility analysis for onshore wind and open-field PV, combining insights from previous relevant research conducted in Europe, particularly in Greece. In order to assess land suitability, a comprehensive list of land constraints was formulated, encompassing social, technological, environmental, and economic factors. By imposing exclusion land constraints using GLAES, the areas eligible for constructing wind turbines and PV power plants were obtained. However, it is important to note that the land eligibility analysis, based solely on the current policies and previous empirical studies, can only provide indicative results. The overall process of constructing renewable energy system entails complex aspects that must be considered. These include assessing the potential of RESs, evaluating construction costs, and analyzing topographic and geological conditions specific to the target area. Further investigations can be carried out to examine the discrepancy between the estimated suitability areas for wind turbines and the wind turbines actually built (see Figure 5). Especially on mountain tops, the wind potential is systematically underestimated when using averaged wind speed data, as shown in the study by Hu et al. [60]. In a future study, the effect of topology at the sub-grid level will be included in the land suitability analysis.

Additionally, it is crucial to secure the support and involvement of local residents and stakeholders, among other relevant considerations. Consequently, the successful implementation of wind turbines and PV power plants necessitates not only model simulations but also thorough field investigations, which together enable optimizing electric power generation efficiency and minimizing costs to ensure the utmost effectiveness of renewable energy generation.

With the phasing out of lignite mines in Greece, there is a significant opportunity to construct renewable energy systems in decommissioned lignite mines. However, there are certain limitations in constructing wind turbines and PV power plants in these mining areas. Geotechnical stability, slopes, and landslides, for Example, should be considered during construction [61]. Moreover, this study compared the potential of wind energy and solar PV energy in Greece, specifically focusing on decommissioned lignite mines. The analysis mainly utilized the WPD and GHI maps of Greece for discussion. It should be noted that there is no direct relationship between WPD and GHI values; rather, they serve as reference indices for estimation proposes. The electric power output of wind energy and solar PV predominantly depends on their respective power capacities. Therefore, this work used a reference wind turbine and a reference PV panel to estimate the number of constructable turbines and the area for PV parks so that the energy potential can be calculated under ideal conditions. Finally, it was found that the PV energy potential significantly surpassed that of wind energy in the decommissioned lignite mines under study. It is essential to recognize that the estimated potential represents a technically possible maximum. The exploitable potential may be substantially lower due to practical, ecological, or technological limitations (which cannot be addressed in this work). Detailed site-specific studies are imperative for a realistic assessment.

Furthermore, a comprehensive calculation of electric generation is an intricate process that necessitates the consideration of numerous factors. In the case of wind energy, the power generation capacity is determined by climatic factors, such as air density and temperature, as well as physical factors, like wind speed and the spatial and temporal variations in wind patterns. Similarly, the power generation capacity of solar PV is influenced by climatic parameters, like air quality and sunshine hours, in addition to physical factors, such as solar irradiance in combination with atmospheric conditions. Overall, the calculation of wind energy and solar PV energy capacities for the whole territory of Greece is an intricate and challenging task, thus representing a limitation.

5. Conclusions

In the context of Greece's energy transition dominated by wind and solar energy, this work assessed the land eligibility of onshore wind turbines and open-field PV power plants to provide insights of reference for their suitable construction. Meanwhile, the electric generation potential of the two RESs in the decommissioned lignite mines of Megalopoli and Ptolemeida were specifically compared to further discuss the most suitable RESs for Greece, either wind or solar. This work concluded that solar (PV) energy in Greece has a greater potential of electricity generation than that of wind energy. Considering only land eligibility during optimal conditions, Greece's 132,032 km^2 of suitable areas for PV could generate an electricity of 205 TWh, significantly surpassing the country's total electricity consumption of 52.8 TWh in 2020 provided by International Energy Agency. And in decommissioned lignite mines, while wind turbines appear unsuitable for electricity generation, PV systems have significant potential. Moreover, it is worth mentioning that a joint venture between the German utility RWE and Public Power Corp. (PPC) of Greece is already constructing solar projects, named Amynteo Cluster I and II, with a total capacity of 210 MWp and 280 MWp, respectively, in the former Amynteo open-pit lignite mine of Western Macedonia [62].

Through the land eligibility analysis of onshore wind turbines and open-field PV power plants using GLAES, it was found that Greece has significant land potential for both compared to the current state. It is worth noting that the construction locations of wind turbines modeled using GLAES is quite different from the current installations. Most of the current wind turbines are located at high altitudes with a high wind energy potential, which is contrary to the predicted results. Additionally, although the eligible area for wind turbines (12.16%) is much larger than that of PV power plants (4.67%), a significant proportion of the predicted eligible land for wind turbines is located in low electric generation potential areas, especially in the studied lignite mining areas. Meanwhile, this study made a start to analyze the land eligibility for both the onshore wind and open-field PV of Greece and compared the electric energy potential of wind and solar power. In general, the successful construction of wind turbines and PV power plants not only needs to be based on model predictions considering socio-politics, technology, the environment, and economy, but also requires a thorough field investigation of the target construction areas. Moreover, the most direct way to compare the potential of wind and solar energy sources is to calculate their electric power capacity. However, due to the complexity of the calculation, which needs to consider many climatic factors, this work did not conduct the calculation, thus remaining a limitation. Overall, this work provided meaningful references for the construction of wind turbines and open-field PV power plants for Greek decarbonization. Notably, part of the wider significance of this study lies in the estimation of the potential of renewable energy available by using the areas of three decommissioned lignite mines, which is in the order of magnitude of the electricity produced by gas in 2020 in all of Greece.

Author Contributions: Conceptualization, Q.W., F.C.K., E.G. and M.S.; methodology, Q.W., M.S., E.G. and P.S.; software, Q.W., P.S. and M.S.; validation, Q.W., E.G., P.S. and M.S.; formal analysis, Q.W.; investigation, Q.W., E.G., M.S., F.C.K. and V.L.; resources, F.C.K.; data curation, Q.W. and M.S.; writing—original draft preparation, Q.W.; writing—review and editing, Q.W., M.S., F.C.K., P.S. and E.G.; visualization, Q.W.; supervision, M.S., E.G., F.C.K. and P.S.; project administration, F.C.K. All authors have read and agreed to the published version of the manuscript.

Funding: This research received no external funding.

Data Availability Statement: Data are contained within the article.

Conflicts of Interest: Author Martin Spiller was employed by the ISATEC GmbH. The remaining authors declare that the research was conducted in the absence of any commercial or financial relationships that could be construed as a potential conflict of interest.

Abbreviations

DEM	Digital Elevation Model
GLAES	Geospatial Land Availability for Energy Systems
GHI	Global Horizontal Irradiation
GIS	Geographic Information System
NECP	National Energy and Climate Plan
PV	Photovoltaic
RES	Renewable Energy Source
WPD	Wind Power Density
SFSPSD-RES	Specific Framework for Spatial Planning and Sustainable Development for Renewable Energy Sources

References

1. United Nations. *Paris Agreement*; United Nations: Paris, France, 2015.
2. Kaldellis, J.K.; Kapsali, M.; Katsanou, E. Renewable Energy Applications in Greece—What Is the Public Attitude? *Energy Policy* **2012**, *42*, 37–48. [CrossRef]
3. IEA. Energy Policy Review Greece 2023. Available online: https://iea.blob.core.windows.net/assets/5dc74a29-c4cb-4cde-97e0-9e218c58c6fd/Greece2023.pdf (accessed on 26 June 2023).
4. Central Intelligence Agency Greece Country Summary. Available online: https://www.cia.gov/the-world-factbook/countries/greece/#energy (accessed on 27 June 2023).
5. IEA 2023. Greece—Countries & Regions. Available online: https://www.iea.org/countries/greece (accessed on 30 December 2023).
6. Mourmouris, J.C.; Potolias, C. A Multi-Criteria Methodology for Energy Planning and Developing Renewable Energy Sources at a Regional Level: A Case Study Thassos, Greece. *Energy Policy* **2013**, *52*, 522–530. [CrossRef]
7. Tsoutsos, T.; Tsitoura, I.; Kokologos, D.; Kalaitzakis, K. Sustainable Siting Process in Large Wind Farms Case Study in Crete. *Renew. Energy* **2015**, *75*, 474–480. [CrossRef]
8. Latinopoulos, D.; Kechagia, K. A GIS-Based Multi-Criteria Evaluation for Wind Farm Site Selection. A Regional Scale Application in Greece. *Renew. Energy* **2015**, *78*, 550–560. [CrossRef]
9. Bili, A.; Vagiona, D.G. Use of Multicriteria Analysis and GIS for Selecting Sites for Onshore Wind Farms: The Case of Andros Island (Greece). *Eur. J. Environ. Sci.* **2018**, *8*, 5–13. [CrossRef]
10. Karamountzou, S.; Vagiona, D.G. Suitability and Sustainability Assessment of Existing Onshore Wind Farms in Greece. *Sustainability* **2023**, *15*, 2095. [CrossRef]
11. Vagiona, D.G. Comparative Multicriteria Analysis Methods for Ranking Sites for Solar Farm Deployment: A Case Study in Greece. *Energies* **2021**, *14*, 8371. [CrossRef]
12. Greece. Available online: https://www.britannica.com/place/Greece (accessed on 26 June 2023).
13. National Archive of Monuments. Available online: https://www.arxaiologikoktimatologio.gov.gr/ (accessed on 26 June 2023).
14. Natura 2000 Viewer. Available online: https://natura2000.eea.europa.eu/ (accessed on 26 June 2023).
15. Regions of Greece. Available online: https://geodata.gov.gr/en/dataset/periphereies-elladas (accessed on 26 June 2023).
16. Baltas, A.E.; Dervos, A.N. Special Framework for the Spatial Planning & the Sustainable Development of Renewable Energy Sources. *Renew. Energy* **2012**, *48*, 358–363. [CrossRef]
17. Ryberg, D.S.; Tulemat, Z.; Robinius, M.; Stolten, D. Geospatial Land Availability for Energy Systems (GLAES). Available online: https://github.com/FZJ-IEK3-VSA/glaes (accessed on 23 June 2023).
18. European Evironment Agency Digital Elevation Model over Europe. Available online: https://www.eea.europa.eu/data-and-maps/data/eu-dem (accessed on 28 June 2023).
19. Global Wind Atlas. Available online: https://globalwindatlas.info/en/ (accessed on 26 June 2023).
20. Global Solar Atlas. Available online: https://globalsolaratlas.info/global-pv-potential-study (accessed on 26 June 2023).
21. RAE Geospatial Map for Energy Units & Requests. Available online: https://geo.rae.gr/?tab=viewport_maptab (accessed on 27 June 2023).
22. Iqbal, M.; Azam, M.; Naeem, M.; Khwaja, A.S.; Anpalagan, A. Optimization Classification, Algorithms and Tools for Renewable Energy: A Review. *Renew. Sustain. Energy Rev.* **2014**, *39*, 640–654. [CrossRef]
23. Ryberg, D.S.; Robinius, M.; Stolten, D. Methodological Framework for Determining the Land Eligibility of Renewable Energy Sources. *arXiv* **2017**, arXiv:1712.07840.
24. *European Environment Agency Europe's Onshore and Offshore Wind Energy Potential: An Assessment of Environmental and Economic Constraints*; Publications Office: Luxembourg, 2009.
25. McKenna, R.; Hollnaicher, S.; Ostman, V.D.; Leye, P.; Fichtner, W. Cost-Potentials for Large Onshore Wind Turbines in Europe. *Energy* **2015**, *83*, 217–229. [CrossRef]
26. Eurek, K.; Sullivan, P.; Gleason, M.; Hettinger, D.; Heimiller, D.; Lopez, A. An Improved Global Wind Resource Estimate for Integrated Assessment Models. *Energy Econ.* **2017**, *64*, 552–567. [CrossRef]

27. Ryberg, D.S.; Tulemat, Z.; Stolten, D.; Robinius, M. Uniformly Constrained Land Eligibility for Onshore European Wind Power. *Renew. Energy* **2020**, *146*, 921–931. [CrossRef]
28. Ryberg, D.S. Generation Lulls from the Future Potential of Wind and Solar Energy in Europe. 2020, pp. 96–109. Available online: https://publications.rwth-aachen.de/record/805445 (accessed on 12 June 2023).
29. GISCO Corine Land Cover V2012 2012. Available online: https://land.copernicus.eu/pan-european/corine-land-cover/clc-2012 (accessed on 12 July 2023).
30. GISCO Clusters: Urban 2011 2016. Available online: http://ec.europa.eu/eurostat/web/gisco/geodata/reference-data/elevation/hydrography-laea (accessed on 12 July 2023).
31. OpenStreetMap Contributors Planet Dump 2017. Available online: https://planet.osm.org (accessed on 12 July 2023).
32. UNEP-WCMC and IUCN Protected Planet: The World Database on Protected Areas (WDPA) 2016. Available online: https://www.protectedplanet.net/en/search-areas?filters[db_type][]=wdpa&geo_type=region (accessed on 12 July 2023).
33. Baseer, M.A.; Rehman, S.; Meyer, J.P.; Alam, M.M. GIS-Based Site Suitability Analysis for Wind Farm Development in Saudi Arabia. *Energy* **2017**, *141*, 1166–1176. [CrossRef]
34. Sánchez-Lozano, J.M.; García-Cascales, M.S.; Lamata, M.T. Identification and Selection of Potential Sites for Onshore Wind Farms Development in Region of Murcia, Spain. *Energy* **2014**, *73*, 311–324. [CrossRef]
35. Ayodele, T.R.; Ogunjuyigbe, A.S.O.; Odigie, O.; Munda, J.L. A Multi-Criteria GIS Based Model for Wind Farm Site Selection Using Interval Type-2 Fuzzy Analytic Hierarchy Process: The Case Study of Nigeria. *Appl. Energy* **2018**, *228*, 1853–1869. [CrossRef]
36. Höfer, T.; Sunak, Y.; Siddique, H.; Madlener, R. Wind Farm Siting Using a Spatial Analytic Hierarchy Process Approach: A Case Study of the Städteregion Aachen. *Appl. Energy* **2016**, *163*, 222–243. [CrossRef]
37. Tlili, O.; Mansilla, C.; Robinius, M.; Ryberg, D.S.; Caglayan, D.G.; Linssen, J.; André, J.; Perez, Y.; Stolten, D. Downscaling of Future National Capacity Scenarios of the French Electricity System to the Regional Level. *Energy Syst.* **2022**, *13*, 137–165. [CrossRef]
38. Réfet de la Région Nord-Pas-de-Calais Annexe: 'Schéma Régional Eolien'. Available online: https://www.hauts-de-france.developpement-durable.gouv.fr/IMG/pdf/sre_et_srs_npdc-2.pdf (accessed on 16 August 2023).
39. Maestre, V.M.; Ortiz, A.; Ortiz, I. Transition to a Low-Carbon Building Stock. Techno-Economic and Spatial Optimization of Renewables-hydrogen Strategies in Spain. *J. Energy Storage* **2022**, *56*, 105889. [CrossRef]
40. Colak, H.E.; Memisoglu, T.; Gercek, Y. Optimal Site Selection for Solar Photovoltaic (PV) Power Plants Using GIS and AHP: A Case Study of Malatya Province, Turkey. *Renew. Energy* **2020**, *149*, 565–576. [CrossRef]
41. Günen, M.A. Determination of the Suitable Sites for Constructing Solar Photovoltaic (PV) Power Plants in Kayseri, Turkey Using GIS-Based Ranking and AHP Methods. *Environ. Sci. Pollut. Res.* **2021**, *28*, 57232–57247. [CrossRef]
42. Kocabaldır, C.; Yücel, M.A. GIS-Based Multicriteria Decision Analysis for Spatial Planning of Solar Photovoltaic Power Plants in Çanakkale Province, Turkey. *Renew. Energy* **2023**, *212*, 455–467. [CrossRef]
43. Tercan, E.; Saracoglu, B.O.; Bilgilioğlu, S.S.; Eymen, A.; Tapkın, S. Geographic Information System-Based Investment System for Photovoltaic Power Plants Location Analysis in Turkey. *Environ. Monit. Assess.* **2020**, *192*, 297. [CrossRef] [PubMed]
44. Perpiña Castillo, C.; Batista E Silva, F.; Lavalle, C. An Assessment of the Regional Potential for Solar Power Generation in EU-28. *Energy Policy* **2016**, *88*, 86–99. [CrossRef]
45. Yousefi, H.; Hafeznia, H.; Yousefi-Sahzabi, A. Spatial Site Selection for Solar Power Plants Using a GIS-Based Boolean-Fuzzy Logic Model: A Case Study of Markazi Province, Iran. *Energies* **2018**, *11*, 1648. [CrossRef]
46. *National Energy and Climate Plan*; HELLENIC REPUBLIC Ministry of the Environment and Energy: Athens, Greece, 2019.
47. Argyropoulos, D. Coal, on Its Way out—Greece's Plans to Phase out Lignite Are Boosted by the Pandemic. 2020. Available online: https://energytransition.org/2020/09/coal-on-its-way-out-how-the-greek-plans-to-phase-out-lignite-are-boosted-by-the-pandemic/ (accessed on 5 July 2023).
48. Liaggou, C. PPC Extends Operation of Seven Lignite Units to Ensure Supply 2021. Available online: https://www.ekathimerini.com/economy/1174648/ppc-extends-operation-of-seven-lignite-units-to-ensure-supply/ (accessed on 5 July 2023).
49. Chatzitheodoridis, F.; Kolokontes, A.D.; Vasiliadis, L. Lignite Mining and Lignite-Fired Power Generation in Western Macedonia of Greece: Economy and Environment. *J. Energy Dev.* **2023**, *33*, 267–282.
50. 2022 Annual Statistics of Wind Farms 2023. Available online: https://resinvest.gr/2023/02/01/2022-annual-statistics-hellenic-wind-energy-association-2/ (accessed on 7 July 2023).
51. Petrova, V. Greece Adds Record 1.4 GW of Fresh Solar in 2022—Report 2023. Available online: https://renewablesnow.com/news/greece-adds-record-14-gw-of-fresh-solar-in-2022-report-818214/ (accessed on 7 July 2023).
52. Suri, M.; Betak, J.; Rosina, K.; Chrkavy, D.; Suriova, N.; Cebecaue, T.; Caltik, M.; Erdelyi, B. *Global Photovoltaic Power Potential by Country (English)*; Energy Sector Management Assistance Program (ESMAP): Washington, DC, USA; World Bank Group: Washington, DC, USA, 2020.
53. 30 Years of Policies for Wind Energy: Lessons from Greece; International Renewable Energy Agency. Available online: https://www.irena.org/publications/2013/Jan/30-Years-of-Policies-for-Wind-Energy-Lessons-from-12-Wind-Energy-Markets (accessed on 9 July 2023).
54. Greek PV Market Investment Opportunities. Hellenic Association of Photovoltaic Companies. Available online: https://helapco.gr/pdf/Greek_PV_Market_Opportunities_Aug2019.pdf (accessed on 10 January 2024).

55. Schmidt, O.; Melchior, S.; Hawkes, A.; Staffell, I. Projecting the Future Levelized Cost of Electricity Storage Technologies. *Joule* **2019**, *3*, 81–100. [CrossRef]
56. Tassenoy, R.; Couvreur, K.; Beyne, W.; De Paepe, M.; Lecompte, S. Techno-Economic Assessment of Carnot Batteries for Load-Shifting of Solar PV Production of an Office Building. *Renew. Energy* **2022**, *199*, 1133–1144. [CrossRef]
57. Aposporis, H. Solar Power Capacity in Greece Overtakes Wind for First Time 2022. Available online: https://balkangreenenergyne com/solar-power-capacity-in-greece-overtakes-wind-for-first-time/ (accessed on 10 July 2023).
58. Creutzig, F.; Hilaire, J.; Nemet, G.; Müller-Hansen, F.; Minx, J.C. Technological Innovation Enables Low Cost Climate Change Mitigation. *Energy Res. Soc. Sci.* **2023**, *105*, 103276. [CrossRef]
59. Kougioumoutzis, K.; Kokkoris, I.; Panitsa, M.; Kallimanis, A.; Strid, A.; Dimopoulos, P. Plant Endemism Centres and Biodiversity Hotspots in Greece. *Biology* **2021**, *10*, 72. [CrossRef]
60. Hu, W.; Scholz, Y.; Yeligeti, M.; Bremen, L.V.; Deng, Y. Downscaling ERA5 Wind Speed Data: A Machine Learning Approach Considering Topographic Influences. *Environ. Res. Lett.* **2023**, *18*, 094007. [CrossRef]
61. Antoniadis, A.; Roumpos, C.; Anagnostopoulos, P.; Paraskevis, N. Planning RES Projects in Exhausted Surface Lignite Mines—Challenges and Solutions. In Proceedings of the International Conference on Raw Materials and Circular Economy, Athens, Greece, 5–9 September 2021; p. 93.
62. Knauber, S. RWE and PPC to Build Solar Projects with 280 Megawatts of Capacity in Greece. Available online: https://www.rwe.com/en/press/rwe-renewables-europe-australia/23-07-31-rwe-and-ppc-to-build-solar-projects-with-280-mw-of-capacity-in-greece/ (accessed on 10 December 2023).

Article

Empowering Active Users: A Case Study with Economic Analysis of the Electric Energy Cost Calculation Post-Net-Metering Abolition in Slovenia

Eva Tratnik * and Miloš Beković

Faculty of Electrical Engineering and Computer Science, University of Maribor, 2000 Maribor, Slovenia
* Correspondence: eva.tratnik@um.si; Tel.: +386-(2)-22-07-086

Abstract: This paper addresses the issue of the abolition of annual net metering in Slovenia and compares the electric energy costs for the studied active user after the abolition. The article also provides an exploration of the role played by an aggregator, which serves as a central entity that enables individuals to participate in the electric energy market. An analysis of the case study of an active user was made, where an analysis was made of the measurements of household consumption and photovoltaic plant production for the year 2022. This article presents an economic analysis with and without net metering and an analysis of the aggregator involvement strategies. In addition, a battery energy storage system was also considered in the analysis. An important part of the article is the identification of the flexibility potential for shiftable loads, which enable an aggregator to acquire insight into the energy consumption profile and energy production profile of active users. The following indicators were used to compare the strategies: annual electric energy cost and the indicators including self-sufficiency, self-consumption, and grid dependency. The findings indicate that, even in the absence of annual net metering, the active user can lower their costs for electric energy with the help of an aggregator.

Keywords: active user; battery storage system; case study; economic analysis; net metering; photovoltaic system

Citation: Tratnik, E.; Beković, M. Empowering Active Users: A Case Study with Economic Analysis of the Electric Energy Cost Calculation Post-Net-Metering Abolition in Slovenia. *Energies* **2024**, *17*, 1501. https://doi.org/10.3390/en17061501

Academic Editor: Marco Merlo

Received: 26 February 2024
Revised: 15 March 2024
Accepted: 18 March 2024
Published: 21 March 2024

1. Introduction

The flexibility potential of active users is ever increasing due to the growth in the use of the following power-consuming and/or power-producing appliances in households: heat pumps (HPs) for heating, electric vehicles (EVs), battery energy storage systems (BESSs), and renewable energy sources (RESs) such as solar power plants. An active user is an individual or a group of end-users who provide services for adjusting consumption and/or production. The integration of home appliances with home energy management systems (HEMSs) does present some challenges for households due to the complexity of the underlying processes and technologies [1,2], but also helps to improve the energy efficiency in households. The lack of awareness among active users about the sustainable use of electric energy decelerates the usage of HEMSs in households [3].

The EU countries have a variety of programs to encourage households to install photovoltaic systems (PVs) in their homes, as discussed in [4–6]. One of the programs that encourage households to install PV plants in Slovenia is net metering, which is described later. In [7], a method for the sizing of a PV plant and the environmental perspective are presented for a case study in Slovenia. The support laws and regulations of countries that encourage PV plants are some of the key factors in the economic feasibility of investing in a PV plant. An overview of the supporting policies in five EU countries through investment profitability is provided in [8]. This paper offers a detailed analysis of the support policies for photovoltaic installations in the residential sector across major European markets such as Flanders (Belgium), Germany, Italy, Spain, and France. It evaluates the economic viability of a household investment in a photovoltaic installation through a model based on the discounted cash flows of the installation over its lifetime. The study suggests that Italy's support system has been the most profitable among the countries studied since 2010. Despite decreasing support levels, residential installations are still profitable in most cases under the current support policies, except for Spain. The study further highlights the potential of self-consumption to increase profits, especially in Spain and Germany. However, Flanders' policy has no impact on levels of self-consumption. Finally, a comparison of past and present policies indicates varying levels of success in keeping the profitability of investments stable over the years, depending on the efficiency of the support policy. Germany's support system stands out as the most balanced one over the last five years. The results of the study show that the supporting policies have a significant impact on investments' profitability. Household users can also sell their electric energy and have different options, as discussed in [9], which compares the following three electric energy sales mechanisms: feed-in tariffs, net metering, and net purchase with sale. Each mechanism has been described using a simple microeconomic model. The mechanisms have been compared in terms of social welfare and the retail electric energy rate.

Annual net metering allows the solar power plant to cover the household's electric energy needs while surpluses and deficits are regulated by the grid. When production exceeds consumption, the power plant sends the excess energy to the grid, while, during lower production time, the household receives electric energy from the grid. The annual calculation is made only once a year, at the end of the calendar year, and is based on the difference between the supplied and received electric energy. The decision as to whether any excess energy that arises after the end of the billing period will be handed over to the supplier for a fee or free of charge is a matter of agreement between the customer and the supplier. Several studies have been performed to optimize the size and operation of PV plants. In [10], the researchers were focused on the optimal design of PV systems in light of the specificity of Polish regulations. Since the annual net metering was valid, the annual consumption was relevant in the dimensioning of the PV plant. With the abolition of net metering, the methods for sizing PV plants will change.

The Directive 2019/944 EU [11] requires that the network fee must be non-discriminatory, regardless of whether the household user is included in the net metering policy. With the new regulation, the household user will be required to pay a network fee for all the electric energy delivered from the distribution grid [12]. Therefore, Slovenia coordinated its legalization with the EU legislation and adopted the Act on Promoting the Use of renewable energy sources [13] and the regulation on self-supply with electric energy from renewable energy sources [14], which will enter into force in 2024. Legislations will bring major changes in Slovenia's network fee tariffs and abolish the concept of annual net metering, and cause many households in Slovenia to wonder about the profitability of investments in a PV plant. Households with installed PV plants need to consider their energy management strategies.

There is always a discrepancy between produced energy and household consumption. Therefore, for a reliable power supply, it is essential to connect the PV plant to the distribution grid or energy storage system. With the development and massive integration of BESS, many studies have been conducted to address the advantages of BESS in PV applications. A study performed by [15] is being carried out to provide household and commercial users with the optimal size of a PV plant based on their yearly consumption. In [16], the researchers proposed a multi-objective optimal sizing of a grid-connected household with a PV plant and BESS to minimize the total electric energy cost and grid dependency. In [17], the optimal size for a PV-BESS system is considered by the depth of discharge value and the optimal tilt angle of the PV panels. The BESS degradation was ignored. In [18], a framework is presented of a two-stage optimization model for the planning and operation for a household with a PV-BESS system. As a result, it is necessary to choose the optimal size of the PV and BESS to achieve the minimum cost of the system.

Active users can vary their electric energy consumption according to their needs and preferences, taking advantage of dynamic pricing and generation from renewable energy sources. Price-based programs are based on dynamic tariffs, where the prices of the tariff values change depending on the electric energy market circumstances and network conditions. Price-based programs based on time and power demand can be classified as the following [19]: time-of-use pricing, real-time pricing, and critical-peak pricing. In [20], a self-scheduling model for a HEMS is presented considering time-of-use pricing and real-time pricing, aiming to reduce the users' electric energy costs. Additionally, various scenarios with the inclusion of BESSs are studied, to examine their impact on optimal performance. A HEMS optimization model for scheduling home appliances effectively under a time-of-use pricing tariff is presented in [21]. The HEMS optimizes the performance of multiple household appliances, including BESSs, EVs, air conditioners, and water heaters, to minimize electric energy costs.

The HEMS allows active users to manage their consumption and/or production actively, and to participate in new energy services and new opportunities in flexibility markets [22,23]. New energy services are implemented by combining consumers and producers. A more active involvement of active users will also require a new concept of the flexibility market. In [24], the trends in the new flexibility market were selected and compared, where the results showed that the new market platforms present promising models regarding technical and economic justification. Aggregator platforms for local trading between active users, like promoting peer-to-peer trading, were also analyzed in the article.

Demand-side flexibility (DSF) mechanisms can be classified into implicit and explicit DSF mechanisms [25]. Under the implicit DSF mechanism, active users increase or decrease consumption or production in response to electric energy price signals. Under the explicit DSF mechanism, active users increase or decrease consumption or production in response to the aggregator signal, which is based on energy market needs and ancillary products. The aggregator is a retailer of flexibility and has different roles: they aggregate, and coordinate flexibility provided by users, trade with flexibility, and conclude contracts with active user and flexible users (Flex Requesting parties), the transmission system operator, the distribution system operator, and the balance responsible party (BRP). In response, new aggregator platforms and flexibility market models are being developed and analyzed in [24]. The regulatory framework for the aggregator role is still under discussion, and is implemented differently in several EU countries. The status of explicit DSF for active users in the 26 EU Member States was examined in [26].

Consumption in households consists of different types of power-consuming appliances. In [27], home appliances are categorized as shiftable, interruptible, weather-based, and non-manageable. Shiftable appliances have flexible delays with specific energy consumption profiles, e.g., washing machines, dishwashers, etc. Interruptible appliances have fixed energy consumption when they are switched on and non-energy consumption when they are switched off, e.g., water heaters, refrigerators, etc. Weather-based appliances depend upon the weather and dimensions of the premises, e.g., heating, ventilation, and air conditioning appliances. Non-manageable appliances do not have the option to adjust their consumption, e.g., TVs, lights, etc. The home appliances in [28] are divided into two groups, shiftable and non-shiftable, and this division of home appliances is used in the following.

The HEMS requires the development of a framework for handling the energy needs and demands of an active user without compromising the comfort level and without greater involvement of the active user, so the process must be as automated as possible. The increasing use of RESs and BESSs in households can have significant benefits to the operations of HEMSs. The advantage of using a BESS is to store produced energy with the RES, and to use energy when the electric energy price is high. In [29], a HEMS with a RES and BESS is presented, where the findings obtained by simulating the problem using household data thoroughly confirm the efficiency of the provided HEMS in lowering the electric energy cost while keeping the appropriate level of consumer discomfort.

This work aims to examine the possibilities for an active user after the abolition of net metering and explore the economic analysis of households with solar PV plants in Slovenia, comparing the cost-effectiveness of households with and without net metering policies. An integral aspect of the article involves calculating the flexibility potential of an active user, which is technologically available using algorithms for detecting flexible loads. The focus is on a case study of a single-family home with an installed PV plant. In the study, a comprehensive analysis is carried out with and in the absence of net metering, as well as investigating strategies with aggregator involvement. The self-sufficiency indicator (SS), self-consumption (SC) indicator, grid-dependency indicator (GD), and annual electric energy costs (P_{EB}) are calculated to determine the economic viability of each strategy. Two main contributions are provided in this work. Firstly, the identification algorithms for identifying shiftable loads are included, which offers a robust methodology for the recognition of such loads. Secondly, it presents an analysis encompassing scenarios with and without net metering alongside strategies involving aggregators.

The paper is organized as follows. An introduction to the discussed problem is given in the first section. The case methodology is discussed in the second section. In the third section, a case study description is presented, whereas the fourth section presents and discusses the obtained results. The fifth section gives a conclusion, where the outline is presented for future work.

2. Methodology

2.1. Electric Energy Cost Calculation Model

Understanding the electric energy cost can help active users to make informed decisions about their energy consumption and to optimize their energy use to reduce costs. Electric energy costs consist of the following: the market price of electric energy, the network charge (transmission network charge and distribution network charge), contributions (a contribution for a market operator (C_1), a contribution for energy efficiency (C_2), a contribution to support the production of electric energy from renewable energy sources (RES), and high-efficiency co/generations (C_3)), and the excise duties on electric energy (C_4). The network charge represents the amount that an active user of the transmission system must pay for utilizing the system. Its purpose is to cover the costs incurred by the electricity system operators in each year of the regulatory period. The Energy Agency establishes the network charge for the electricity system in accordance with the Legal Act on the methodology for determining the regulatory framework for electricity system operators. The prices set by the Energy Agency for contributions and charges, which were valid in 2022, were used. In Slovenia, there is a two-rate tariff system, which is shown Figure 1. The low tariff (LT) is applied during off-peak hours, which last from 00:00 to 5:59 and from 22:00 to 23:59. The LT is also used on holidays and weekends. The high tariff (HT) is applied during peak hours, which last from 6:00 to 21:59.

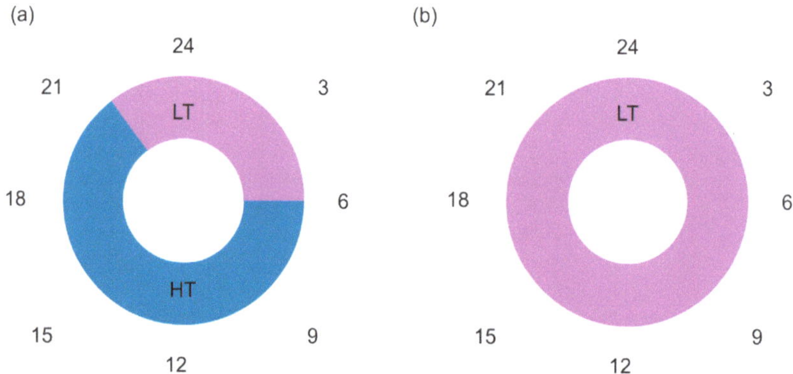

Figure 1. LT and HT on (**a**) weekdays (Monday to Friday) and (**b**) weekends (Saturday and Sunday) and national holidays.

The prices for LT (P_{LT}) and price HT (P_{HT}) are shown in Table 1. The network charge for LT (NC_{LT}) is lower than the network charge for HT (NC_{HT}). The capacity charge (CC) is a fixed charge based on the contracted capacity of the discussed household user. The capacity charge must be paid monthly. The rated power is determined by the current limiter, while the connection power and billing power are determined by the network regulation. The billing power (P_B) of the user is determined by the contract with the user based on the network regulation, according to which the billing power may also be lower or higher than the connection power. The billing power determines the amount of two charges on the electric energy bill: C_C and C_3. The C_C and C_3 factors are paid monthly and are not related to the electric energy consumption, but rather to the contracted power. Table 1 presents prices for tariffs and different charges for calculating the final price of electric energy without VAT.

Table 1. Charges and prices that are used to calculate the final price of electric energy for Slovenian households.

P_{LT} [EUR/kWh]	P_{HT} [EUR/kWh]	NC_{LT} [EUR/kWh]	NC_{HT} [EUR/kWh]	C_C [EUR/kW]	C_1 [EUR/kWh]	C_2 [EUR/kWh]	C_3 [EUR/kW]	C_4 [EUR/kWh]
0.118	0.082	0.03215	0.04182	0.77417	0.00013	0.0008	0.36948	0.00153

The annual electric energy costs (P_E) for a user are calculated using Equation (1), where factor 12 is used for the C_C and C_3 factors because we are calculating the consumption for the whole year. To allocate E_L^c to the different tariff rates accurately, it is necessary to segregate the total load consumption based on the time of the day. The E_G^c data were divided into corresponding periods to determine the energy consumption under LT (E_G^{cLT}) and HT (E_G^{cHT}), and can be written by (1).

$$E_G^c(t) - E_G^{cLT}(t) + E_G^{cHT}(t) \tag{1}$$

$$P_E = E_G^{cLT} \cdot P_{LT} \cdot C_{LT} + E_G^{cHT} \cdot P_{HT} \cdot C_{HT} + 12 \cdot P_B \cdot C_c + \\ + E_G^c \cdot C_1 + E_G^c \cdot C_2 + 12 \cdot P_B \cdot C_3 + E_G^c \cdot C_4 \tag{2}$$

2.2. Calculation of the SS, SC, and GD Indicators

The self-sufficiency (*SS*), self-consumption (*SC*), the grid-dependency (*GD*) indicators are calculated and can be used to assess the level of energy independence of a system. The *SS* indicator is used to estimate how much energy a PV plant can provide for its own needs without relying on external resources. The *SS* indicator describes the relationship between the energy produced from the PV plant for the observed time period and the total consumption of the household for an observed time period, and can be calculated using (3). The observed time period (*T*) can be a day, a month, a year, or any period of time.

$$SS = \frac{\sum\limits_{t=1}^{T} E_{PV}^P(t)}{\sum\limits_{t=1}^{T} E_L^c(t)} \cdot 100 \tag{3}$$

The *SC* indicator is used to estimate how much electric energy generated from the PV plant is actually used in a 15 min time interval. The *SC* indicator describes the relationships between the energy actually consumed from the PV plant for the observed time period and the total consumption of the load for the observed time period, and can be written by (4).

$$SC = \frac{\sum\limits_{t=1}^{T} E_{PV}^{c}(t)}{\sum\limits_{t=1}^{T} E_{L}^{c}(t)} \cdot 100 \tag{4}$$

When the BESS is included in the household, the *SC* indicator is used to estimate how much electric energy is generated from the PV plant, and the BESS is actually used in a 15 min time interval. Equation (5) is used to calculate the *SC* indicator with the inclusion of the BESS in the household.

$$SS = \frac{\sum\limits_{t=1}^{T} E_{PV}^{p}(t) + \sum\limits_{t=1}^{T} E_{BESS}^{c}(t)}{\sum\limits_{t=1}^{T} E_{L}^{c}(t)} \cdot 100 \, [\%] \tag{5}$$

The difference between the *SS* in the *SC* indicators is that the *SS* indicator is used to calculate a ratio between the total energy produced from the PV plant and the total consumption of load for the observed time. The *SC* indicator is used to calculate how much energy produced electric energy from a PV plant is actually consumed in a 15 min interval by a household user. The *GD* indicator is used to estimate the level of grid dependency. It is defined by (6) and describes the ratio between the consumption from the grid and the total load consumption.

$$GD = \frac{\sum\limits_{t=1}^{T} E_{G}^{c}(t)}{\sum\limits_{t=1}^{T} E_{L}^{c}(t)} \cdot 100 \, [\%] \tag{6}$$

The mentioned parameters, particularly the *SS* and *GD* indicators, assume a crucial role under the new accounting system, as the network no longer serves as virtual storage for produced energy. In ideal scenarios, our goal is absolute independence from the grid (*SS* = 100%). However, achieving this would necessitate an economically unjustifiably large BESS capable of maintaining such autonomy, especially during the winter months. Conversely, we aim to minimize the *GD* indicator, implying a shift in the load profile. Active users are encouraged to align their consumption with daily production, thereby reducing the *GD* indicator. Both parameters, therefore, wield significant influence in the decision-making process for individuals when determining the placement of the RES and BESS.

2.3. Description of the Aggregator Role

Household consumers and active users are becoming more equipped (PV plants and advanced metering) and have the potential to provide new services of flexibility. Flexibility is described as a movement in the active user's load profile caused by changes in the load profile. Although the flexibility potential of household active users is limited, an effective grouping of active users may be sufficient to keep the power system balanced. A new role—the aggregator—is required in the energy value chain for the consumers and prosumers to have access to this flexibility market. The aggregator connects numerous modest flexibility resources into a usable flexibility volume by working between active suppliers and active users. With the participation of active users in the energy value chain, an understanding of new relationships, services, and programs will be required. Any legal or natural person who wishes to participate actively in the electric energy market will have to join the Balance Scheme. Balance Schemes and electric energy market operators are connected through a set of procedures, regulations, and market mechanisms that ensure an efficient and reliable supply of electric energy. This connection is critical for maintaining the balance between electric energy generation and consumption in the power grid. The electric energy market in Slovenia is organized hierarchically into a Balance Scheme, which is managed by the electric energy market operator, Borzen, and d.o.o.

Document [30] addresses several possible aggregator implementation models that can be used to implement the aggregator role in current energy markets. The document presents six aggregator models, and one of these models was adopted for further research. The relationships between an active user, a new entity in the future electric energy market aggregator, and supplier are shown in Figure 2. The aggregator establishes all the contractual relations necessary to participate in the flexibility market and establishes connections with customers who own flexible resources. Active users will be compensated according to the degree of asset flexibility they provide. In order to adhere to the size and temporal requirements of particular flexibility products, the aggregator constructs a portfolio of assets. A balance responsible party (BRP) is in charge of balancing demand and supply actively for its portfolio. The aggregator is in charge of activating flexibility, while the active user is in charge of energy change. It is important to emphasize that, in order to participate in market mechanisms where active users can offer their flexibility potential, the active user must/can have separate contracts with the electric energy supplier and the aggregator. Aggregators empower active users to participate actively in DSF mechanisms to obtain incentives to reduce the load during peak periods. It is crucial to emphasize that aggregators play a key role in optimizing energy consumption. By managing and coordinating the various resources in their portfolios actively, aggregators can help active users implement energy conservation strategies, demand response initiatives, and redeployment practices.

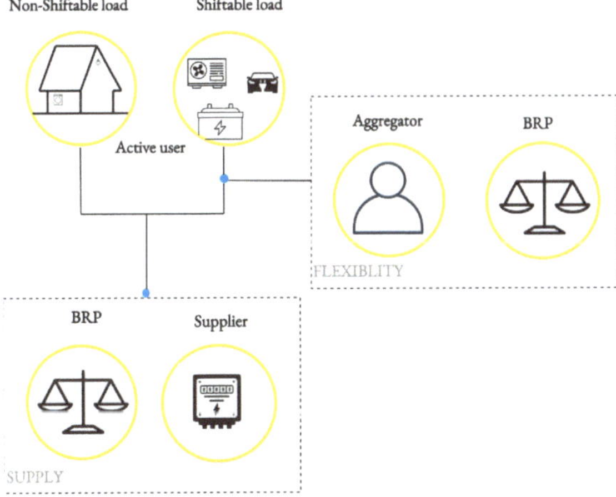

Figure 2. The assumed relationship between active user, new entity aggregator, and supplier in the future electric energy market.

Electric energy trading involves buying and selling electric energy in various forms, such as power contracts, futures, options, and swaps. This trading can occur between different market participants, including active users, aggregators, and energy service providers. Electric energy markets are structured differently in different regions and countries, but they all aim to provide an efficient and reliable electric energy supply at a reasonable price. Generally, electric energy markets operate based on supply and demand, where the price is set by the intersection of the supply curve (i.e., the amount of electric energy producers are willing to generate at a given price) and the demand curve (i.e., the amount of electric energy consumers are willing to buy at a given price).

Electric energy trading can occur in various markets, including wholesale markets, balancing markets, capacity markets, and futures markets. Wholesale markets are used for large-scale transactions between electric energy producers and consumers, while spot markets allow for the real-time trading of electric energy at current prices. Futures markets allow market participants to trade contracts for the future delivery of electric energy.

An electric energy market model requires crucial data from previous trading periods. Patterns and trends in electric energy demand and supply, as well as the prices at which electric energy is traded on the market, can be identified by analyzing the data. This information can then be utilized to develop a model that simulates the behavior of the electric energy market under different conditions. Energy exchanges such as the BSP South Pool Energy Exchange serve as an important source of data for data acquisition. The BSP South Pool Energy Exchange allows electric energy trading in the Slovenian and Serbian electric energy on the day-ahead, intraday, and balancing markets. Furthermore, the BSP Energy Exchange offers various other services and products, such as capacity reserve auctions, balancing market services, and the trading of guarantees of origin certificates. These products and services aim to provide a comprehensive platform for market participants to manage their electric energy portfolios and risks efficiently. BSP stands for the "BSP Regional Energy Exchange", which is a Slovenian energy exchange company that operates a trading platform for trading electric energy. A market model for electric energy trading was developed based on day-ahead trading annual data. The hourly trading price (T_p) data were employed in our analysis. This information was used to develop a model that simulates the behavior of the day-ahead electric energy market. Figure 3 offers a detailed overview of the hourly trading price fluctuations that occurred throughout 2022.

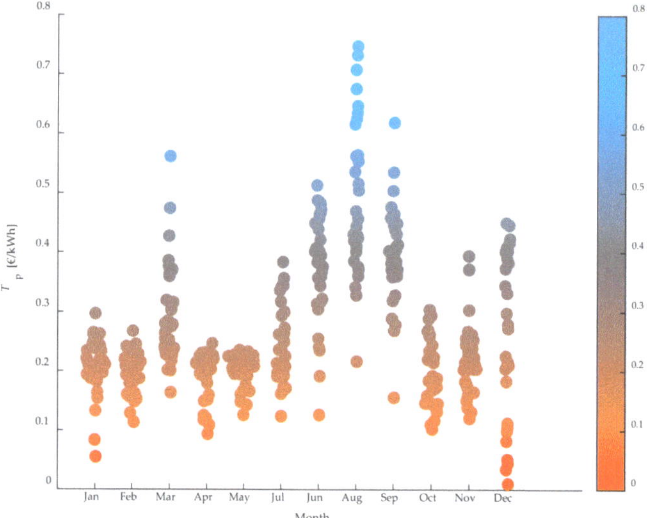

Figure 3. Hourly trading price on the day-ahead market for electric energy.

The patterns in the prices at which electric energy is traded on the day-ahead market were identified by analyzing the data. The analysis of the trading price on the day-ahead market revealed that the peak trading price of electric energy on the day-ahead market was observed in August, while the minimum trading price of electric energy on the day-ahead market was noted in December.

2.4. Identification of Flexibility Potential for a Shiftable Load

By defining the flexibility potential, an active user's appliances and sources can be utilized optimally. This enables the aggregator to gain insight into the active user's energy consumption profile and energy production profiles. The ability of an active user to consume or produce electric energy in response to an external signal is referred to as positive and negative flexibility. The positive flexibility potential represents the capability to increase energy consumption or decrease electric energy production as needed. The negative flexibility potential relies on the ability to decrease electric energy consumption or increase electric energy production as needed. Both types of flexibility potential are crucial and can be offered to an aggregator. The flexibility potential of the active user is determined by evaluating the flexibility potential for each shiftable household load. Shifting the active user's appliances and sources can help reduce the cost of electric energy. However, these changes require some level of compromise in terms of daily electricity pattern usage routines. This compromise can result in discomfort, which can be defined as a loss or impairment of energy comfort. The literature has identified three types of discomfort that have been discussed in this context [31]: delay or waiting time due to shifting time-flexible appliances, temperature deviation due to thermostat settings of temperature-based appliances, and power deviation resulting from reduced power of instant power-based appliances. To ensure the active user's comfort in strategy 2, a strategy was implemented where shiftable loads were manipulated only once a day, with each activation lasting just 30 min. The approach aimed to minimize the active user's discomfort while still achieving a reduction in cost for electric energy.

The flexibility potential of EVs is defined by the EV charging habits that are analyzed throughout the year. The flexibility potential of EVs is defined with (7), where E_{EV}^c is the electric energy consumption of the EV. The assumptions considered in defining the flexibility potential of the EV are as follows: the EV is charged every day and the EV is available every day after 16:00. The positive flexibility potential of the EV is defined by allowing the reduction and rescheduling of EV charging during charging periods, and changing the positive flexibility potential on an annual basis, as can be seen in Figure 4a. The negative flexibility potential of the EV includes identifying instances when the EV has not yet been fully charged within the day, and if it is past 16:00, the charging can be rescheduled. The negative flexibility potential of the EV throughout the year can be seen in Figure 4b.

$$E = \begin{cases} E_+(t) = E_{EV}^c(t), \; if \; t > 16:00 \, \& \, \sum_{t=0}^{t} E_{EV}^c(t) = 0 \\ E_-(t) = -E_{EV}^c(t), \end{cases}$$

(7)

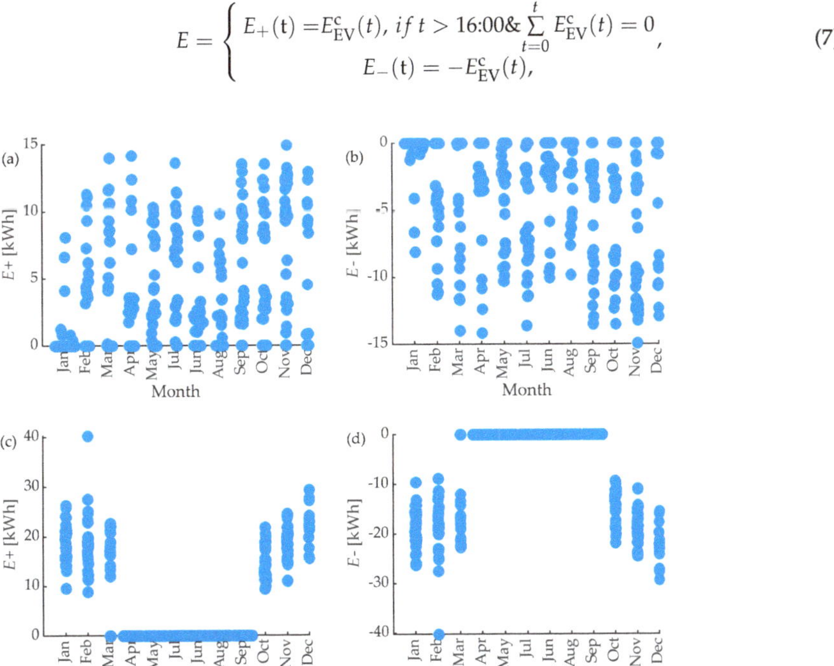

Figure 4. Positive and negative flexibility potential for the EV (**a**,**b**) and HP (**c**,**d**).

The flexibility potential of the HP is defined by (8), where E_{HP}^c is the electric energy consumption of the HP. The negative flexibility potential of the HP on an annual basis can be seen in Figure 4b, while the positive flexibility potential of the HP on an annual basis can be seen in Figure 4d. The f_{HP} weighting factor is a parameter that typically ranges between 0 and 1, determining the degree to which we want to adjust the power of the HP.

$$E = \begin{cases} E+(t) = E_{HP}^c(t) \cdot f_{HP} \\ E-(t) = -E_{HP}^c(t) \cdot f_{HP} \end{cases} \tag{8}$$

The flexibility potential of the PV is defined with (9), where E_{PV}^p is the electric energy produced from the PV and E_L^c is the consumption of the load. The negative flexibility potential of the PV size is defined as the difference between the energy produced from the PV and the consumed energy, and can be seen in Figure 5a. The positive flexibility potential of the PV is defined as zero, because PV plant systems are dependent on solar irradiance, which cannot be controlled by an active user, and can be seen in Figure 5b. The amount of produced energy from the PV plant is determined primarily by the environmental conditions.

$$E = \begin{cases} E+(t) = 0 \\ E-(t) = E_{PV}^p(t) - E_L^c(t) > 0 \end{cases} \tag{9}$$

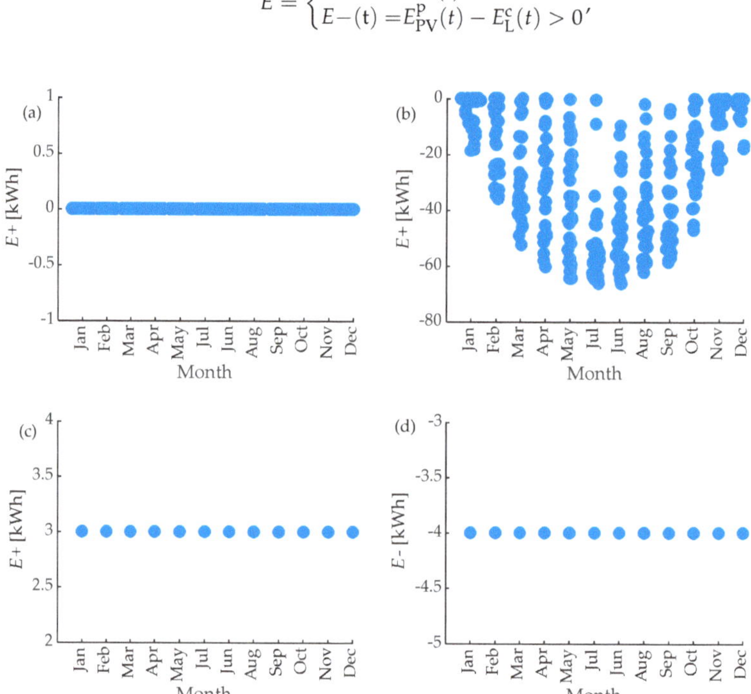

Figure 5. Positive and negative flexibility potential for a PV (**a,b**) and a BESS (**c,d**).

The flexibility potential of a BESS can be quantified based on its capacity to store and release energy at specific times, its efficiency in these processes, and its ability to respond to various grid events and signals. The operation and dynamics of the charging and discharging of the BESS are different for different strategies, which are described below. The flexibility potential of the BESS is defined by (10), where SOC_{min} and SOC_{max} are changing according to the strategy and different modes. The flexibility potential of the BESS is offered to an aggregator only in strategy 2. The negative flexibility potential of the

BESS on an annual basis can be seen in Figure 5c, while the positive flexibility potential of the BESS on an annual basis can be seen in Figure 5d.

$$E = \begin{cases} E+(t) = \dfrac{(SOC(t)-SOC_{\max}) \cdot C_{BESS}}{100} \\ E-(t) = \dfrac{(SOC(t)-SOC_{\min}) \cdot C_{BESS}}{100} \end{cases}, \tag{10}$$

2.5. Definition of Strategies

The research aims to explore the possibilities for active users after the abolition of net metering, and two strategies were designed for this purpose. Engaging in the electric energy market is a viable option for an active user. Nonetheless, since direct participation in the electric energy market is not possible, the active user needs to establish a connection with an aggregator. Two strategies were formulated to evaluate the advantages for an active user in collaborating with an aggregator. Given the recent approval of subsidies by the state for the construction of a BESS, there has been a notable increase in their adoption in households. Consequently, the upcoming analysis focuses on testing the strategy on two active user cases: one with the PV plant and another with both the PV plant and BESS. The operation for strategy 1 involves offering only excess energy from the PV plant to the aggregator. Strategy 2 offers a shiftable household load in addition to excess energy from the PV plant to the aggregator.

2.5.1. Strategies for an Active User with a PV Plant Only

In Figure 6a, the presentation shows strategy 1, which shows the scenario where the engaged active user provided excess energy from the PV plant to the aggregator. In Figure 6b, the presentation shows strategy 2, which shows the scenario where the engaged active user provided excess energy from the PV plant and shiftable load to the aggregator. Shiftable loads in strategy 2 are identified using the identification algorithm for shiftable loads presented in Section 2.3. Strategy 2 offers the flexibility potential of the HP and the flexibility potential of the EV to an aggregator. By incorporating shiftable loads, the active user's flexibility potential can be increased, providing the aggregator with more flexibility potential.

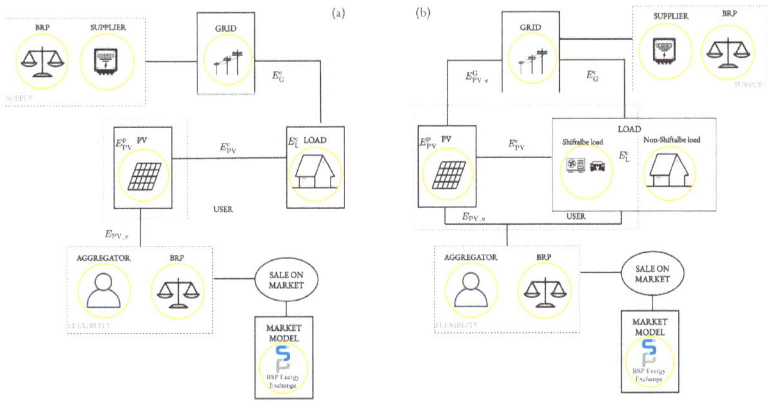

Figure 6. Strategy 1 (**a**) and strategy 2 (**b**) for an active user with a PV plant only.

Operation during strategy 1 for an active user with a PV plant only: The electric energy produced from the PV plant (E_{PV}^p) is divided into two energies: the energy consumed directly by the load (E_{PV}^c) and the excess energy (E_{PV_e}) which is offered to an aggregator, and, further, to the electric energy market, and can be written with (11).

$$E_{PV}^p(t) = E_{PV}^c(t) + E_{PV\_e}^A(t) \tag{11}$$

The load consumption E_L^c is equal to the energy taken from the grid (E_G^c) and the energy produced with the PV plant that is actually consumed directly by the load (E_{PV}^c), and can be written by (12).

$$E_L^c(t) = E_G^c(t) + E_{PV}^c(t) \tag{12}$$

In strategy 2, the following three modes of operation are distinguished: operation before activation, operation during activation, and operation after activation. Operation during activation is the same as in strategy 1, as the active user's flexibility potential is offered to an aggregator. Operation before and after activation will be presented in the following. The energy produced from the PV plant (E_{PV}^p) is divided into the following two energies: the energy that supplies the load directly (E_{PV}^c) and the excess energy ($E_{PV\_e}^G$) which is sent to the grid, and can be written by (13). The distribution of the load consumption (E_L^c) remains the same as in (13).

$$E_{PV}^p(t) = E_{PV}^c(t) + E_{PV\_e}^G(t) \tag{13}$$

2.5.2. Strategies for an Active User with a PV and BESS

In Figure 7a, the presentation shows strategy 1, which shows the scenario where the engaged active user provided excess energy from the PV plant that remained after charging the BESS to the aggregator. In Figure 7b, the presentation shows strategy 2, which shows the scenario where the engaged active user provided excess energy from the PV plant, BESS, and shiftable load to the aggregator. The shiftable loads in strategy 2 are identified the same as in the previous example.

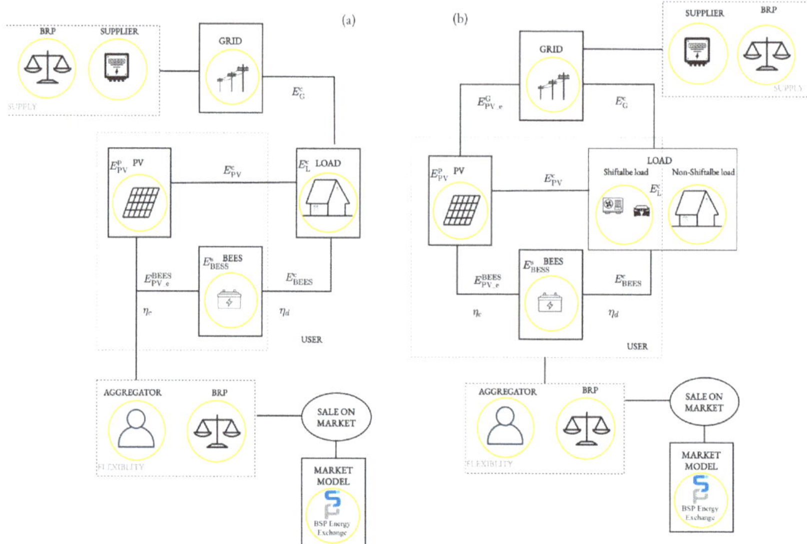

Figure 7. Strategy 1 (**a**) and strategy 2 (**b**) for an active user with a PV plant and BESS.

Operation during strategy 1 for an active user with a PV and BESS: the energy produced from the PV plant (E_{PV}^p) is divided into the following two energies: the energy that supplies the load directly (E_{PV}^c) and the excess energy (E_{PV_e}) which is offered to an aggregator, and can be written by (14).

$$E_{PV}^p(t) = E_{PV}^c(t) + E_{PV\_e}(t) + E_{PV\_e}^{BESS}(t) \tag{14}$$

The load consumption (E_L^c) is equal to the energy taken from the grid (E_G^c) and the energy produced with the PV plant that is actually consumed directly by the load (E_{PV}^c), and can be written by (15).

$$E_L^c(t) = E_G^c(t) + E_{PV}^c(t) + E_{BESS}^c(t) \tag{15}$$

Excess energy from the PV plant E_{PV_e} is equal to the energy sent to the aggregator ($E_{PV\_e}^A$) and the energy that charges the BESS ($E_{PV\_e}^{BESS}$).

$$E_{PV\_e}(t) = E_{PV\_e}^A(t) + E_{PV\_e}^{BESS}(t) \tag{16}$$

Three modes of operation are discussed for operation during strategy 2 for an active user with a PV and BESS. The operation before and after activation will be presented in the following. The energy produced from the PV plant (E_{PV}^P) is divided into two energies: the energy that supplies the load directly (E_{PV}^c) and the excess energy (E_{PV_e}) which is offered to an aggregator, and can be written by (17). The distribution of the load consumption (E_L^c) remains the same as in (15).

$$E_{PV}^P(t) = E_{PV}^c(t) + E_{PV\_e}^G(t) + E_{PV\_e}^{BESS}(t) \tag{17}$$

The excess energy from the PV plant E_{PV_e} is equal to the energy sent to the aggregator ($E_{PV\_e}^A$) and the energy that charges the BESS ($E_{PV\_e}^{BESS}$).

$$E_{PV\_e}(t) = E_{PV\_e}^G(t) + E_{PV\_e}^{BESS}(t) \tag{18}$$

Operation of the BESS during strategy 1: The efficiency of charging (η_c) and discharging (η_d) are considered when energy is stored in the BESS. The BESS is charged when there is excess energy available. The energy stored in the BESS (E_{BESS}^s) is positive in the case of charging the BEES, and is calculated according to (19).

$$E_{BESS}(t) = E_{PV\_e}^{BESS}(t) \cdot \eta_c \tag{19}$$

Discharge begins when the energy from the PV plant is not sufficient to consume the load. E_{BESS}^c is negative in the case of discharging the BEES and is calculated according to (20).

$$E_{BESS}(t) = E_{BESS}^c(t) \cdot \eta_d \tag{20}$$

The energy stored in the BESS E_{BESS}^s is calculated using (21), where $E_{BESS}^s(t-1)$ is the energy stored at the previous moment and the E_{BESS}.

$$E_{BESS}^s(t) = E_{BESS}^s(t-1) + E_{BESS}(t) \tag{21}$$

The state of charge (SOC) is the ratio between the energy stored in the BESS and the size of the BESS, and is calculated using (22).

$$SOC(t) = \frac{E_{BESS}^s(t)}{C_{BESS}} \cdot 100 \tag{22}$$

The operation of the BESS during strategy 1 can be described using (23), and operates between the lower limit (SOC_{min}), which is 20%, and the upper limit (SOC_{max}), which is 90%.

$$SOC_{min} \leq SOC(t) \leq SOC_{max} \tag{23}$$

2.5.3. Elaboration on the Distinction in the Operation of the BESS between Strategy 1 and Strategy 2

In strategy 1, the BESS operates within the range of the lower limit of the SOC_{min} set to 20% and the upper limit of the SOC_{max} set to 90%, as shown in Figure 8a. The operation of the BESS in strategy 2 is divided into the following 3 modes: before the operation, during the operation, and after the operation of the BESS. Before the activation, the SOC_{max} is set to 60%, as seen in Figure 8b. Therefore, the BESS remains available for charging or discharging during the activation mode. During the activation, the BESS undergoes charging or discharging based on the aggregator signal. The BESS is operated during the activation within the range of the lower limit of the SOC_{min} set to 10% and the upper limit of the SOC_{max} set to 90%. After the activation mode, two cases can be distinguished based on the action during activation. The BESS was discharged when a signal for negative flexibility was received from the aggregator. Subsequently, after the activation, the BESS underwent recharging from the grid. In a case where a signal for positive flexibility was received from the aggregator, and the BESS was charged from the grid; subsequently, after the activation, the BESS underwent discharging to the grid. A more detailed presentation of the BESS performance during and after activation will be elucidated in the Results section.

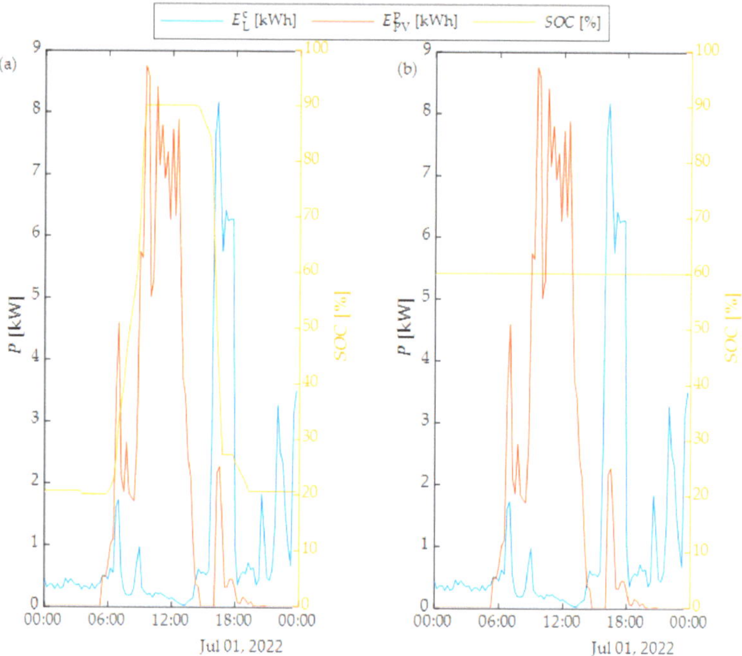

Figure 8. Performance of the BESS for strategy 1 (**a**) and strategy 2 (**b**).

2.6. Profit Determination

The profit from the sale of excess energy from the PV plant, P_{PV_e} is calculated using (24), where $E^G_{PV\_e}$ is the excess energy from the PV plant, T_p is the trading price, and f_A is a weighting factor that indicates how much the aggregator reduces profit relative to the T_p. Factor f_A is in the range of 0 to 1. If factor f_A is equal to 1, it signifies that the aggregator compensates in alignment with the prevailing electric energy prices on the day-ahead market.

$$P_{PV\_e}(t) = E^G_{PV\_e}(t) \cdot T_p(t) \cdot f_A(t) \; [€] \tag{24}$$

The difference between the electric energy payment P_{EB} and the profit from the sale of excess energy P_{PV_e} is calculated using (25), which is represented as P_d. P_d signifies the actual costs for electric energy that the active user is obligated to pay subsequent to the sale of excess energy.

$$P_d = P_{EB} - P_{PV\_e} \tag{25}$$

3. Case Study Description

The case study involved an active user in a single house with an installed PV plant of 11 kW. The house was equipped with an HP for central heating and a charging station for an EV. The research process consisted of several critical stages, as shown in Figure 9, starting with data acquisition. Information on electric energy consumption and energy produced from the PV plant in the 15 min interval was obtained on the "Moj elektro" platform. The data from the platform were exported in Excel and processed further in Matlab. The next step was data analysis, focusing on the consumption and production patterns. Following that, the identification of shiftable loads was conducted, and, additionally, a BESS was included. Important data-driven results were derived as a result of this comprehensive analysis. The results were divided into the following 3 categories: the analysis with net metering, the analysis without net metering, and the analysis to evaluate the advantages of a collaboration between an active user and an aggregator, where two strategies were tested.

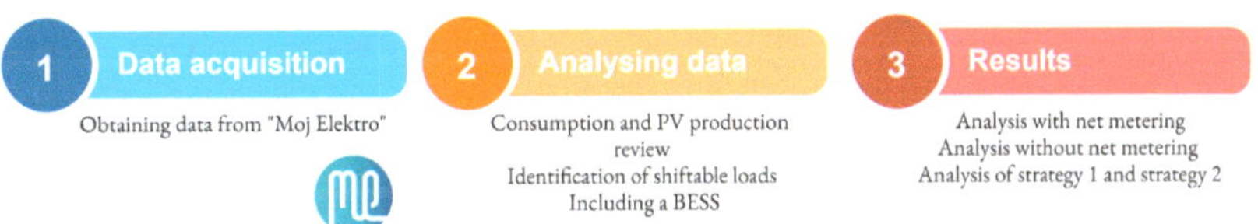

Figure 9. Process of the research.

3.1. Analyzing the Data

The analysis was conducted by examining the daily energy consumption and PV plant production records from the previous years. An analysis was carried out of the active user electric energy consumption and PV production for the year 2022. The daily E_L^c of each month is presented in Figure 10a. It can be seen how E_L^c changed depending on the season. The average daily E_L^c was 31.26 kWh, with a total yearly E_L^c of 11.17 MWh. In Figure 10b, the average E_{PV}^p of each month is presented, with higher production during the summer months. The average daily E_{PV}^p was 29.81 kWh, with a total yearly E_{PV}^p of 11.97 MWh. Based on the data, it can be inferred that the winter consumption is higher than the summer consumption due to the use of an HP for heating. Additionally, PV production is at its peak during the summer and at its lowest during the winter.

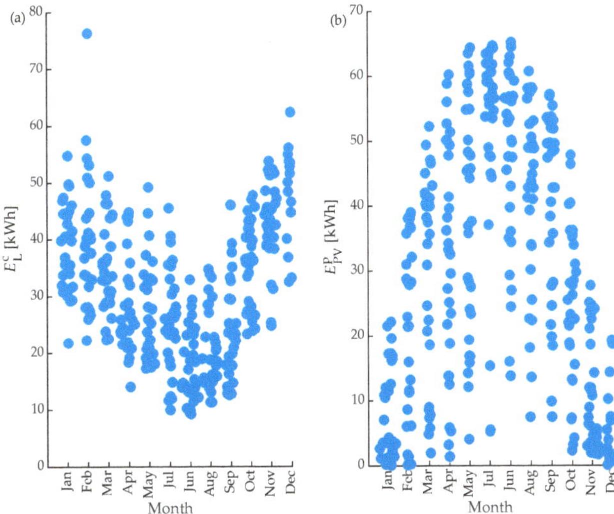

Figure 10. Daily load consumption (**a**) and PV production distributed (**b**) by month.

3.2. Algorithm for Shiftable Load Identification

Active users must estimate their flexibility potential to provide flexibility services. The identification of flexible loads in an active user profile is key in the estimation of the flexibility potential. It is also necessary to determine the degree of flexibility of the associated loads with a weighting factor. The algorithm for the identification of the EV and HP is presented below. The Algorithm 1: Identification of EV is designed to identify the moment with the highest electric energy consumption within each day. After the moment with the highest consumption is identified, the consumption for four moments before and after are examined to determine the EV consumption.

Algorithm 1: Identification of EV

Load input parameter (E_L^c, t)
for $t = 1,...,t_{end}(365) \rightarrow$ days in year
    if $t == t_{LT} \rightarrow$ max E_L^c for that day
    $\left[E_{Lmax}^c, t_{EV}\right] = \max(E_L^c(t)) \rightarrow$ find max consumption and index for that moment
    for every day
    End
calculate the consumption of EV if $t_{EV} - 4 \leq t \leq t_{EV} + 4$
$E_{EV}(t) = E_L^c(t) \cdot c_1$
calculate the consumption of EV $1 < t_{EV} - 4, t_{EV} + 4 > t_{end}$ $E_{EV}(t) = 0$
$E_{EV}(t)$ is a vector of EV consumption for every 15 min for a whole year

---

The Algorithm 2: Identification of HP was designed to compare the consumption between summer and winter days. The first step is to deduct the consumption of the EV from the total consumption. The average consumption for the summer dates is calculated next. Then, the algorithm subtracts the average summer consumption from the consumption in the heating season to determine the HP consumption. The summer season was defined from 21 June 2022 to 23 September 2022 and the heating season was defined from 1 January 2022 to 10 March 2022 and from 1 October 2022 to 31 December 2022.

Algorithm 2: Identification of HP

Load input parameter (E_L^c, t)
$E_{LW}^c(t) = E_L^c(t) - E_{EV}(t) \rightarrow$ EV consumption is subtracted from the consumption of all loads
Find summer dates (21 June 2022 till 23 September 2022) $\rightarrow t_{sd}$
For $t = t_{sd}$
    $\overline{E}_{LW}^c(t) = \text{mean}(E_{LW}^c(t)) \rightarrow$ average summer profile for every 15 min
End
Find dates for the heating season (21 June 2022 till 23 September 2022) $\rightarrow t_{hd}$
for $t = 1,...,t_{hd} \rightarrow$ days in year
    $E_{HP}^c(t) = E_{LW}^c(t) - \overline{E}_{LW}^c(t)$
End
calculate consumption of HP $E_{HP}(t) = E_{HP}^c(t) \cdot c_1$
$E_{HP}(t)$ is a vector of EV consumption for every 15 min for a whole year

---

The results of the algorithm for the identification of shiftable loads are presented in Figure 11, providing a comprehensive overview of the energy consumption patterns. The results of the EV recognition algorithm can be observed in Figure 11a,c. The highest point of total consumption during the day was located and is marked in Figure 11a,c with a gray dashed line. There is no noticeable difference between the EV consumption during the summer months in Figure 11a and the winter months in Figure 11c. The performance of the HP recognition algorithm is observed in Figure 11b,d. On the selected summer day, there was no HP consumption, while, on the selected winter day, the HP consumption was detected.

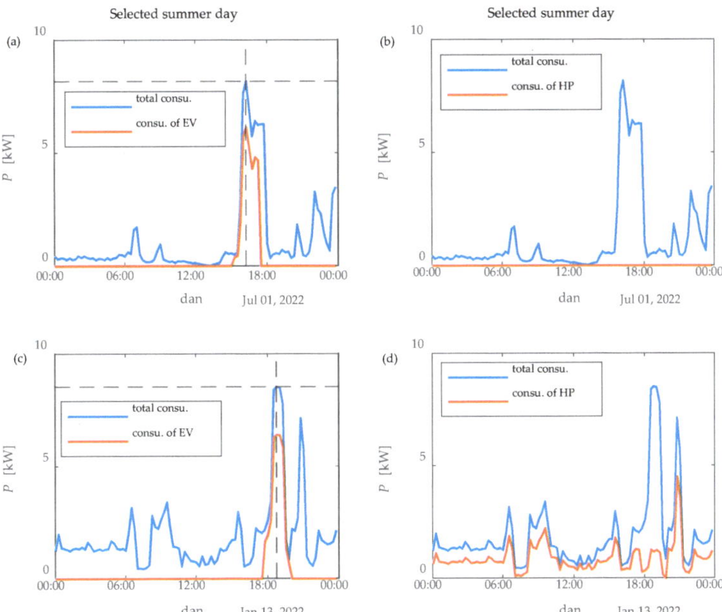

Figure 11. Results of the EV (**a,c**) and HP (**b,d**) recognition algorithm for the selected summer and winter days.

With the help of the identification algorithms, the consumption was divided into the following three categories: EV consumption and HP consumption, which both can be referred to as the consumption of shiftable loads and non-shiftable loads, as seen in Figure 12, where the consumption is presented for each month. The annual consumption of the EV amounted to 2 MWh, with the electric energy consumption remaining relatively consistent throughout the year, with a slight increase during the winter months due to the cold weather impacting to the travel range of the EV. The annual consumption of the HP amounted to 3 MWh, while the annual consumption of the non-shiftable household load amounted to 6 MWh.

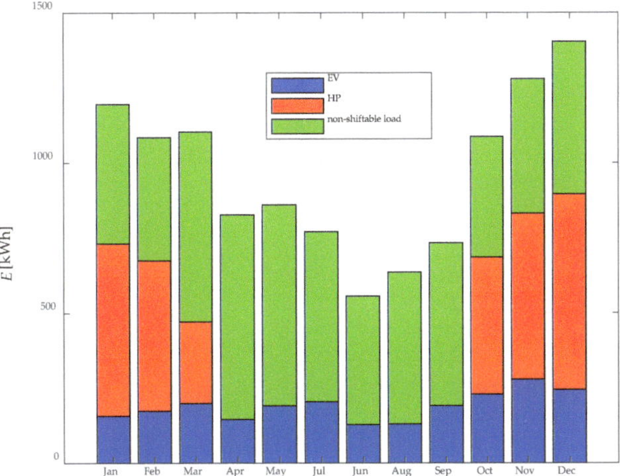

Figure 12. Load consumption for the shiftable (EV in HP) and non-shiftable household loads for every month.

4. Results and Discussion

4.1. Analysis with Net Metering

The analysis carried out using net metering revealed that the active user involved in net metering only has to pay the difference between the consumed and produced electric energy if the user consumed more energy than produced. Based on the analyzed data, the annual load consumption of E_c^L was 11.17 MWh, while the energy produced from the PV E_{PV}^P was 11.97 MWh. Since the active user produced as much energy as they consumed, they only needed to pay the monthly fixed fees, which include the capacity charge and contribution C_3. As a result, the active user involved in net metering paid EUR 96.07 without VAT for P_E, which is calculated using (26).

$$P_E = 12 \cdot P_B \cdot C_c + 12 \cdot P_B \cdot C_3 \tag{26}$$

Despite its proven benefits, the net metering policy in Slovenia is set to be abolished. This upcoming change is expected to transform the financial landscape for active users. As a result, it has become crucial to evaluate the economic feasibility of PV plants in the absence of net metering and explore the benefits of the active user's collaboration with an aggregator, where two strategies were investigated.

4.2. Analysis without Net Metering

With the upcoming abolition of net metering, the active user will no longer be enrolled in the net metering policy, which means that annual accounting will no longer be available. Therefore, an analysis of the consumed and produced energy was conducted in 15 min time intervals without net metering, which yielded the results shown in Table 2. The analysis of 15 min time intervals showed that the E_{PV}^c was only 1.92 MWh, which means that minimal electric energy was consumed in the 15 min interval. Without involvement in the net metering policy, the active user with a PV plant would only have to pay EUR 1401.2 for the P_{EB}, and the P_{PV_e} would be EUR 0, which means the P_d would be EUR 1401.2.

Table 2. Results of analysis without net metering.

Strategy	P_d [EUR]	C_{BESS} [kWh]	SS [%]	SC [%]	GD [%]
Without	1401.2	-	107	17	83
Without	1116.0	10	107	36	64

The grid is often perceived as a vast storage system due to the implementation of net metering policies. However, this perception fails to acknowledge the inherent limitations of the grid in storing excess energy generated by PV plants effectively, particularly during peak production periods in the summer. As a result, mismatches between energy production and consumption arise, notably during the winter months, when the demand escalates due to increased use of heat pumps for heating purposes. Such discrepancies pose challenges to system stability. This article has introduced indicators such as SS, GD, and SC to provide a more accurate representation of the level of self-sufficiency exhibited by active users and their dependence on the power grid. The SS indicator was 107%, indicating that the PV plant produces more electric energy annually than the active user consumes. However, only 17% of the electric energy produced by the PV plant was consumed, resulting in an SC indicator of 17%. The GD indicator was 83%, indicating that the majority of the electric energy consumed was from the grid. The GD indicator was very high because the produced energy from the PV was not consumed at the same 15 min interval. In the absence of net metering, the results of the indicators remained unchanged when compared to the results obtained through strategy 1. Therefore, no distinct outcomes were mentioned for strategy 1. Figure 13a–c represent the variations in the SS, GD, and SC indicators over the entire year. The SS indicator was highest in the summer months, where the values reached up to 300%, which means that the PV plant produced more electric energy than the active user consumed. However, in the winter months, particularly in January and December, the amount of the *SS* indicator decreased drastically to 20%, which was the lowest point throughout the year. The *GD* indicator was higher in the winter months and lower in the summer months, while the *SC* indicator was lower in the summer months and higher in the winter months. In the month of August, the *SS* indicator showed that the active user had produced more electric energy than they had consumed, with a percentage of 300%. However, taking a closer look at the *GD* indicator for the same month, it was observed that it was only 67%, while the *SC* indicator was 33%. This implies that, despite producing more electric energy than it consumed on a monthly basis, the active user still relied heavily on the grid for the majority of the 15 min intervals.

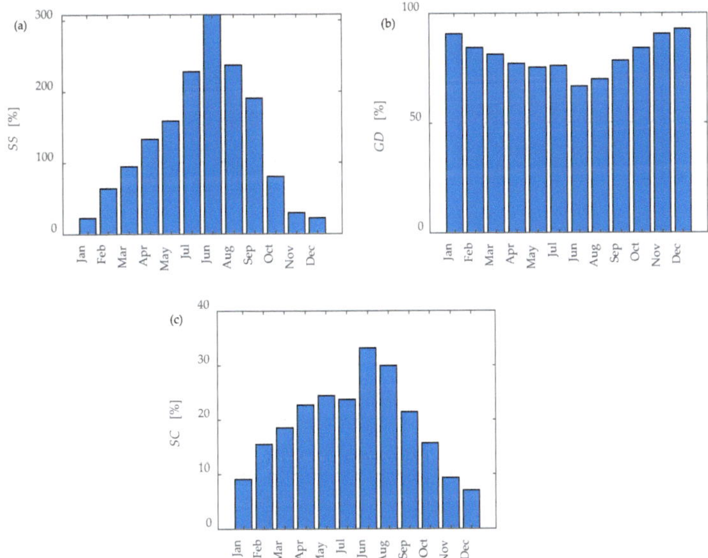

Figure 13. Analysis of the *SS* (**a**), *SC* (**b**), and *GD* (**c**) indicators for every month without the net metering policy and strategy 1 for an active user with a PV only.

An active user with a PV plant and BESS would have to pay less for the P_{EB}, which can be seen in results displayed in Table 2. The active user with a PV plant and BESS paid EUR 285.2 less for the P_{d}, and the $P_{\text{PV\_e}}$ was EUR 0 because the excess electric energy was not sold. The *SC* indicator was higher than in the case with the PV only, since a certain part of the excess energy from the PV plant was stored in the BESS and then consumed when needed. Therefore, the *GD* indicator was lower in the case with the PV plant and BESS.

Figure 14 shows the energy exchange with the grid for each day of the year. The blue color indicates the case without the included BESS and the red color indicates the case with the included BESS. Figure 14a shows how much energy is taken from the grid for each day, where it can be observed that, in the case without the BESS, more energy from the grid was taken compared to the case with the BESS. For the E_{G}^{c} for the whole year without the BESS, it amounts to 9.25 MWh, and for the case with the BESS included, it amounts to 7.35 MWh. Figure 14b shows how much excess energy from the PV plant is sent to the grid, where can be observed that, in both cases, more excess energy is sent to the grid during the summer months. In the case without the BESS, more excess energy is sent to the grid. For the whole year without the BESS included, it amounts to 10.05 MWh, and for the case with the BESS included, it amounts to 7.95 MWh.

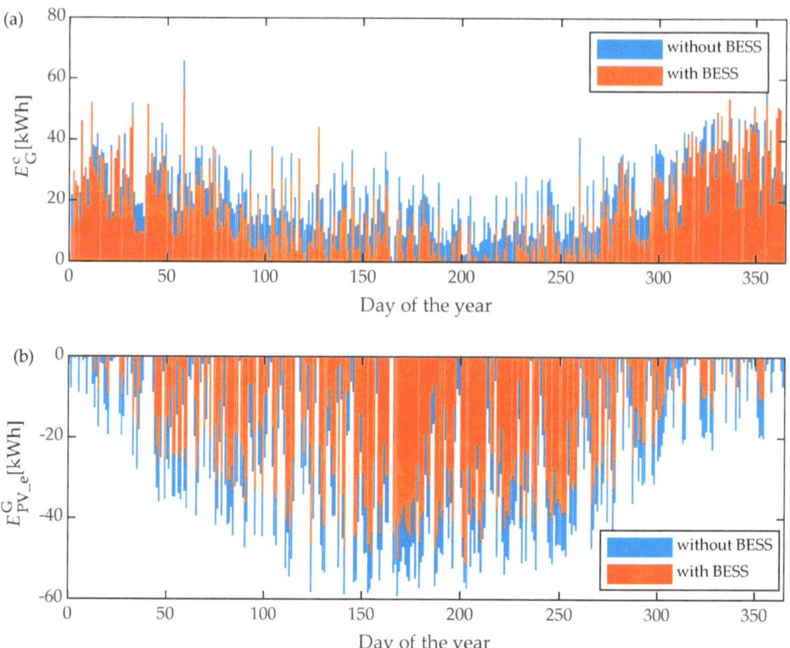

Figure 14. Energy exchange with the grid: (**a**) energy taken from the grid and (**b**) energy sent to the grid.

To evaluate the potential benefits of combining an active user and an aggregator, the following two strategies were taken into account for two different active users: an active user with a PV only and an active user with a PV and BESS. Through this analysis, the costs for electric energy and the benefits of these strategies can be evaluated, and the optimal strategy can be determined for an active user. The results of implementing both strategies are presented as follows: a comparison of the results obtained with the analysis for the active user with a PV plant only with the results obtained with the analysis for the active user with a PV plant and BESS for each strategy separately. Finally, a comparison of the results is presented in terms of the price, profit, and indications between the results obtained without using net metering and the results obtained from implementing the strategies.

4.3. Analysis of Strategy 1

Figure 15 presents a comparison of the flexibility potential between an active user with a PV plant only and an active user with a PV plant and BESS for strategy 1. The positive flexibility potential was equal to zero, since only excess energy was offered to the aggregator. A comparison between the flexibility potential of two active users, one with a PV plant and the other with a PV plant and BESS, is presented in Figure 15. Figure 15a shows the daily flexibility potential for the selected summer day, while Figure 15c shows the corresponding flexibility potential for the winter day. The active user with a PV plant and BESS had a lower negative flexibility potential of the PV plant than the active user with the PV plant only, because the excess energy generated by the PV must first charge the BESS before being offered to the aggregator. Figure 15b shows a comparison of the sum of the flexibility potential for each month. For an active user with a PV plant, the annual negative flexibility potential was 10.05 MWh. However, for an active user with both a PV plant and BESS, the annual negative flexibility potential was lower, and it was 7.68 MWh.

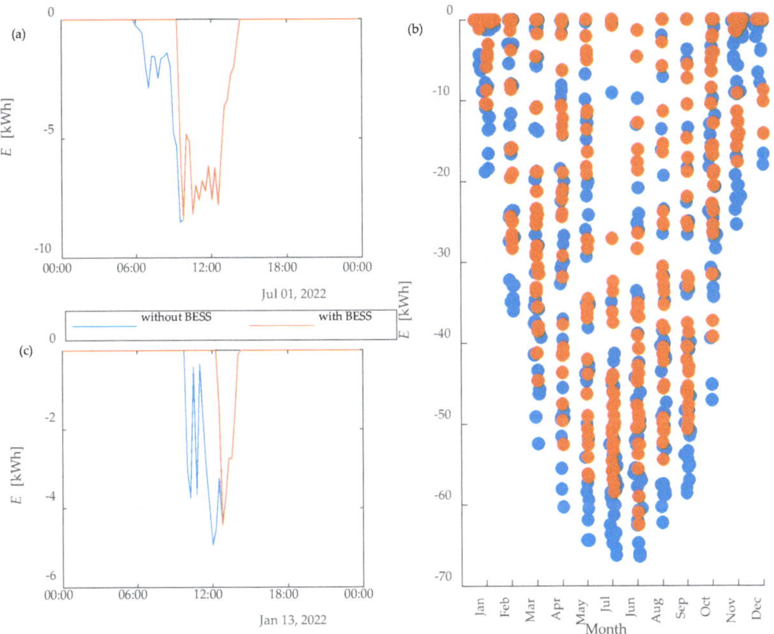

Figure 15. Comparison of the flexibility potential between an active user with a PV plant only and an active user with a PV plant and BESS for strategy 1: (**a**) for the selected summer day, (**b**) annual monthly flexibility potential, and (**c**) for the selected winter day.

An analysis of strategy 1 was performed, yielding the results shown in Table 3. For an active user with a PV plant, only the P_{PV_e} was greater than in the case with an active user with a PV and BESS, because an active user with a PV plant has more excess energy that can be sold to the aggregator. The P_{PV_e} was dependent on the factor f_A and agreed with the aggregator, and the higher the factor is, the greater the P_{PV_e} that can be obtained.

Table 3. Results from strategy 1 analysis

Strategy	C_{BESS} [kWh]	f_A	P_{PV_e} [EUR]	P_d [EUR]
1	-	0.2	599.7	801.5
1	-	0.4	1199.5	201.7
1	-	0.6	1.799.2	−397.9
1	10	0.2	459.8	656.1
1	10	0.4	919.7	196.3
1	10	0.6	1379.5	−263.54

4.4. Analysis of Strategy 2

Figure 16 presents a comparison of the flexibility potential between an active user with a PV plant only and an active user with a PV plant and BESS for strategy 2. Figure 16a shows the daily negative potential for the selected summer day, while Figure 16c shows the positive potential for the same day. The flexibility potential for an active user with a PV plant consists of the sum of individual flexibility potentials: the flexibility potential of the PV plant, the flexibility potential of the EV, and the flexibility potential of the HP. In addition to the previous ones, the potential for an active user with a PV plant and BESS also contains the potential of the BESS. Figure 16b shows a comparison of the sum of the negative flexibility potential for each month, while 16d shows a comparison of the sum for the positive flexibility potential for each month. The annual negative potential for an active user with a PV plant was 15.34 MWh, while the annual positive potential was 4.4 MWh. The annual negative potential for an active user with a PV plant and BESS was 66 MWh, while the annual positive potential was 22 MWh. The flexibility potential represents the capacity for active users to adjust their electric energy consumption patterns, especially in response to external signals. However, the extent to which this flexibility potential is harnessed and offered to aggregators varies from one individual to another, depending on their willingness to adapt and optimize their energy consumption patterns. Therefore, this flexibility potential serves as the basis for assessing the amount of potential that the active user can offer to the aggregator.

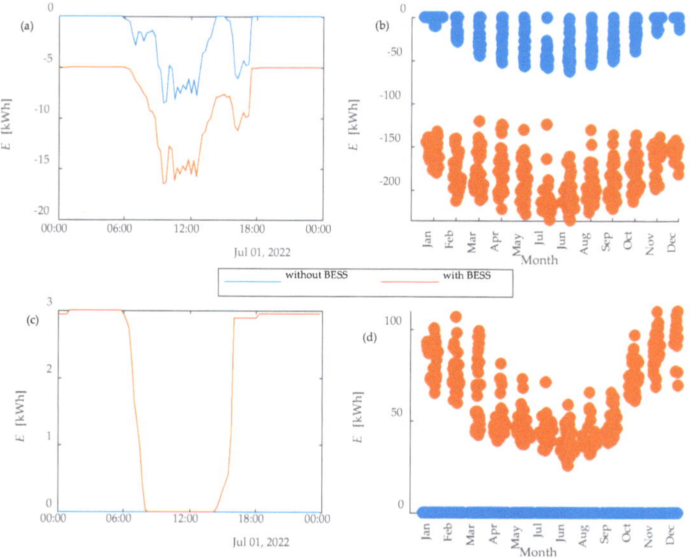

Figure 16. Comparison of the flexibility potential between an active user with a PV plant only and an active user with a PV plant and BESS for strategy 2: (**a**) negative flexibility potential for the selected summer day, (**b**) annual monthly negative flexibility potential, (**c**) positive flexibility potential for the selected summer day, and (**d**) annual monthly positive flexibility potential.

Figures 17 and 18 were created to provide a clearer understanding of how the flexibility potential was composed. Figure 17 shows the composition of the flexibility potential for an active user with a PV only. Figure 17a shows the negative flexibility potential for the selected summer day, while Figure 17b shows the negative flexibility potential for the selected winter day. The negative flexibility potential for active user with the PV consists of the negative flexibility potential of the PV, the negative flexibility potential of the EV, and the negative flexibility potential of the HP during the heating season. The negative flexibility potential of the PV was evident during the electric energy generation. The negative flexibility potential of the HP was present during the heating season, and the power of the HP can be manipulated. The negative flexibility potential of an EV occurs when the EV is charging and can be rescheduled. Figure 17c shows the positive flexibility potential for the selected summer day, while Figure 17b shows the positive flexibility potential for the selected winter day. The positive flexibility potential of the PV is not present, since the input power of the PV plant cannot be manipulated. As a result, active users with PVs do not have any positive flexibility potential. However, during the heating season, the positive flexibility potential of the HP arises, as its power can be manipulated.

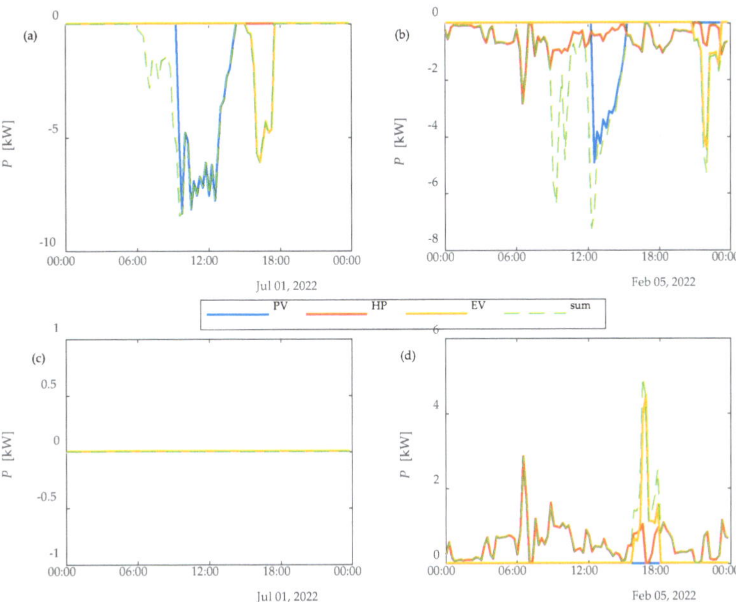

Figure 17. Composition of the flexibility potential for an active user with a PV: (**a**) the negative flexibility potential for the selected summer day, (**b**) the negative flexibility potential for selected winter day, (**c**) the positive flexibility potential for selected summer day, (**d**) the flexibility potential for selected winter day.

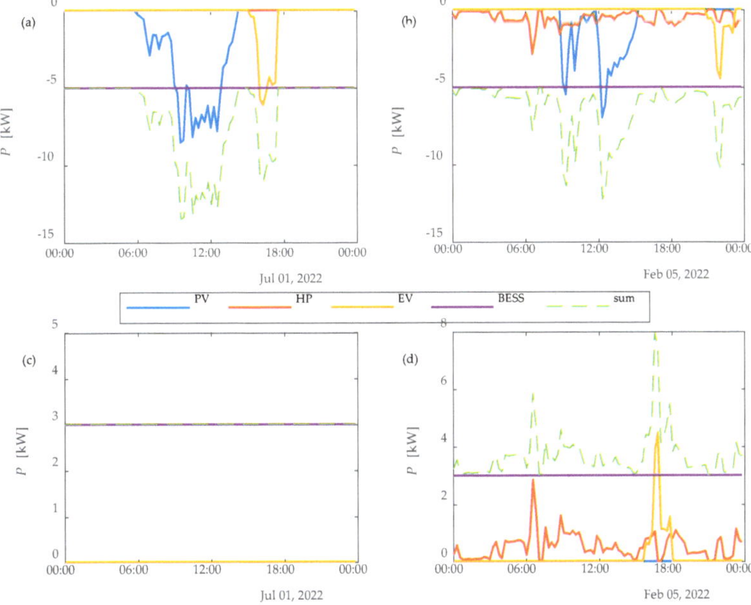

Figure 18. Composition of the flexibility potential for an active user with a PV and BESS (**a**) the negative flexibility potential for the selected summer day, (**b**) the negative flexibility potential for selected winter day, (**c**) the positive flexibility potential for selected summer day, (**d**) the flexibility potential for selected winter day.

Figure 18 shows the composition of the flexibility potential for an active user with a PV and BESS. Figure 18a shows the negative flexibility potential for the selected summer day, while Figure 18b shows the negative flexibility potential for the selected winter day. The negative flexibility potential for an active user with a PV and consists of the same negative flexibility potential as an active user with a PV only with the addition of the negative flexibility potential of the BESS. The BESS possesses negative flexibility potential, which can be utilized by discharging the BESS from 60% to 10% of the *SOC*. Figure 18c shows the positive flexibility potential for the selected summer day, while Figure 18b shows the positive flexibility potential for the selected winter day. The positive flexibility potential of the BESS is present, and it can be charged from 60% to 90% of the *SOC*. The main difference between the flexibility potential of an active user with a PV and an active user with a PV and BESS is that the active user with the BESS constantly has access to the BESS, which is available for the aggregator's use.

The following results are presented to provide a clearer understanding of what occurs during the activation mode/aggregation signal. Figure 19 shows the load consumption for the selected day during two responses to the aggregator signal. In Figure 19, activation 1 was marked at 7:15, when the aggregator sent a signal for positive flexibility. Similarly, activation 2 was marked at 17:15, when the aggregator sent a signal for negative flexibility. Figure 19 displays the changes during and after activation 1 in the box on the left corner, where the load consumption changed by 0.50 kWh during activation 1. After the activation, the load consumption for activation 1 was lower than in the case without activation, due to the fact that, during activation, the house was heated more than normal, so, after activation, the heating of the house was reduced. Changes during and after activation 2 are shown in Figure 19 in the box in the middle. The load consumption changed by 3.73 kWh during the activation.

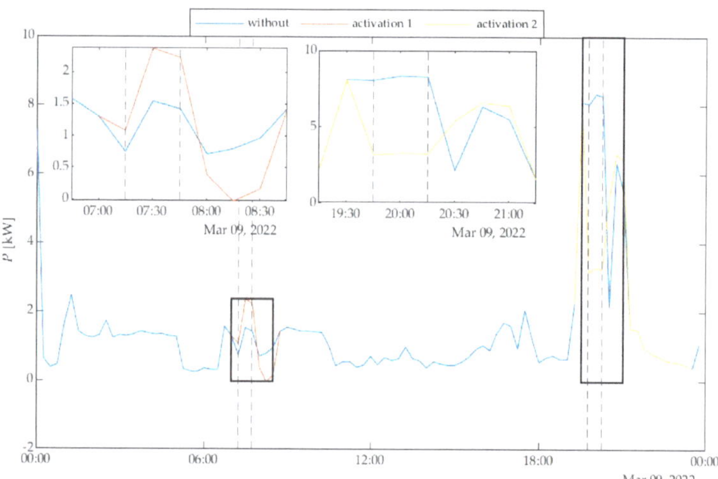

Figure 19. Changes in load consumption during the aggregation signal, where activation 1 is marked with first black box and activation 2 is marked with the second black box.

The consumption from the grid for the selected day during two responses to the aggregator signal is displayed in Figure 20. The changes during and after activation 1 are shown in Figure 20 in the box in the left corner. The change in consumption from the grid during the activation 1 was 3.34 kWh. The changes during and after activation 2 are shown in Figure 20 in the box in the middle. The change in consumption from the grid during the activation 2 was 6.17 kWh.

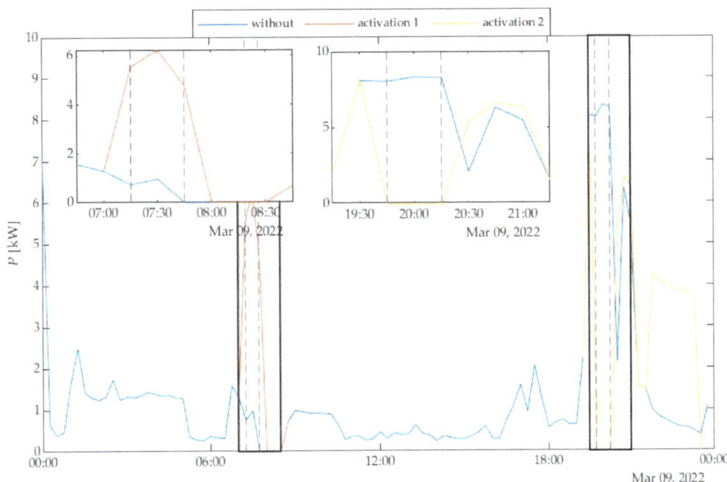

Figure 20. Consumption from the grid during the aggregator signal for positive and negative flexibility signals, where activation 1 is marked with first black box and activation 2 is marked with the second black box.

Figure 21 shows the excess energy sent to the grid, and Figure 22 shows the *SOC* for the selected day during two responses to the aggregator signal. The BESS was charged to 90% *SOC* during activation 1, which means that, as part of its operational requirements, it must subsequently be discharged down to 60% *SOC*. This ensures that the BESS operates within the specified *SOC* range and is ready for activation. Therefore, after activation 1, the excess energy from the BESS is sent to the grid. During activation 2, the BESS is discharged from 60% to 20% *SOC*. Hence, not only the load consumption was reduced during activation 2, but also excess energy was sent to the grid.

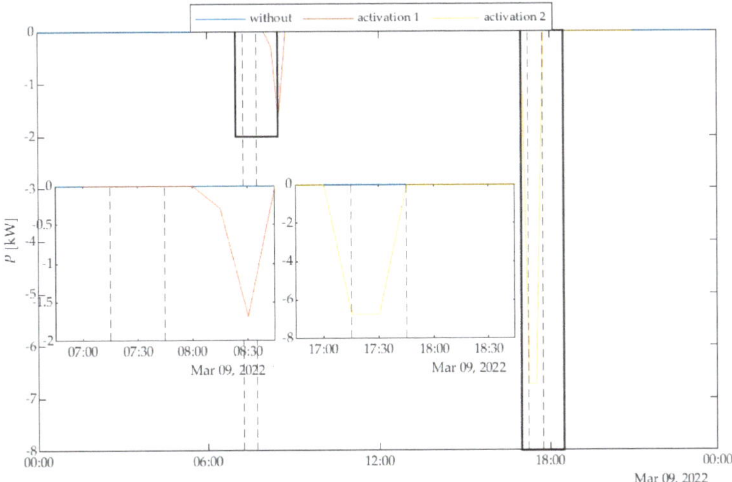

Figure 21. Excess energy sent to the grid during the aggregator signal for positive and negative flexibility, where activation 1 is marked with first black box and activation 2 is marked with the second black box.

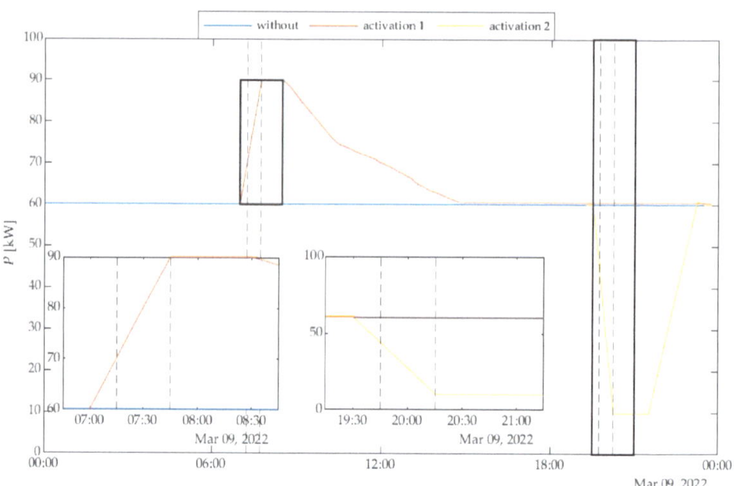

Figure 22. *SOC* during the aggregator signal for positive and negative flexibility, where activation 1 is marked with first black box and activation 2 is marked with the second black box.

An analysis of strategy 2 was performed, yielding the results shown in Table 4. For a case with an active user with a PV plant and BESS, the $P_{\mathrm{PV\_e}}$ was 56% higher than in the case of an active user without the BESS. It is important to consider that an amount of profit was generated through 365 activations, each lasting only 30 min. After the activations, there was still excess energy left. In the case of an active user with a PV plant, the excess energy was higher than in the case of an active user with a PV and BESS. In the case of strategy 2, the P_{d} for the active customer was lower as compared to strategy 1. However, strategy 2 involved offering significantly less excess energy than in strategy 1. Excess energy was still left post-activation and represents the potential for the sale of excess energy. Active users have an exciting opportunity to make the most of their excess energy, either by selling it or exploring alternative ways to use it effectively. This not only helps them to maximize their renewable energy investments, but also presents an intriguing prospect to contribute towards a sustainable future. The *SS* indicator remained the same. The *GD* indicator increased by 2% compared to the previous strategy, because the operation of the BESS changed during activation.

Table 4. Results for strategy 2.

Strategy	C_{BESS} [kWh]	f_{A}	$P_{\mathrm{PV\_e}}$ [EUR]	P_{d} [EUR]	$E_{\mathrm{PV\_e}}^{\mathrm{G}}$ [kWh]	*SS* [%]	*SC* [%]	*GD* [%]
2	-	1	26.6	1346.0	8.29	107	15	85
2	10	1	148.8	967.2	6.01	107	32	68

4.5. Limitations and Usability

The presented results cater to two primary audiences: potential active users on the one hand and potential aggregators (flexibility providers) on the other. For the first group, active users, it is crucial to have a clear and transparent understanding of the implications of their potential inclusion in the flexibility system. By doing so, they can actively engage in the flexibility market using their investment inputs (PVs and BESSs), thereby attaining economic benefits through relatively minor adjustments in energy consumption. On the active user's side, limitations arise from the permitted connection power of the PV system and BESS, along with financial constraints. We also highlight that active users can engage in multiple daily activations in both energy transmission and reception directions, contingent upon their capacities.

There is compelling potential for flexibility and the activation of energy needs among a larger number of active users. This collective energy could prove valuable for balance groups, the distribution network, or the transmission network. The essence of this segment lies in securing adequate capacity, enabling the aggregator to present a compelling offering to the flexibility market on behalf of active users.

As the European flexibility market aims for uniformity across the entire European Union, the outcomes of the case at hand are valuable not only for the current users but also for other potential users of this service.

5. Conclusions

This study focused on investigating the economic feasibility of households with PV plants in Slovenia in light of the abolition of net metering policies. One of the important parts of this article was conducting a preliminary analysis based on a case study. The highest electric energy consumption is in the winter season and the highest PV production is in the summer season. Assuming that the active user is generating more electric energy from their PV plant than they consume, without net metering they would be forced to sell any excess electric energy to the grid for a relatively low price and buy electric energy from the grid when they need it at a higher price. This can result in a significantly higher cost of electric energy. To explore options for active users after the abolition of net metering, two strategies were designed, including engaging in the electric energy market. The recent approval of subsidies for the BESS led to the testing of two active users' cases: an active user with a PV plant only and an active user with a PV and BESS. Collaboration with the aggregator is necessary, since direct participation in the electric energy market for active users is not possible.

In order to provide flexibility services, active users need to estimate their flexibility potential, which involves identifying shiftable loads. The identification algorithm outlined in the study for identifying the EV and HP emerges as a pivotal component, offering a robust methodology for discerning such loads. This aspect of shiftable load identification assumes paramount importance within the broader context of resource management and sustainability initiatives, as it lays the groundwork for optimizing energy usage and facilitating the integration of renewable energy sources. Moreover, shiftable load identification plays a crucial role in DRP and grid flexibility initiatives.

We highlight various economic aspects comparing scenarios with and without net metering, with the latter carrying an exceptional cost burden. The economic advantage of using battery energy storage systems (BESSs) becomes apparent, particularly when coupled with collaboration with an aggregator.

Without net metering, we introduced two strategies to address high electricity bills. Strategy 1 focuses on maximizing revenue by selling excess energy generated by renewable sources, such as a PV plant. Strategy 2 involves activations through an aggregator, presenting a potential for profit despite offering less excess energy. In strategy 1, profit depends on the agreed factor fA and is higher for active users with a PV plant.

In the article, new indicators such as *SS*, *GD*, and *SC* have been introduced to provide a more precise representation of the level of self-sufficiency shown by active users and their reliance on the grid. The *SS* indicator remained the same, regardless of the chosen strategy, indicating that the PV plant generated more electric energy annually than the active user consumed. The *GD* indicator was higher, and, therefore, the *SC* indicator was lower in the case of strategy 2 due to the modified operation of the BESS.

This work contributes to the understanding of the potential of renewable energy sources and the BESS as an alternative to traditional energy sources. The results of this analysis can guide active users in making informed decisions regarding their investment in PV systems and energy management strategies. Future work will be based on production and consumption forecasts and detailed economic analysis, with the inclusion of the price of the investment.

Author Contributions: Conceptualization, E.T. and M.B.; methodology, E.T.; software, E.T.; validation, E.T. and M.B.; formal analysis, E.T.; investigation, E.T.; resources, E.T.; data curation, E.T.; writing—original draft preparation, E.T. and M.B.; writing—review and editing, E.T.; visualization, E.T.; supervision, M.B.; project administration, M.B.; funding acquisition, M.B. All authors have read and agreed to the published version of the manuscript.

Funding: This research received no external funding.

Data Availability Statement: The original contributions presented in the study are included in the article, further inquiries can be directed to the corresponding author.

Conflicts of Interest: The authors declare no conflict of interest.

Abbreviations

C_1	A contribution for a market operator [EUR]
C_2	A contribution for energy efficiency [EUR]
C_3	A contribution to support the production of electric energy from renewable energy sources and high-efficiency co-generations [EUR]
C_4	Excise duties on electric energy [EUR]
C_{BESS}	BESS capacity [kWh]
C_c	Billing power [kW]
E_{BESS}	Instantaneous energy in the BESS [kWh]
E_{BESS}^c	Energy consumed from the BESS [kWh]
E_G^c	Energy taken from the grid [kWh]
E_{PV}^c	Energy that supplies the household load directly from the PV [kWh]
E_L^c	Load consumption energy [kWh]
E_{PV}^p	Energy produced from the PV plant [kWh]
$E_{\text{PV\_e}}^G$	Excess energy sent to the grid [kWh]
$E_{\text{PV\_e}}^{\text{BESS}}$	Excess energy from the PV plant that supplies the BESS [kWh]
E_{BESS}^s	Energy stored in the BESS [kWh]
E_G^{cLT}	Energy used under LT [kWh]
E_G^{cHT}	Energy used under HT [kWh]
f_A	Weighting factor that represents the reduction in profit, and it must be agreed upon with the aggregator
f_{HP}	Weighting factor that determines the extent to which we need to modify the power of the HP
GD	Grid-dependency indicator [%]
N_{LT}	Network charge for LT [EUR/kWh]
N_{HT}	Network charge for HT [EUR/kWh]
P_{EB}	Initial costs for electric energy [EUR]
P_d	Costs for electric energy after deduction of the costs for $P_{\text{PV\_e}}$ [EUR]
$P_{\text{PV\_e}}$	Profit from selling the excess energy [EUR]
P_{LT}	Cost for LT [EUR/kWh]
P_{HT}	Cost for HT [EUR/kWh]
SC	Self-consumption indicator [%]
SOC	State of charge [%]
SOC_{min}	Minimum state of charge [%]
SOC_{max}	Maximum state of charge [%]
SS	Self-sufficiency indicator [%]
T_p	Trading price on the day-ahead market [EUR/kWh]
η_c	Charging efficiency
η_d	Discharging efficiency

References

1. Burbano, A.M.; Martín, A.; León, C.; Personal, E. Challenges for citizens in energy management system of smart cities. In Proceedings of the 2017 Smart City Symposium Prague (SCSP), Prague, Czech Republic, 25–26 May 2017; pp. 1–6. [CrossRef]
2. Prakash, K.B.; Padmanaban, S.; Mitolo, M. 4 Security Challenges in Smart Grid Management. In *Smart and Power Grid Systems—Design Challenges and Paradigms*; River Publishers: Aalborg, Denmark, 2022; pp. 117–134.
3. Mahapatra, B.; Nayyar, A. Home energy management system (HEMS): Concept, architecture, infrastructure, challenges and energy management schemes. *Energy Syst.* **2019**, *13*, 643–669. [CrossRef]
4. López Prol, J.; Steininger, K.W. Photovoltaic self-consumption regulation in Spain: Profitability analysis and alternative regulation schemes. *Energy Policy* **2017**, *108*, 742–754. [CrossRef]
5. Ramirez, F.J.; Honrubia-Escribano, A.; Gomez-Lazaro, E.; Pham, D.T. Combining feed-in tariffs and net-metering schemes to balance development in adoption of photovoltaic energy: Comparative economic assessment and policy implications for European countries. *Energy Policy* **2017**, *102*, 440–452. [CrossRef]
6. Vilaça Gomes, P.; Knak Neto, N.; Carvalho, L.; Sumaili, J.; Saraiva, J.T.; Dias, B.H.; Miranda, V.; Souza, S.M. Technical-economic analysis for the integration of PV systems in Brazil considering policy and regulatory issues. *Energy Policy* **2018**, *115*, 199–206. [CrossRef]
7. Virtič, P.; Kovačič Lukman, R. A photovoltaic net metering system and its environmental performance: A case study from Slovenia. *J. Clean. Prod.* **2019**, *212*, 334–342. [CrossRef]
8. De Boeck, L.; Van Asch, S.; De Bruecker, P.; Audenaert, A. Comparison of support policies for residential photovoltaic systems in the major EU markets through investment profitability. *Renew. Energy* **2016**, *87*, 42–53. [CrossRef]
9. Yamamoto, Y. Pricing electricity from residential photovoltaic systems: A comparison of feed-in tariffs, net metering, and net purchase and sale. *Sol. Energy* **2012**, *86*, 2678–2685. [CrossRef]
10. Górnowicz, R.; Castro, R. Optimal design and economic analysis of a PV system operating under Net Metering or Feed-In-Tariff support mechanisms: A case study in Poland. *Sustain. Energy Technol. Assess.* **2020**, *42*, 100863. [CrossRef]
11. European Union. Directive (EU) 2019/944 of the European Parliament and of the Council of 5 June 2019 on common rules for the internal market for electricity. *Off. J. Eur. Union* **2019**, *158*, 125–211.
12. Uradni list Republike Slovenije, Uredba o Samooskrbi z Električno Energijo iz Obnovljivih Virov Energije. 2015. Available online: http://www.pisrs.si/Pis.web/pregledPredpisa?id=URED7867 (accessed on 5 September 2023).
13. Uradni List Republike Slovenije, Zakon o Spodbujanju Rabe Obnovljivih Virov Energije (ZSROVE). 2021. Available online: http://www.pisrs.si/Pis.web/pregledPredpisa?id=ZAKO8236 (accessed on 5 September 2023).
14. Official Journal of European Union, Directive (EU) 2018/ 2001 of the European Parliament and of the Council of 11 December 2018 on the Promotion of the Use of Energy from Renewable Sources z Električno Energijo. 2018. Available online: https://eur-lex.europa.eu/legal-content/EN/TXT/?uri=uriserv:OJ.L_.2018.328.01.0082.01.ENG&toc=OJ:L:2018:328:TOC (accessed on 5 September 2023).
15. Rauf, A.; Al-Awami, A.T.; Kassas, M.; Khalid, M. Optimal Sizing and Cost Minimization of Solar Photovoltaic Power System Considering Economical Perspectives and Net Metering Schemes. *Electronics* **2021**, *10*, 2713. [CrossRef]
16. Ndwali, K.; Njiri, J.G.; Wanjiru, E.M. Multi-objective optimal sizing of grid connected photovoltaic batteryless system minimizing the total life cycle cost and the grid energy. *Renew. Energy* **2020**, *148*, 1256–1265. [CrossRef]
17. Alramlawi, M.; Li, P. Design Optimization of a Residential PV-Battery Microgrid With a Detailed Battery Lifetime Estimation Model. *IEEE Trans. Ind. Appl.* **2020**, *56*, 2020–2030. [CrossRef]
18. Bhamidi, L.; Sivasubramani, S. Optimal Sizing of Smart Home Renewable Energy Resources and Battery Under Prosumer-Based Energy Management. *IEEE Syst. J.* **2021**, *15*, 105–113. [CrossRef]
19. Meng, Q.; Li, Y.; Ren, X.; Xiong, C.; Wang, W.; You, J. A demand-response method to balance electric power-grids via HVAC systems using active energy-storage: Simulation and on-site experiment. *Energy Rep.* **2021**, *7*, 762–777. [CrossRef]
20. Javadi, M.S.; Nezhad, A.E.; Nardelli, P.H.J.; Gough, M.; Lotfi, M.; Santos, S.; Catalão, J.P.S. Self-scheduling model for home energy management systems considering the end-users discomfort index within price-based demand response programs. *Sustain. Cities Soc.* **2021**, *68*, 102792. [CrossRef]
21. Munankarmi, P.; Wu, H.; Pratt, A.; Lunacek, M.; Balamurugan, S.P.; Spitsen, P. Home Energy Management System for Price-Responsive Operation of Consumer Technologies Under an Export Rate. *IEEE Access* **2022**, *10*, 50087–50099. [CrossRef]
22. Lopez, S.R.; Gutierrez-Alcaraz, G.; Javadi, M.S.; Osório, G.J.; Catalão, J.P.S. Flexibility Participation by Prosumers in Active Distribution Network Operation. In Proceedings of the 2022 IEEE International Conference on Environment and Electrical Engineering and 2022 IEEE Industrial and Commercial Power Systems Europe (EEEIC/I&CPS Europe), Prague, Czech Republic, 28 June–1 July 2022; pp. 1–6.
23. Khorasany, M.; Shokri Gazafroudi, A.; Razzaghi, R.; Morstyn, T.; Shafie-khah, M. A framework for participation of prosumers in peer-to-peer energy trading and flexibility markets. *Appl. Energy* **2022**, *314*, 118907. [CrossRef]
24. Valarezo, O.; Gómez, T.; Chaves-Avila, J.P.; Lind, L.; Correa, M.; Ulrich Ziegler, D.; Escobar, R. Analysis of New Flexibility Market Models in Europe. *Energies* **2021**, *14*, 3521. [CrossRef]
25. D'Ettorre, F.; Banaei, M.; Ebrahimy, R.; Pourmousavi, S.A.; Blomgren, E.M.V.; Kowalski, J.; Bohdanowicz, Z.; Łopaciuk-Gonczaryk, B.; Biele, C.; Madsen, H. Exploiting demand-side flexibility: State-of-the-art, open issues and social perspective. *Renew. Sustain. Energy Rev.* **2022**, *165*, 112605. [CrossRef]

26. Nouicer, A.; Meeus, L.; Delarue, E. The Economics of Explicit Demand-Side Flexibility in Distribution Grids: The Case of Mandatory Curtailment for a Fixed Level of Compensation. European University Institute. 2020. Available online: https://cadmus.eui.eu/handle/1814/67762 (accessed on 5 September 2023).
27. Qayyum, F.A.; Naeem, M.; Khwaja, A.S.; Anpalagan, A.; Guan, L.; Venkatesh, B. Appliance Scheduling Optimization in Smart Home Networks. *IEEE Access* **2015**, *3*, 2176–2190. [CrossRef]
28. Javadi, M.S.; Gough, M.; Lotfi, M.; Esmaeel Nezhad, A.; Santos, S.F.; Catalão, J.P.S. Optimal self-scheduling of home energy management system in the presence of photovoltaic power generation and batteries. *Energy* **2020**, *210*, 118568. [CrossRef]
29. Javadi, M.S.; Nezhad, A.E.; Gough, M.; Lotfi, M.; Anvari-Moghaddam, A.; Nardelli, P.H.J.; Sahoo, S.; Catalão, J.P.S. Conditional Value-at-Risk Model for Smart Home Energy Management Systems. *e-Prime* **2021**, *1*, 100006. [CrossRef]
30. USEF Foundation. Workstream on Aggregator Implementation Models. 2017. Available online: https://www.usef.energy/app/uploads/2016/12/Recommended-practices-for-DR-market-design.pdf (accessed on 5 September 2023).
31. Alıç, O.; Filik, Ü.B. A multi-objective home energy management system for explicit cost-comfort analysis considering appliance category-based discomfort models and demand response programs. *Energy Build.* **2021**, *240*. [CrossRef]

Article

Evaluation of Technological Configurations of Residential Energy Systems Considering Bidirectional Power Supply by Vehicles in Japan

Jun Osawa

Department Management Systems, College of Informatics and Human Communication, Kanazawa Institute of Technology, Nonoichi 921-8501, Japan; j-osawa@neptune.kanazawa-it.ac.jp

Abstract: To reduce CO_2 emissions in the residential and transportation sectors, distributed energy technologies, such as photovoltaic power generation (PV), stationary storage batteries (SBs), battery electric vehicles (BEVs), and vehicle-to-home (V2H) systems, are expected to be introduced. The objective of this study was to analyze the impact of the installed capacity of PV and SB, the type of vehicle, and their combination on the economic and environmental performance of the total energy consumption of residences and vehicles. Thus, this study developed a model to optimize the technological configuration of residential energy systems, including various vehicle types and driving patterns. The simulation results showed that it is more economically and environmentally efficient to install a BEV and a V2H system in households with longer parking times at the residence and to install an SB in addition to these technologies in households with shorter parking times at the residence. Furthermore, comparing a gasoline vehicle and an SB, the most economical combination, with a BEV and a V2H system and with a BEV, a V2H system, and an SB, estimated the carbon tax rate necessary for cost equivalence. The result indicated that the carbon tax rate needs to be increased from its current level.

Keywords: clean energy vehicle; optimization; photovoltaic; residential energy system; vehicle-to-home

Citation: Osawa, J. Evaluation of Technological Configurations of Residential Energy Systems Considering Bidirectional Power Supply by Vehicles in Japan. *Energies* **2024**, *17*, 1574. https://doi.org/10.3390/en17071574

Academic Editors: Boštjan Polajžer, Davood Khodadad, Younes Mohammadi and Aleksey Paltsev

Received: 24 February 2024
Revised: 22 March 2024
Accepted: 23 March 2024
Published: 26 March 2024

1. Introduction

As floods, droughts, extreme heat, torrential rains, and other weather disasters have occurred in many parts of the world in recent years, climate change has become an important issue of concern. To mitigate the negative impacts of climate change, Japan and many other countries are supporting the development of new technologies to achieve carbon neutrality. In the transportation sector, in addition to improving the fuel efficiency of gasoline vehicles (GVs), clean energy vehicles such as hybrid electric vehicles (HEVs), plug-in hybrid electric vehicles (PHEVs), battery electric vehicles (BEVs), and fuel cell vehicles (FCVs) are being developed and introduced to the market. Furthermore, the use of distributed energy technologies, such as decentralized power sources and power storage devices, is expected in the residential sector as a measure to improve energy use efficiency and reduce CO_2 emissions. Photovoltaic power generation (PV) and stationary storage batteries (SBs) have already begun to be introduced into residences. The introduction of BEVs as an energy storage technology into residences has been attracting attention in recent years. Moreover, the introduction of a vehicle-to-home (V2H) system using a dedicated power conditioner that connects the BEV to the distribution board of the residence will enable a bidirectional power supply between the BEV and the residence, allowing the BEV to be used as a residential power storage system. This is expected to reduce CO_2 emissions through more efficient energy use in the residential and transportation sectors.

Several studies have evaluated the effects of PVs, BEVs, and V2H systems on the efficiency of electricity use in residences. Osawa et al. [1] calculated and compared the self-sufficiency rate, self-consumption rate, and CO_2 emissions in five cases, i.e., GVs,

BEVs, GVs and SBs, BEVs and SBs, and BEVs and V2H systems. The self-sufficiency rate indicates the percentage of energy consumption of the residence that is covered by the amount of PV electricity generated, whereas the self-consumption rate indicates the percentage of PV electricity generated that is consumed by the residence. The comparison results indicated that the introduction of a BEV and V2H system has the potential to reduce CO_2 emissions compared with the case of BEVs alone. In addition, Akimoto et al. [2] investigated SB charging, BEV charging, and daytime operation of heat pump water heaters as operational methods for the self-consumption of PV surplus electricity and compared the effects of installing each device. The results showed that households with less frequent use of vehicles and shorter driving distances have higher self-consumption of PV surplus power and lower CO_2 emissions and annual costs, whereas households with more frequent use of vehicles and longer driving distances cannot fully consume PV surplus power. Kobashi et al. [3] assumed a cost decline in PVs, SBs, and BEVs until 2030 in Kyoto, Japan, and Shenzhen, China, and evaluated the economic and environmental advantages of PVs and V2H systems over PVs and SBs in Japan in the future. Nishioeda et al. [4] surveyed the PV generation capacity, private vehicle usage patterns, and electricity demand of residences in Tagajo, Japan, through questionnaires and on-site surveys. Then, based on those results, they estimated the amount of PV-generated surplus electricity and the amount of surplus electricity charged to BEVs.

Several studies have examined the optimal configuration of equipment installed in residences to improve the efficiency of residential energy systems. Higashitani et al. [5,6] optimized residential equipment configuration and operation, such as with SBs and gas water heaters, to minimize the annual costs in each case of GVs, BEVs, and combined BEVs and V2H installation. The results showed that it is more economical to combine V2H systems and SBs in households where BEVs are absent for long periods of time. New installations of heat pump water heaters were also shown to be effective. Erdinc et al. [7] developed a model to optimize PV and SB capacities in a residence with a BEV and V2H system. Wu et al. [8] optimized the SB size and battery operation strategy in a household with PVs and a BEV. The results showed that V2H systems have the potential to reduce daily electricity costs.

As described above, existing studies have evaluated the effectiveness of BEVs and V2H systems from various perspectives. However, these studies have focused on BEVs and have yet to analyze the implications of the introduction of other types of vehicles, such as HEVs, PHEVs, and FCVs. Moreover, vehicle types have not been included as design variables for optimization. A comprehensive study to optimize the technological configuration of vehicle types and PV and SB installed capacities, with or without V2H installation, considering the variety of vehicle types and the future price declines in these technologies has not yet been conducted. Various powertrains are being developed with the goal of achieving carbon neutrality and satisfying diverse consumer preferences, and consideration of the characteristics of these various vehicle types is important for the design of future energy systems.

Several previous studies have focused solely on the transportation sector, without considering the residential sector, and have examined optimal plans for the introduction of clean energy vehicles [9–13]. These studies have shown that the characteristics of each vehicle type, such as the energy source, CO_2 emissions, and cost, affect future installation plans. On the other hand, the introduction and operational effects of clean energy vehicles in conjunction with the transportation and residential sectors, such as V2H systems, have not been considered. Assuming a wide variety of vehicle options, there is currently no comprehensive discussion about whether they should be implemented alone or in combination with SBs or V2H systems. Such information would be valuable for developing future scenarios for vehicles and residential energy systems for more efficient energy use.

In this study, a model was developed to evaluate and optimize the technological configuration of residential energy systems that include various vehicle types. The objective of this study was to analyze the impact of the installed capacity of PVs and SBs, the type of vehicle, and their combination on the economic and environmental performance of the total energy consumption of residences and vehicles. A unique feature of this study is that it considered various types of vehicles with low environmental impacts in addition to the introduction of PVs, SBs, and V2H systems into residences. Each vehicle type has different battery capacities and energy sources, resulting in different operational effectiveness and optimal PV and SB capacities. Moreover, the cost and CO_2 emissions of the residential energy system for each technological configuration were calculated, and a multi-objective optimization model considering both economic and environmental aspects was developed. In recent years, consumer preferences have diversified, and some consumers are concerned not only with economic efficiency but also with environmental efficiency. In addition, the government of Japan needs to reduce CO_2 emissions to achieve carbon neutrality. When considering future environmental and energy policies for the diffusion of distributed energy technologies, it would be beneficial to develop a multi-objective optimization method that considers both environmental and economic aspects and provides results. Furthermore, multiple patterns of vehicle usage were assumed. The study examined how the time of vehicle absence during the day would affect the effectiveness of V2H systems and SB implementation.

Section 2 of this paper describes the new model for evaluating and optimizing the technological configuration of residential energy systems. Section 3 elaborates on the various prerequisite data for the model. Section 4 presents and explains the simulation results generated by the model, and Section 5 presents the conclusions of this study and discusses future prospects based on the discoveries made herein.

2. Methods

2.1. Framework

An overview of the analytical model constructed in this study is shown in Figure 1. The study consists of three categories of inputs: residential data, vehicle data, and equipment data. To define the residential data, the first step is to select the target region. In this study, the Kanto region of Japan was selected as a case study. Based on the selected target region, residential data such as electricity demand, solar radiation, and electricity prices were established. In addition, vehicle data and equipment data such as price, charge/discharge capacity, and durability years were used. Because vehicle absence time is considered an important parameter that can affect the economic and environmental performance of V2H systems, three patterns were established. Then, based on these inputs, the charging and discharging of electricity and the total energy consumption of the residence and the vehicle were calculated. As there are various possible configurations of vehicle type, PV capacity, SB capacity, and V2H installations, it is important to consider which types of technological configurations will be required in the future based on economic and environmental aspects. It is then necessary to provide information that will be valuable in making policy decisions for improving energy efficiency. Therefore, this study developed an optimization model to calculate the optimal configuration of vehicle type, PV capacity, SB capacity, and V2H installation with two objective functions: minimization of annual costs and minimization of annual CO_2 emissions. In this study, the genetic algorithm (NSGA-II) of Pymoo (version 0.6.1) [14] was used to perform multi-objective optimization. The mathematical formulas for the optimization model are described in the following section.

A flow chart of the residential energy system in the analytical model developed in this study is shown in Figure 2. In this model, electricity demand at residences and vehicle driving demand were considered. Several configurations of PV capacity, SB capacity, and vehicle type were assumed. The model allows for the analysis of changes in electricity demand and supply for each configuration. The time resolution of the model is 1 h. Operation on a total of 48 representative days by month, weekday/holiday, and weather (sunny/other) is considered, and this is multiplied by the number of days in each category for each region to calculate values for one year. Considering the vehicle types currently on the market that are compatible with the V2H system, we set two vehicle types capable of using the V2H system: PHEVs and BEVs. The electricity supply from these two vehicle types to the residence was assumed to be rechargeable and dischargeable only when the vehicles were at the residence.

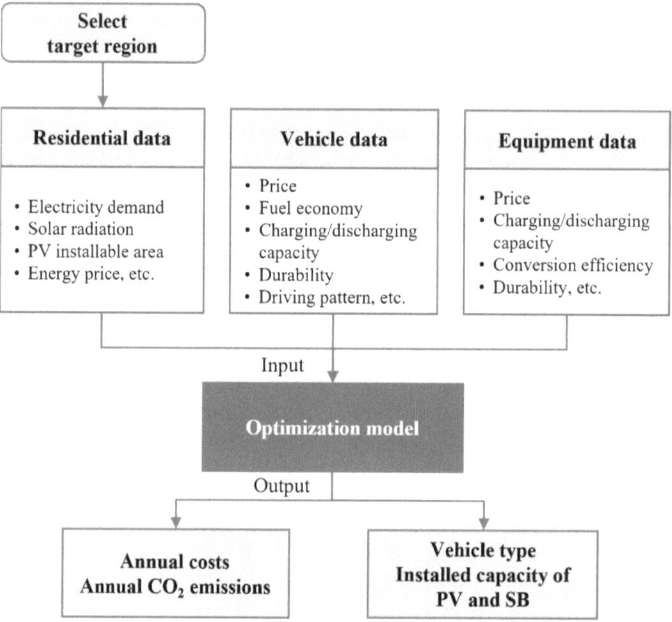

Figure 1. Overview of the analytical model.

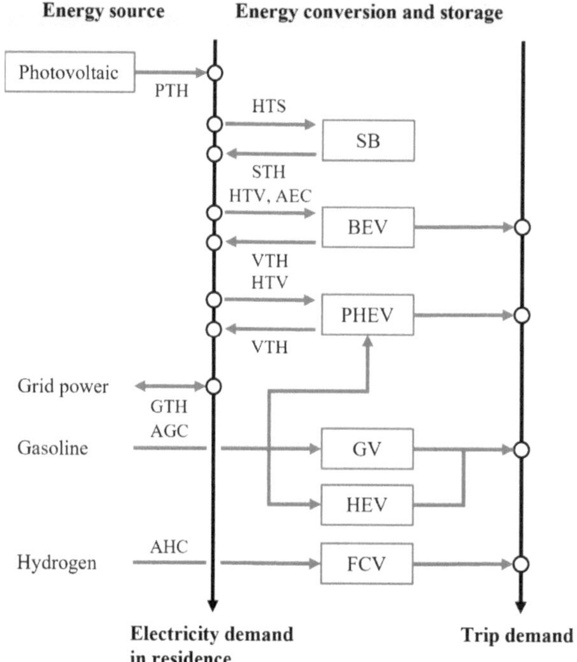

Figure 2. Flow chart of the modeled residential energy system.

2.2. Optimization Model

2.2.1. Objective Function

In this study, two objective functions were established, i.e., annual costs and annual CO_2 emissions (Equations (1) and (2), respectively).

Annual costs consisted of three costs: the installation cost of vehicles and equipment, the energy cost, and the carbon cost for CO_2 emissions (Equation (1)). The installation cost of vehicles and equipment, which is a fixed cost, was converted to a one-year cost by using the durable years and the interest rate (Equation (3)). The interest rate was set at 3%. The energy cost was defined as the sum of the annual purchase costs of electricity, gasoline, and hydrogen (Equation (4)). The electricity cost was defined as the difference between the cost of purchasing grid power and the profit from the sale of surplus power generated by PVs. The time resolution of the model is 1 h. Based on the operation on a total of 48 representative days by month, weekday/holiday, and weather (sunny/other), the values were multiplied by the number of days in each category and calculated for one year (Equations (4)–(8)). Electricity prices were set as amounts for daytime (7–22) and nighttime (23–6), respectively. In addition, electricity prices were set according to the energy mix of the Kanto region, based on the business scope of the major electric power companies in Japan. Many countries have introduced or are considering carbon taxes to achieve carbon neutrality. In Japan, a carbon tax has been partially introduced as a "tax for global warming countermeasures" [15]. Therefore, the carbon tax rate was multiplied by the CO_2 emissions in each case of technological configuration and added to the annual cost.

Annual CO_2 emissions were defined as the annual purchases of grid power, gasoline, and hydrogen multiplied by their respective CO_2 intensity (Equations (9)–(12)). The CO_2 intensity of grid power was set based on the energy mix of the Kanto region. The sale of surplus electricity to the grid was not considered a CO_2 reduction effect (Equation (10)).

$$min\ AC_k(T_{ik}, PVA_{ik}, SBC_{ik}) = \sum_i T_{ik}(TC_{ik} + EC_{ik} + CC_{ik}) \tag{1}$$

$$min\ ACE_k(T_{ik}, PVA_{ik}, SBC_{ik}) = \sum_i T_{ik}(ELCE_{ik} + GACE_{ik} + HYCE_{ik}) \tag{2}$$

$$TC_{ik} = VTC_{ik}\frac{r(1+r)^{vtd_i}}{(1+r)^{vtd_i}-1} + PVA_{ik}ID\frac{r(1+r)^{pvd}}{(1+r)^{pvd}-1} + SBC_{ik}\frac{r(1+r)^{sbd}}{(1+r)^{sbd}-1} \tag{3}$$

$$EC_{ik} = \sum_{m,d,w} ND_{kmdw}(ELC_{ikmdw} + GAC_{ikmdw} + HYC_{ikmdw}) \tag{4}$$

$$ELC_{ikmdw} = \sum_h ELU_{ikhmdw}ELP_{kh} \tag{5}$$

$$ELU_{ikhmdw} = GTH_{ikhmdw} + AEC_{ikhmdw} \tag{6}$$

$$GAC_{ikmdw} = \sum_h AGC_{ikhmdw}GP_k \tag{7}$$

$$HYC_{ikmdw} = \sum_h AHC_{ikhmdw}HP_k \tag{8}$$

$$CC_{ik} = CP(ELCE_{ik} + GACE_{ik} + HYCE_{ik}) \tag{9}$$

$$ELCE_{ik} = ELCI_k \sum_{h,m,d,w} ND_{kmdw}\left(\begin{cases} ELU_{ikhmdw}, & GTH_{ikhmdw} \geq 0 \\ AEC_{ikhmdw}, & GTH_{ikhmdw} < 0 \end{cases}\right) \tag{10}$$

$$GACE_{ik} = GACI \sum_{h,m,d,w} ND_{kmdw}AGC_{ikhmdw} \tag{11}$$

$$HYCE_{ik} = HYCI_k \sum_{h,m,d,w} ND_{kmdw} AHC_{ikhmdw} \tag{12}$$

In the above, i is the technological configuration type [GV, HEV, PHEV, BEV, FCV, GV_SB_PV, HEV_SB_PV, PHEV_SB_PV, BEV_SB_PV, FCV_SB_PV, PHEV_V2H_PV, BEV_V2H_PV, PHEV_V2H_SB_PV, BEV_V2H_SB_PV]; k is the target year [2025, 2030]; h is the hour [0–23]; m is the month [1–12]; d is the day category [weekday, holiday]; w is the weather category [sunny, other]; r is the discount rate [%]; vtd is the durable years of each vehicle type [years]; pvd is the durable years of PV power [years]; sbd is the durable years of the SB [years]; T is the sales volume of vehicle type [units]; PVA is the PV installation area [m^2]; SBC is the storage capacity [kWh]; AC is the annual costs [million yen]; ACE is the annual CO_2 emissions [t-CO_2]; TC is the installation cost of vehicles and equipment [million yen]; EC is the energy cost [million yen]; CC is the carbon cost for CO_2 emissions [million yen]; $ELCE$ is the CO_2 emissions from electricity [t-CO_2]; $GACE$ is the CO_2 emissions from gasoline [t-CO_2]; $HYCE$ is the CO_2 emissions from hydrogen [t-CO_2]; VTC is the installation cost of each vehicle type [million yen]; ID is the installation density [kW/m^2]; ND is the number of days in each category [days]; ELC is the electricity cost [million yen]; GAC is the gasoline cost [million yen]; HYC is the hydrogen cost [million yen]; ELU is the electricity usage [MJ]; ELP is the electricity price [yen/MJ]; AEC is the amount of electricity for additional charging while driving [MJ]; AGC is the consumption of gasoline used for driving [MJ]; GP is the gasoline price [yen/MJ]; AHC is the consumption of hydrogen used for driving [MJ]; HP is the hydrogen price [yen/MJ]; CP is the carbon tax rate for CO_2 emissions [yen/t-CO_2]; $ELCI$ is the CO_2 intensity of electricity [g-CO_2/MJ]; $GACI$ is the CO_2 intensity of gasoline [g-CO_2/MJ]; $HYCI$ is the CO_2 intensity of hydrogen [g-CO_2/MJ]; and GTH is the amount of electricity purchased or sold between the residence and the grid power [MJ].

Equation (13) shows the energy balance. The hourly residential electricity demand and supply coincide. Considering residential electricity consumption, PV generation, vehicle charging/discharging, and SB charging/discharging, the system was set up to purchase electricity from the grid when demand exceeds supply and sell electricity to the grid when demand falls below supply. Moreover, based on previous research [16], hourly PV generation was calculated by multiplying the amount of solar radiation by the installed area, module conversion efficiency, temperature loss factor, and integration factor (Equation (14)).

$$GTH_{ikhmdw} = (ELD_{khmdw} + HTV_{ikhmdw} + HTS_{ikhmdw})$$
$$-(PTH_{ikhmdw} + VTH_{ikhmdw} + STH_{ikhmdw}) \tag{13}$$

$$PTH_{ikhmdw} = SR_{khmdw} PVA_{ik} CE TL_m IF \tag{14}$$

In the equations, ELD is the amount of electricity consumed by a residence [MJ]; HTV is the amount of electricity flowing from the residence to the vehicle to charge the vehicle [MJ]; HTS is the amount of electricity flowing from the residence to the SB to charge the SB [MJ]; PTH is the amount of electricity generated by PVs [MJ]; VTH is the amount of electricity flowing from the vehicle to the residence due to the vehicle's discharge [MJ]; STH is the amount of electricity flowing from the SB to the residence due to the SB discharge [MJ]; SR is the amount of solar radiation [MJ/m^2]; CE is the module conversion efficiency [-]; TL is the temperature loss factor [-]; and IF is the integration factor [-].

2.2.2. Constraints

First, constraints were set for V2H and SB operations. Electricity generated by PVs is prioritized for consumption in the residence; if there is a surplus, it is charged to a SB or vehicle. Even after charging a SB or vehicle, if there is a surplus, electricity is sold to the grid. If the consumption in the residence exceeds the electricity generated by PVs, electricity is supplied from a SB or vehicle. If there is still a shortage after power supply from a SB or vehicle, electricity is purchased from the grid. When both V2H systems and SBs are introduced, priority is given to the V2H system. In other words, after charging/discharging from the vehicle by the V2H system, if there is still a surplus/deficiency of electricity, SB charging/discharging is performed. Moreover, the purchase of electricity from the grid and the sale of electricity to the grid cannot occur at the same time. Similarly, it is not possible to charge a SB and vehicle and supply power from a SB and vehicle at the same time. The two types of vehicles that could supply power to the residence were PHEVs and BEVs. The vehicles were designed to be able to charge and discharge electricity only when parked at a residence.

Additionally, the amount of electricity stored in the batteries of SBs, PHEVs, and BEVs must be less than or equal to their storage capacity (Equations (15) and (16)). The system was set to stop discharging to the residence when the amount of charge fell below the lower limit. However, during driving, there is no lower limit of discharge, and the system is set to charge the battery externally if the amount of charge is insufficient. With reference to previous studies [1,17,18], the lower limit of the charge for BEVs and PHEVs was set at 40% of the storage capacity. The lower limit of the charge for SBs was set at 20% of the storage capacity. The lower limit was set with the strategy of not affecting driving in the case of vehicles and of allowing use in an emergency in the case of SBs. Furthermore, the amount of electricity entering and leaving the storage battery each hour coincides with the change in the state of charge (SOC) of the storage battery (Equations (17) and (18)):

$$VSOC_{ikhmdw} \leq VBC_{ik} \tag{15}$$

$$SSOC_{ikhmdw} \leq SBC_{ik} \tag{16}$$

$$VSOC_{ikhmdw} - VSOC_{ikh-1mdw} + (HTV_{ikhmdw} - VTH_{ikhmdw} + AEC_{ikhmdw})/cmk \tag{17}$$

$$SSOC_{ikhmdw} = SSOC_{ikh-1mdw} + (HTS_{ikhmdw} - STH_{ikhmdw})/cmk \tag{18}$$

where $VSOC$ is the SOC of the battery of each vehicle type [kWh]; VBC is the battery capacity of each vehicle type [kWh]; $SSOC$ is the SOC of the SB [kWh]; and cmk is the conversion factor between megajoules and kilowatt–hours [-].

Then, two constraint conditions were set: the number of units and the PV area installed. In this study, the number of vehicles owned per household was assumed to be one (Equation (19)). Furthermore, the area available for PV installation depends on the area of the detached houses, which varies from prefecture to prefecture. Therefore, the area available for installation in the Kanto region was calculated by multiplying the average area of a detached house in each prefecture of the Kanto region (Ibaraki, Tochigi, Gunma, Saitama, Chiba, Tokyo, Kanagawa, and Yamanashi Prefecture) by the installable area ratio. The value was then used as the upper limit for the PV installed area (Equation (20)):

$$\sum_i T_{ik} = 1 \tag{19}$$

$$PVA_{ik} \leq PVUL \tag{20}$$

where $PVUL$ is the upper limit for the installed PV area [m^2].

3. Prerequisite Data

3.1. Technology Parameters

The values of the parameters for vehicles and equipment that do not change from year to year are shown in Table 1.

Considering the vehicle types currently on the market that are compatible with the V2H system, we set two vehicle types capable of using a V2H system: PHEVs and BEVs. Then, based on previous studies [5,6], the charging and discharging capacity was set to 6 kWh when the V2H system was used. Conversely, when charging without the V2H system, the charging capacity was set to 3 kWh.

Previous studies [17,18] have found that many users recharge BEVs when their SOC is approximately 30–35% because of concerns about their driving range. Therefore, 40% was set as the lower limit rate of discharge. If the SOC falls below 40%, discharge is stopped, and the battery is recharged when it is parked at the residence the next time. When driving, there is no lower limit of discharge. If the battery capacity becomes insufficient while driving, the vehicle is recharged externally using electricity or gasoline. In addition, as a response to emergencies, a lower limit discharge rate of 20% was also set for SBs.

Table 1. Technology parameters that do not change from year to year.

Technology	Parameter	
Vehicle	Durability [year]	14
	Charging capacity [kWh]	6 (V2H), 3 (Others)
	Discharging capacity [kWh]	6 (V2H)
	Lower limit rate of discharge [%]	40
	Rate of initial SOC [%]	50
PV	Durability [year]	25
	Installation area [m^2]	6, 11, 16, 21, 26, 31, 36, 41, 46, 48
	Maximum installable area (Kanto region) [m^2]	37
	Module conversion efficiency [-]	0.18
	Temperature loss factor [-]	0.9 (December–February) 0.8 (June–August) 0.85 (Others)
	Integration factor [-]	0.7
SB	Price [million yen/kWh]	0.042
	Durability [year]	15
	Storage capacity [kWh]	4, 6, 8, 10, 12
	Charging and discharging capacity [kWh]	2.5
	Lower limit rate of discharge [%]	20
	Rate of initial SOC [%]	50

The optimal levels of PV installation area and SB capacity are expected to vary depending on the technological configuration, such as which type of vehicle is used and whether a V2H system is used. Therefore, the PV installation area and SB capacity were considered as design variables, consisting of 10 and 5 alternatives, respectively.

Furthermore, distributed energy technologies are still in the early stages of market penetration, and prices tend to be relatively high, but the cost is projected to decrease in the future. Therefore, the prices of each vehicle type and PV generation were set based on the projected values in the target year based on previous research [9,19]. On the other hand, GVs and SBs were considered to have less price volatility than other technologies and thus were assumed to have fixed values [5,9]. Additionally, the fuel economy and battery capacity of each vehicle type were set based on previous studies [9,20], considering changes over time.

3.2. Solar Radiation Data and Electricity Demand Data

Data on the hourly values of the total solar radiation and weather for January–December 2018–2022 were obtained from the Japan Meteorological Agency [21]. Average values were then calculated by month, weekday/holiday, and weather category (sunny/other). Data for the Kanto region were taken from Tokyo, where the headquarters of Tokyo Electric Power Company, one of Japan's major power companies operating in the Kanto region, is located.

For electricity demand and total floor area of houses, we used data actually measured in detached houses from the "Database of Energy Consumption in Houses" created by the Research and Study Committee on Energy Consumption in Houses of the Architectural Institute of Japan [22]. The study used data from nine residences in the Kanto region between October 2002 and March 2005. We also obtained data from the Japan Meteorological Agency [21] on total solar radiation and weather for hourly values at the same time of year when electricity demand was actually measured. The median electricity demand per area by month, weekday/holiday, and weather category was then calculated. The median electricity demand per area was then multiplied by the average of the total floor area in the Kanto region [23] to calculate the electricity demand of the average household by month, weekday/holiday, and weather category. The weather category for each future day was randomly assigned based on the percentage of weather categories for each month in 2018–2022. It was assumed that the incidence of each weather category in the future would be approximately the same as in the past. In this study, the total floor area was considered a characteristic of each detached house, and electricity demand per area was used. On the other hand, from the perspective of considering the overall trend in electricity demand for detached houses, this study did not analyze other characteristics in detail, such as the number of people in the household. Although these data are not necessarily the same as current data, this study used median values that exclude outliers, which we believe is useful for observing overall trends in the effectiveness of the linkage between vehicles and residences. It is also possible to change the input data as the survey data are updated. Thus, we do not believe that this circumstance will affect the usefulness of the model itself.

3.3. Energy Parameters

The energy sources are gasoline for GVs, HEVs, and PHEVs, electricity for BEVs and PHEVs, and hydrogen for FCVs. For PHEVs, electricity is used for BEV driving, and gasoline is used for HEV driving. This study also considered CO_2 emissions during both the driving phase of the vehicle and the production phase of the energy source.

Based on previous studies [9,24], the price of gasoline was calculated based on the price of crude oil plus petroleum, coal, and gasoline taxes and refining and distribution margins. With reference to the study of Osawa [9], the CO_2 intensity of gasoline was set to approximately 85.

The price and CO_2 intensity of hydrogen were set based on the assumption that blue hydrogen will be mainly used until 2029 and that green hydrogen will be mainly used after 2030. The price and CO_2 intensity of hydrogen vary depending on whether it is produced domestically or abroad. However, hydrogen is still a technology in the development and demonstration stages, and the cost and CO_2 intensity of hydrogen produced abroad and transported to Japan will vary greatly depending on the country of production. Therefore, in this study, the price and CO_2 intensity were set assuming domestic production and transportation [25–29].

The purchase price of electricity is affected by the mixture of power sources in the region. Based on previous work [24,30], this study calculated the power supply mixture for the Kanto region in 2020 and multiplied it by the cost of each power source to determine the cost of electricity generation. The cost of electricity generation was then deducted from the current daytime and nighttime electricity purchase prices, and the respective margins were calculated. The future electricity purchase price was then set by multiplying the Kanto region's energy mix by the future cost of each power source and adding each margin. Based on a previous study [6], the sale price of surplus electricity was fixed at 8.5 yen/kWh. The CO_2 intensity of electricity was also calculated by multiplying the energy mix by the CO_2 intensity of each power source, based on previous work [31].

3.4. Vehicle Driving Patterns

It is assumed that the time spent parked at the residence affects the effectiveness of the V2H system. Therefore, based on previous studies [1,5,6], three driving patterns were developed for this study (Table 2). Pattern A assumes that the main purpose of use is shopping and that users make short-distance trips. Pattern B assumes that the main purpose of use is commuting on weekdays and shopping on weekends and holidays. The SOC is assumed to remain unchanged (constant) during the time the vehicle is parked at the office. Pattern C assumes that the main purpose of use is shopping on weekdays and long-distance driving on weekends and holidays, and that the vehicle is used for long-distance travel. The annual mileage values in 2025 calculated from the hourly mileage were approximately 8490 km for driving pattern A, 10,950 km for driving pattern B, and 13,250 km for driving pattern C. Osawa [9] estimated the average annual mileage to be 9601 km, which is close to the mileage values used in this study; thus, the values are considered to be a reasonable setting.

Table 2. Vehicle driving patterns.

Pattern	Day Category	Use	Parking Time at Residence	Mileage [km/h]
A	Weekdays	Shopping	0–9 12–23	10
	Holidays	Shopping	0–9 13–23	
B	Weekdays	Commuting	0–7 18–23	10
	Holidays	Shopping	0–9 13–23	
C	Weekdays	Shopping	0–9 12–23	10
	Holidays	Long-distance driving	0–9 17–23	

4. Results and Discussion

4.1. Optimal Technological Configuration

This section describes the calculation results of the optimal technological configuration for residential energy systems that include a variety of vehicle types.

Figure 3 shows the Pareto optimal solution for each driving pattern in 2025 and 2030. The Pareto optimal solution shown in the figure reveals that the curve shifts to the lower left from 2025 to 2030. This indicates that the reduction potential of annual costs and annual CO_2 emissions improves as the cost of vehicles and PVs declines and their installation becomes easier. By driving pattern, the curve shifts to the upper right for B and C compared to A. This indicates that for driving pattern A, in which vehicles spend more time parked at the residence, a bidirectional power supply could improve both annual costs and annual CO_2 emissions.

Table 3 shows the technological configuration of the optimal solutions that consider both annual costs and annual CO_2 emissions in a balanced manner, which are shown in the rhombus in Figure 3. This optimal solution describes the solution with the lowest sum of values of each objective function of the Pareto optimal solution normalized and multiplied by an equal weight (0.5 each), respectively. For the technological configurations in each optimal solution in Table 3, the combination of BEVs and V2H systems was chosen in both 2025 and 2030 for driving pattern A, which assumes more time parked at the residence. It is assumed that PV power will not be deployed to the maximum extent because of its limited ability to consume and store the amount of electricity it generates. This is attributed to the fact that the BEV is not at the residence for some time. For driving pattern B, which is used for commuting on weekdays, and driving pattern C, which is used for long-distance driving on weekends and holidays, the combination of HEVs and SBs was selected for 2025. On the other hand, the combination of BEVs, V2H systems, and SBs was selected for 2030. This suggests that a V2H system would not be fully effective in 2025 because of the short parking time at the residence and that it would be more beneficial to introduce SBs in addition to HEVs. HEVs also have better fuel economy than GVs, which is advantageous when the driving distance is long. For 2030, BEVs were chosen instead of HEVs, and V2H systems and SBs were introduced. An SB was added as a way to charge the surplus electricity generated by PVs and discharge it for residential electricity consumption when the BEV is not at the residence. Furthermore, the price of PV power will decrease in the future. With the introduction of SBs, the area of PV installation increases, indicating that more energy can be produced and used at the residence.

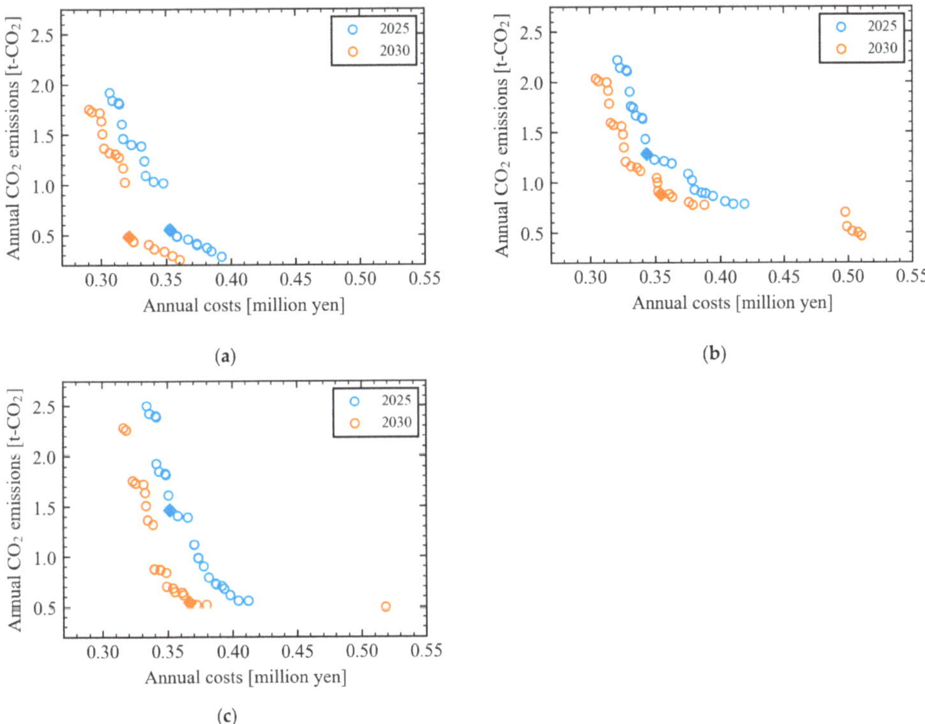

Figure 3. Pareto optimal solution for (**a**) driving pattern A; (**b**) driving pattern B; (**c**) driving pattern C.

Table 3. Optimal technological configurations in 2025 and 2030.

Driving Pattern	Year	Technological Configuration Type	Installation Area of PVs [m^2]	Storage Capacity of an SB [kWh]
A	2025	BEV_V2H_PV	31	-
	2030	BEV_V2H_PV	31	-
B	2025	HEV_SB_PV	21	12
	2030	BEV_V2H_SB_PV	36	4
C	2025	HEV_SB_PV	21	12
	2030	BEV_V2H_SB_PV	31	10

Table 4 shows the technological configurations of the two optimal solutions for driving pattern A in 2030: the case with minimized annual costs and the case with minimized annual CO$_2$ emissions. In the case of minimizing annual costs, a combination of GVs and

SBs was chosen. The area of the PV installation was 26 m^2, and the installed SB capacity was 6 kWh. The system is designed to generate and use only the amount of electricity needed for residential consumption, which is similar to the current residential energy system. On the other hand, in the case of minimizing annual CO_2 emissions, a combination of BEVs, V2H systems, and SBs was chosen. The installed PV area was 36 m^2, and the installed SB capacity was 10 kWh. It is assumed that the maximum amount of electricity is generated at the residence and then stored and used. The introduction of an SB is accompanied by an increase in the installed area of PVs compared with the case of BEVs and V2H systems in driving pattern A in 2030 (Table 3).

Table 4. Optimal technological configurations for driving pattern A in 2030.

Case	Technological Configuration Type	Installation Area of PV [m^2]	Storage Capacity of SB [kWh]
Minimized annual costs	GV_SB_PV	26	6
Minimized annual CO_2 emissions	BEV_V2H_SB_PV	36	10

Figure 4 illustrates the comparison of annual costs and annual CO_2 emissions of GV, HEV, and PHEV_V2H_PV cases with the technological configurations of each optimal solution for driving pattern A in 2030. In all cases, the costs of installing vehicles, PVs, and SBs account for a large proportion of the total costs. The energy cost is lowered significantly by utilizing V2H systems and SBs. Furthermore, the combination of BEVs and V2H systems can significantly reduce annual CO_2 emissions compared with GVs and HEVs. Even when a GV is used, the combination of a GV and an SB results in a significant reduction in annual CO_2 emissions. When a BEV is replaced by a PHEV under the same condition of PV use, a BEV is superior in both annual costs and annual CO_2 emissions.

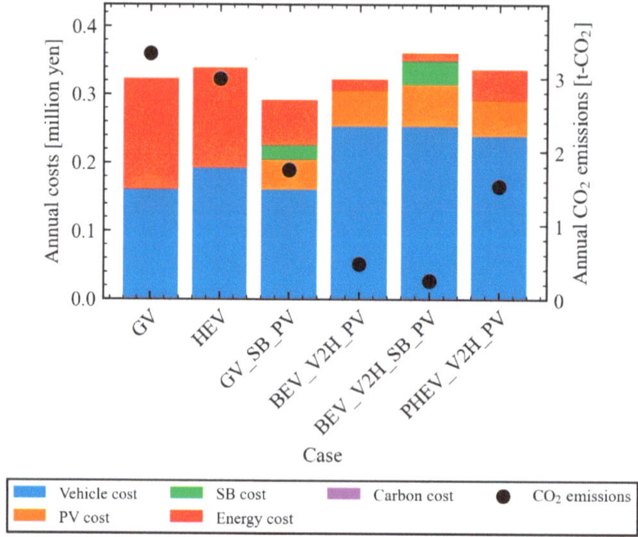

Figure 4. Comparison of annual costs and annual CO_2 emissions between the technological configuration of each optimal solution and GV, HEV, and PHEV_V2H_PV cases (driving pattern A, 2030).

To achieve carbon neutrality, the government of Japan plans to change the energy source of vehicles from gasoline to electricity or hydrogen. However, the result in Figure 4 shows that it is reasonable for consumers to choose a combination of GVs and SBs in addition to PV installation when cost is a priority. To change the energy source, it is necessary to set up a carbon tax that would cost the same amount as the combination of GVs and SBs. The carbon tax rates at which the combinations of BEVs and V2H systems and BEVs, V2H systems, and SBs in driving pattern A in 2030 would cost the same as in the combination of GVs and SBs are approximately 24,000 yen/t-CO_2 and 46,000 yen/t-CO_2, respectively. The current tax rate of the "tax for global warming countermeasures" introduced in Japan is 289 yen/t-CO_2. Furthermore, the carbon tax rate at which the BEVs, V2H systems, and SBs in driving patterns B and C in 2030 would cost about the same amount as the GVs and SBs was more than around 60,000 yen/t-CO_2. It is clear that the operation of vehicles has a significant impact on the design of the carbon tax. Thus, it is essential to reform the existing carbon tax rate.

4.2. Supply and Demand for Electricity

This section describes the daily electricity supply and demand for several optimal solutions and their characteristics.

Figure 5 shows the daily electricity supply and demand in two cases in Tables 3 and 4: a combination of BEVs and V2H systems and a combination of BEVs, V2H systems, and SBs in driving pattern A in 2030. First, in the case of BEVs and V2H systems, it is clear that most of the electricity is supplied by the BEV at night. During the day, the BEV is charged with surplus power according to the amount of PV electricity generated. However, during the time when the vehicle is out of the residence, the PV power generated is not fully consumed and is sold to the grid. In contrast, in the case of BEVs, V2H systems, and SBs, the surplus PV power is charged to the SB, even when the BEV is out of the residence. As a result, the entire amount of PV generation can be consumed in residence or charged to a vehicle or SB. No electricity is purchased from the grid, which contributes to the reduction in CO_2 emissions. On the other hand, the SB charges the surplus PV power generated during the day, becomes fully charged, and then does not operate. This occurs because the BEV discharges at night, utilizing its large-capacity battery to provide the electricity consumed by the residence.

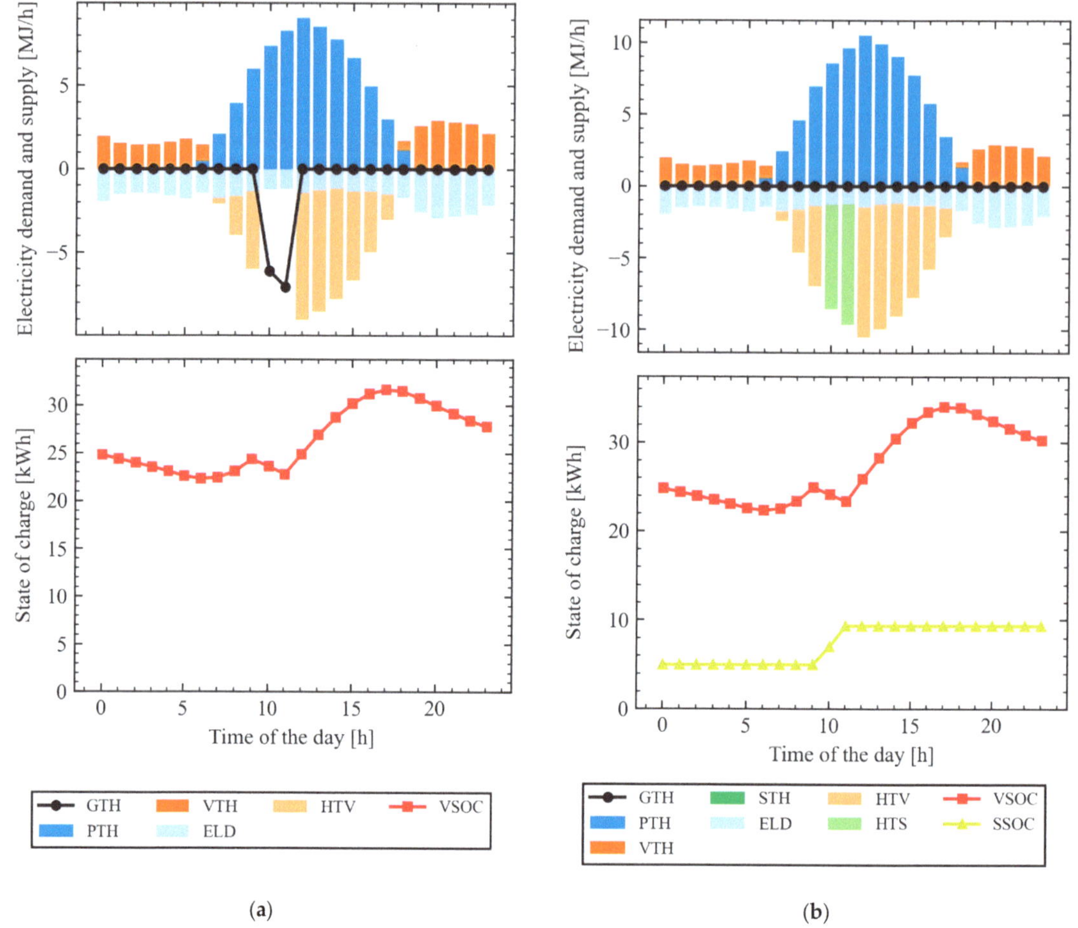

(a) (b)

Figure 5. Calculated daily electricity supply and demand for driving pattern A in 2030 (August, weekday, sunny). (**a**) the case of BEVs and V2H systems; (**b**) the case of BEVs, V2H systems, and SBs.

Figure 6a shows the daily electricity supply and demand for BEVs, V2H systems, and SBs in driving pattern C in 2030. As in driving pattern A, the nighttime electricity consumption is covered by the discharge from the BEV. On the other hand, in driving pattern C, the driver goes for a long drive during the day on holiday, and the SB absorbs the surplus power from the PV generation system. The electricity required to charge the BEV after driving is partly covered by the discharge from the SB, which is charged during the day. It reveals that the economic efficiency of the SB improves when the daytime driving distance is long. On the other hand, due to the upper limit of the SB's discharging capacity, additional electricity is purchased from the grid to recharge the BEV.

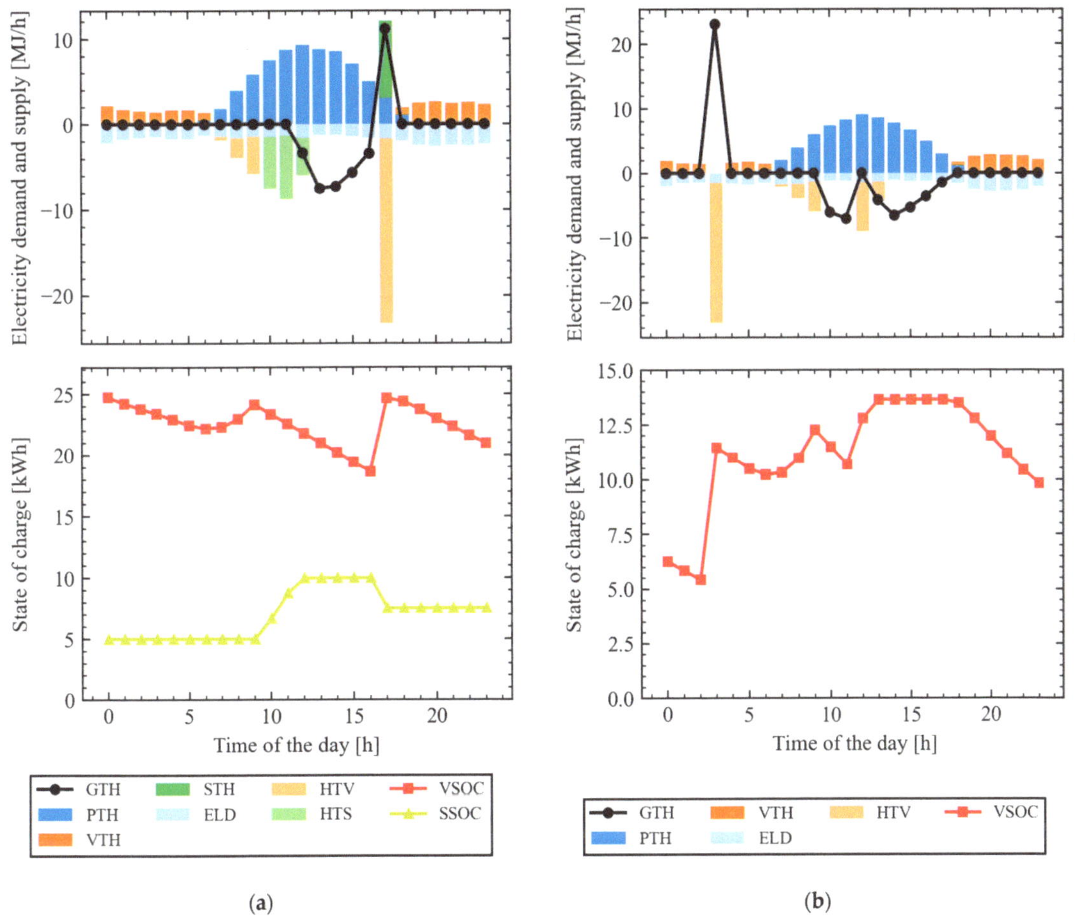

Figure 6. Calculated daily electricity supply and demand in 2030. (**a**) the case of BEVs, V2H systems, and SBs for driving pattern C (August, holiday, sunny); (**b**) the case of PHEVs and V2H systems for driving pattern A (August, weekday, sunny).

In the case of BEVs (Figure 5), there is ample capacity in the storage battery of the BEV. Given that the battery capacity of BEVs is approximately 50 kWh, it is clear that the battery capacity of the BEV is not fully utilized. Therefore, the electricity supply and demand were calculated for the case in which the BEV in Figure 5a is replaced by a PHEV (Figure 6b). As a result, it was found that the battery capacity of the PHEV is approximately 13 kWh, which is insufficient to cover the nighttime electricity demand or absorb all of the surplus PV electricity generated during the daytime. In other words, assuming the residential energy system alone, the battery capacity of the PHEV is insufficient. Thus, increasing the battery capacity of PHEVs would be beneficial for more efficient energy use in residential energy systems. When introducing BEVs, it is also important to consider schemes to utilize their battery capacity more effectively, such as by implementing a bidirectional power supply to not only residences but also offices and commercial facilities where people commute and shop.

5. Conclusions

This study developed a model to evaluate and optimize the technological configuration of residential energy systems, considering a variety of distributed energy technology options. The objective of this study was to use the model to analyze the impacts of the installed capacity of PVs and SBs, the type of vehicle, and their combination on the economic and environmental performance of the total energy consumption of residences and vehicles. The overall results showed that introducing V2H systems and SBs is expected to reduce annual costs and CO_2 emissions. This study underscores the importance of understanding the relationship between vehicle type, the capacity of SBs and PVs, and the effective combination of these technologies. The impact of residents' vehicle driving patterns on choosing optimal technological configurations highlights the necessity for customized energy solutions tailored to individual lifestyles and needs. The extension of the model to households with different geographical and socio-economic backgrounds is expected to broaden our understanding of residential energy system optimization across various contexts. A detailed analysis revealed the following key findings:

- The optimal technological configuration, considering the balance between annual costs and annual CO_2 emissions, varied depending on the vehicle driving patterns. For residences with long parking times, the combination of BEVs and V2H systems was most effective in 2025 and 2030. In contrast, with long times outside the residence, the combination of HEVs and SBs was selected for 2025, and the combination of BEVs, V2H systems, and SBs was selected for 2030. In addition to BEVs, the introduction of SBs has increased the installed PV area and improved energy efficiency.

- When cost is a priority, it is reasonable for consumers to choose a combination of GVs and SBs in addition to PV power. For residences with long parking times, the carbon tax rates at which the combinations of BEVs and V2H systems and BEVs, V2H systems, and SBs in 2030 would cost the same amount as GVs and SBs are approximately 24,000 yen/t-CO_2 and 46,000 yen/t-CO_2, respectively. The current tax rate of the "tax for global warming countermeasures" introduced in Japan is 289 yen/t-CO_2. Increasing the carbon tax rate could be effective.

- Considering the daily electricity supply and demand, when the parking time at the residence is long, a large amount of electricity can be made self-sufficient by installing BEVs and V2H systems, or BEVs, V2H systems, and SBs in addition to PV power. Furthermore, even for households that spend a lot of time outside the residence, by introducing SBs, a portion of the electricity used to charge the BEV after driving could be covered by discharging electricity from the SB, which is charged with surplus PV electricity during the day. The introduction of additional SBs has increased flexibility in electricity supply and demand.

- PHEVs, while cheaper to purchase than BEVs, have less battery capacity, which was found to be insufficient for covering nighttime electricity demands or absorbing all the surplus electricity generated by PV systems during the daytime.

Furthermore, in contrast with previous studies, this study optimized for a variety of vehicle types in addition to PV installed area, SB capacity, and whether a V2H system is used. The additional installation of SBs showed the potential to increase energy efficiency by allowing an increased installed PV area and utilizing its energy for post-drive charging of BEVs. The study also clarified the carbon tax rate at which BEVs and V2H systems and BEVs, V2H systems, and SBs would have cost levels comparable with those of the most economically rational combination, GVs and SBs.

The optimization model of this study provides valuable insights for developing future scenarios for vehicles and residential energy systems for more efficient energy use. The study also contributes to the development of sustainable energy systems in relation to goals 7 and 13 of the Sustainable Development Goals (SDGs) [32]. However, it should be noted that there are uncertainties in the prerequisite data, such as fuel and vehicle prices. In this study, data on demand patterns for electricity were used for a specific region. However, electricity demand patterns may vary from region to region. In addition, although electricity demand per floor area was used as the basis for this study, the number of people in the household and other factors may also affect electricity demand. Vehicle driving patterns may also differ between urban and rural areas. A future issue for this study is to consider the diversity of households. In addition, electricity supply and demand, such as PV generation, and vehicle driving patterns are uncertain, and the model needs to be improved to account for these unexpected fluctuations. It is important to carry out further research to gain a deeper understanding of the impact of the optimal technological configuration of each residence on the electricity system at the local level. Furthermore, raising people's awareness of environmental and energy issues and changing their behavior are also important. Educational programs have been developed at some universities in recent years, and their effectiveness has been confirmed [33–35]. Another challenge is using the findings of this research to educate potential personnel involved in the development of sustainable energy systems.

Funding: This research received no external funding.

Data Availability Statement: Data are contained within the article.

Conflicts of Interest: The author declares no conflicts of interest.

References

1. Osawa, M.; Kuwasawa, Y.; Akimoto, T.; Tajima, D.; Sonoe, H.; Uemori, S.; Komatsu, M. Housing energy performance index including automobile and electricity storage system and effect of reduction of CO_2 emissions by V2H. *J. Jpn. Soc. Energy Resour.* **2019**, *40*, 78–84. [CrossRef]
2. Akimoto, M.; Kim, J.; Tsuneoka, Y.; Oki, R.; Tanabe, S.; Hayashi, Y.; Morito, N. Operational strategies for self-consumption considering the use of an electric vehicle in a net zero energy house. *J. Environ. Eng.* **2020**, *85*, 277–287. [CrossRef]
3. Kobashi, T.; Say, K.; Wang, J.; Yarime, M.; Wang, D.; Yoshida, T.; Yamagata, Y. Techno-economic assessment of photovoltaics plus electric vehicles towards household-sector decarbonization in Kyoto and Shenzhen by the year 2030. *J. Clean. Prod.* **2020**, *253*, 119933. [CrossRef]
4. Nishioeda, T.; Suzuki, M.; Morino, K.; Oka, T.; Takebayashi, Y. Load leveling by using electric vehicle batteries to store surplus energy generated by photovoltaic systems on single-family houses in Japan. *J. Environ. Eng.* **2015**, *80*, 149–158. [CrossRef]
5. Higashitani, T.; Ikegami, T.; Uemichi, A.; Akisawa, A. Evaluation of residential power supply by photovoltaics and electric vehicles. *Renew. Energy* **2021**, *178*, 745–756. [CrossRef]
6. Higashitani, T.; Ikegami, T.; Akisawa, A. Effect of electric vehicles and bidirectional power supplies on economically optimized energy system configurations of residences. *J. Jpn. Soc. Energy Resour.* **2022**, *43*, 140–150. [CrossRef]
7. Erdinc, O.; Paterakis, N.G.; Pappi, I.N.; Bakirtzis, A.G.; Catalão, J.P.S. A new perspective for sizing of distributed generation and energy storage for smart households under demand response. *Appl. Energy* **2015**, *143*, 26–37. [CrossRef]
8. Wu, X.; Hu, X.; Teng, Y.; Qian, S.; Cheng, R. Optimal integration of a hybrid solar-battery power source into smart home nanogrid with plug-in electric vehicle. *J. Power Sources* **2017**, *363*, 277–283. [CrossRef]
9. Osawa, J. Portfolio analysis of clean energy vehicles in Japan considering copper recycling. *Sustainability* **2023**, *15*, 2113. [CrossRef]
10. Romejko, K.; Nakano, M. Portfolio analysis of alternative fuel vehicles considering technological advancement, energy security and policy. *J. Clean. Prod.* **2017**, *142*, 39–49. [CrossRef]
11. Arimori, Y.; Nakano, M. Portfolio optimization for clean energy vehicles in Japan. *Trans. Jpn. Soc. Mech. Eng. Ser. C* **2012**, *78*, 2571–2582. [CrossRef]
12. Uchida, T.; Furubayashi, T.; Nakata, T. Well-to-wheel analysis and a feasibility study of fuel cell vehicles in the passenger transportation sector. *Trans. JSME* **2019**, *85*, 18–00122. [CrossRef]
13. Nakamura, H.; Masaru, N. Scenario analysis for clean energy vehicles in UK considering introduction of renewable energy sources. *Int. J. Autom. Technol.* **2017**, *11*, 592–600. [CrossRef]
14. Blank, J.; Deb, K. Pymoo: Multi-objective optimization in Python. *IEEE Access* **2020**, *8*, 89497–89509. [CrossRef]
15. Ministry of the Environment. Introduction of a Tax to Combat Global Warming. Available online: https://www.env.go.jp/policy/tax/about.html (accessed on 1 February 2024).
16. Sugihara, H.; Yamashita, J.; Ikoma, Y.; Akisawa, A.; Kashiwagi, T. A study on an electricity potential generated photovoltaic systems on whole the houses in Japan using climatic mesh data. *J. Jpn. Sol. Energy Soc.* **2011**, *37*, 41–48.
17. Ueda, Y.; Ota, Y. When do electric vehicle users feel range anxiety and want to recharge their vehicles? *J. Jpn. Soc. Energy Resour.* **2022**, *43*, 84–93. [CrossRef]
18. Pevec, D.; Babic, J.; Carvalho, A.; Ghiassi-Farrokhfal, Y.; Ketter, W.; Podobnik, V. A survey-based assessment of how existing and potential electric vehicle owners perceive range anxiety. *J. Clean. Prod.* **2020**, *276*, 122779. [CrossRef]
19. Fuji Keizai. *Current Status and Future Prospects of Photovoltaic Technologies and Markets, 2021 Edition*; Fuji Keizai: Tokyo, Japan, 2021.
20. Fuji Keizai. *Future Outlook for Energy and Large-Scale Rechargeable Batteries and Materials 2020: Electric Vehicle and Automotive Battery Sector Edition*; Fuji Keizai: Tokyo, Japan, 2020.
21. Japan Meteorological Agency. Past Weather Data Download. Available online: https://www.data.jma.go.jp/gmd/risk/obsdl/index.php (accessed on 1 February 2024).

22. Architectural Institute of Japan. Database of Energy Consumption in Residences. Available online: http://tkkankyo.eng.niigata-u.ac.jp/HP/HP/database/index.htm (accessed on 1 February 2024).

23. Statistics Bureau, Ministry of Internal Affairs and Communications. Housing and Land Survey of Japan. 2018. Available online: https://www.e-stat.go.jp/stat-search/files?page=1&toukei=00200522&tstat=000001127155 (accessed on 1 February 2024).

24. Working Group on Verification of Power Generation Costs, Agency for Natural Resources and Energy. Report for September 2021. Available online: https://www.enecho.meti.go.jp/committee/council/basic_policy_subcommittee/index.html#cost_wg (accessed on 1 February 2024).

25. Nishi, M.; Yamamoto, H.; Takei, K. Cost Analysis of Blue Hydrogen Production in 2030. In *Central Research Institute of Electric Power Industry Report*; Central Research Institute of Electric Power Industry: Tokyo, Japan, 2022. Available online: https://criepi.denken.or.jp/hokokusho/pb/reportDownload?reportNoUkCode=EX21012&tenpuTypeCode=30&seqNo=1&reportId=9840 (accessed on 1 February 2024).

26. Nishi, M.; Yamamoto, H.; Takei, K. Cost Analysis of Hydrogen Produced from Renewable Energy Comparisons of the Production Costs in Japan and Abroad and the Effect of Reducing Capacity of Electrolyzer. In *Central Research Institute of Electric Power Industry Report*; Central Research Institute of Electric Power Industry: Tokyo, Japan, 2020. Available online: https://criepi.denken.or.jp/hokokusho/pb/reportDownload?reportNoUkCode=M19003&tenpuTypeCode=30&seqNo=1&reportId=8945 (accessed on 1 February 2024).

27. Takei, K.; Yamamoto, H.; Mori, M.; Nagata, Y.; Ichikawa, K.; Nishi, M.; Ikegawa, Y. Comparison of the Feasibility of Various Hydrogen Carriers Transported from Overseas for Massive Quantity and Long Distances (FY2018). In *Central Research Institute of Electric Power Industry Report*; Central Research Institute of Electric Power Industry: Tokyo, Japan, 2019. Available online: https://criepi.denken.or.jp/hokokusho/pb/reportDownload?reportNoUkCode=Q18005&tenpuTypeCode=30&seqNo=1&reportId=8886 (accessed on 1 February 2024).

28. The Institute of Applied Energy. Research Report on Economic Evaluation of Synthetic Methane by Methanation—Domestic Delivery. 2018. Available online: https://www.iae.or.jp/wp/wp-content/uploads/2018/10/metanation_201810_v1.1_r.pdf (accessed on 1 February 2024).

29. Mizuho Research & Technologies. Life Cycle Greenhouse Gas Emissions Analysis Report of Hydrogen Supply Chain. 2016. Available online: https://www.mizuho-rt.co.jp/solution/improvement/csr/lca/pdf/jisseki02_ghg1612.pdf (accessed on 1 February 2024).

30. Agency for Natural Resources and Energy. Electric Power Research Statistics. Available online: https://www.enecho.meti.go.jp/statistics/electric_power/ep002/ (accessed on 1 February 2024).

31. Imamura, E.; Iuchi, M.; Bando, S. Comprehensive Assessment of Life Cycle CO_2 Emissions from Power Generation Technologies in Japan. In *Central Research Institute of Electric Power Industry Report*; Central Research Institute of Electric Power Industry: Tokyo, Japan, 2016. Available online: https://criepi.denken.or.jp/hokokusho/pb/reportDownload?reportNoUkCode=Y06&tenpuTypeCode=30&seqNo=1&reportId=8713 (accessed on 1 February 2024).

32. United Nations. Take Action for the Sustainable Development Goals. Available online: https://www.un.org/sustainabledevelopment/sustainable-development-goals/ (accessed on 19 March 2024).

33. Oltra-Badenes, R.; Guerola-Navarro, V.; Gil-Gómez, J.-A.; Botella-Carrubi, D. Design and implementation of teaching–learning activities focused on improving the knowledge, the awareness and the perception of the relationship between the SDGs and the future profession of university students. *Sustainability* **2023**, *15*, 5324. [CrossRef]

34. de la Torre, R.; Onggo, B.S.; Corlu, C.G.; Nogal, M.; Juan, A.A. The role of simulation and serious games in teaching concepts on circular economy and sustainable energy. *Energies* **2021**, *14*, 1138. [CrossRef]

35. Suzuki, K.; Shibuya, T.; Kanagawa, T. Effectiveness of a game-based class for interdisciplinary energy systems education in engineering courses. *Sustain. Sci.* **2021**, *16*, 523–539. [CrossRef]

Stochastic Flow Analysis for Optimization of the Operationality in Run-of-River Hydroelectric Plants in Mountain Areas

Raquel Gómez-Beas [1,2,*], Eva Contreras [1,3], María José Polo [1,3] and Cristina Aguilar [1,2,*]

1   Fluvial Dynamics and Hydrology Research Group, Andalusian Institute for Earth System Research, University of Cordoba, 14071 Cordoba, Spain; econtreras@uco.es (E.C.); mjpolo@uco.es (M.J.P.)
2   Department of Mechanics, School of Engineering Science, University of Cordoba, 14071 Cordoba, Spain
3   Department of Agronomy, Unit of Excellence María de Maeztu (DAUCO), University of Cordoba, 14071 Cordoba, Spain
*   Correspondence: rgbeas@uco.es (R.G.-B.); caguilar@uco.es (C.A.)

Abstract: The highly temporal variability of the hydrological response in Mediterranean areas affects the operation of hydropower systems, especially in run-of-river (RoR) plants located in mountainous areas. Here, the water flow regime strongly determines failure, defined as no operating days due to inflows below the minimum operating flow. A Bayesian dynamics stochastic model was developed with statistical modeling of both rainfall as the forcing agent and water inflows to the plants as the dependent variable using two approaches—parametric adjustments and non-parametric methods. Failure frequency analysis and its related operationality, along with their uncertainty associated with different time scales, were performed through 250 Monte Carlo stochastic replications of a 20-year period of daily rainfall. Finally, a scenario analysis was performed, including the effects of 3 and 30 days of water storage in a plant loading chamber to minimize the plant's dependence on the river's flow. The approach was applied to a mini-hydropower RoR plant in Poqueira (Southern Spain), located in a semi-arid Mediterranean alpine area. The results reveal that the influence of snow had greater operationality in the spring months when snowmelt was outstanding, with a 25% probability of having fewer than 2 days of failure in May and April, as opposed to 12 days in the winter months. Moreover, the effect of water storage was greater between June and November, when rainfall events are scarce, and snowmelt has almost finished with operationality levels of 0.04–0.74 for 15 days of failure without storage, which increased to 0.1–0.87 with 3 days of storage. The methodology proposed constitutes a simple and useful tool to assess uncertainty in the operationality of RoR plants in Mediterranean mountainous areas where rainfall constitutes the main source of uncertainty in river flows.

Keywords: Run-of-River (RoR) hydropower plant; river flow; energy production; operationality; failure; uncertainty; WiMMed

Citation: Gómez-Beas, R.; Contreras, E.; Polo, M.J.; Aguilar, C. Stochastic Flow Analysis for Optimization of the Operationality in Run-of-River Hydroelectric Plants in Mountain Areas. *Energies* **2024**, *17*, 1705. https://doi.org/10.3390/en17071705

Academic Editors: Boštjan Polajžer, Davood Khodadad, Younes Mohammadi and Aleksey Paltsev

Received: 20 February 2024
Revised: 27 March 2024
Accepted: 29 March 2024
Published: 2 April 2024

1. Introduction

Hydropower is one of the cheapest renewable energy sources that can be generated without toxic waste [1,2] and has very low operation and maintenance costs [2,3]. Of the total renewable energy production in Europe, the majority was generated from hydropower, accounting for 425.8 TWh [4]. Most hydropower plants in use today are traditional or conventional hydropower plants designed with a dam, a lake, a penstock, and a powerhouse. In contrast, non-conventional hydropower plants, such as run-of-river (RoR) power plants, constitute a more environmentally friendly option [1,2,5]. In RoR plants, a fraction of the stream flow is diverted through penstocks to a powerhouse and then returned to the stream. The reservoir is frequently absent, or it is a tiny pond or chamber.

The contribution of small hydropower plants (SHPs) to the worldwide electrical capacity is at a more similar scale to the other renewable energy sources (1–2% of the total capacity) [3]. Europe is the market leader in small-scale hydropower technology, with Spain, Italy, France, Germany, and Sweden being the main producers [3]. Small SHPs and, among them, RoR plants are especially useful thanks to their low administrative and executive costs and short construction time compared to projects with storage reservoirs of a similar power capacity [5].

The operation of hydroelectric power plants is subject to different conditioning factors, including the hydraulic conditions of the riverbed, the specific operating conditions of the plant's instrumentation, and certain restrictions, such as compliance with the environmental flow regime established in the corresponding legislation. One of the main drawbacks of RoR plants is the high degree of uncertainty of the available river flows upstream of the plant owing to meteorological fluctuations, resulting in an unpredictable power generation capacity [5,6], which is even more prevalent in the current climate change scenario [2,7,8].

Some of the main energy infrastructures affected by climate change are hydropower plants located in snow-covered mountainous areas, where climate change is expected to result in a later and shorter snow season and less snow coverage [8,9]. Mountains are considered "water towers" since they provide water for both ecosystems and anthropogenic demands in downstream areas [10]. In Mediterranean mountains with both alpine and semi-arid conditions, the variability in the climate enhances the complexity of the hydrological regime. These areas have particularly extreme conditions, in which the high variability in the annual and seasonal climate regimes is usually propagated and amplified by the river flow [10]. Moreover, the highly variable snowpacks, in both time and space, and the presence of several accumulation and melting cycles during the snow season lead to a strong seasonality of the streamflow response in headwater catchments [11]. Thus, RoR plants in Mediterranean mountains operate with even more irregular production subject to the run-of-river flow, which depends on the highly variable forcing agents of the rainfall–runoff processes and snow cover dynamics. Consequently, these plants often have to cease operation when the flow drops below the turbine operating level or rises above the maximum allowed by the turbines, thus affecting power generation and, therefore, plant performance [2].

RoR systems are subject to several uncertainties in both operation and management [12]. Thus, the challenge of the operation of RoR hydroelectric plants for water resource managers is to quantify how much water will be available for power generation. Especially in the Mediterranean snow-dominated mountains where river flows follow a strong seasonal pattern with significant interannual variability, having a seasonal forecasting system with limited uncertainty and sufficient reliability for decision-making would be a very useful tool for their seasonal and annual planning and would reduce opportunity costs due to the lack of such a forecast.

Thus, the main objective of this study was to obtain a simple and versatile stochastic flow-forecasting structure from the significant forcing agents that allow for anticipating the regime of river inflows to run-of-river hydroelectric power plants and the number of days of failure for operational purposes at time scales of interest in hydrological planning. In addition, the effect of possible storage in a load chamber was included to optimize the operation of the chamber by minimizing the run-of-river power plant's dependence on the river flow. The most critical component of river flows in Mediterranean areas are low flow periods because of the mild winter temperatures in combination with long, dry, sunny periods [13]. Thus, the minimum regime as the most limiting variable in the operation of power plants constitutes the basis of this study.

Stochastic models are often applied in hydrology to generate different samples of meteorological and hydrological data that are equally likely with respect to the observed series [14,15]. The stochastic forecasting structure developed in this study is based on scientific knowledge of rainfall–runoff processes and a rigorous analysis of the stochastic relationships of their main descriptor variables. Therefore, the main forcing agents that determine the increase in humidity in the contributing basin were first identified. Secondly, an analysis was carried out to identify the significant relationships between the forcing agents and the target variables at the different temporal scales established. Then, a Bayesian analysis of the probability of occurrence of the inflow rates was made based on the antecedent hydrological conditions. Finally, a scenario analysis was performed to assess the effect of using a small storage chamber to optimize the management of hydroelectric power RoR plants with maximum use of the natural fluvial contributions in the study area.

2. Materials and Methods

2.1. Study Site and Data Sources

The study site was the Poqueira system in Sierra Nevada (37° N; −3.3° W), which is a national park and biosphere reserve located in a mountainous area in southern Spain, where the presence of snow has a great effect on the hydrology of the downstream areas [8]. The contributing basin to the hydropower plants has an area of 38.4 km^2, with an average slope of 23° and an elevation ranging from 3453 m a.s.l. to 449 m a.s.l., with a mean value of 2161 m a.s.l. There is great variability in the precipitation regime due to the interaction of both the alpine and Mediterranean climates of the region, with the accumulated annual precipitation rate reaching 1200 mm in wet years and 220 mm in dry years within the period of 1961–2015 [2]. Above 2500 m a.s.l., the presence of snow is persistent, although it is commonly found at altitudes above 1000 m a.s.l. from November to May, with a very heterogeneous spatial distribution of the snow cover [2,11].

Three consecutive RoR hydroelectric plants, HP1, HP2, and HP3 (Figure 1), belonging to an important Spanish company in the energy sector and with a capacity between 10 and 12 MW, are located in the basin [2]. The hydroelectric plant located at the highest altitude is Poqueira (HP1), which is supplied from a load chamber located at 2100 m a.s.l. The operation of this system is as follows: the turbined flow in the upstream plant (HP1) is carried through a pipeline that connects the load chambers to the next plant (HP2). In the same way, from the latter, it reaches the HP3 plant. For the study of the operation of these mini power plants in a concatenated series, only the case of HP1 has been analyzed since the turbine flow in this plant determines the operation of the other two plants downstream. For this reason, ecological flow restrictions must be applied to the HP1 plant since it is the first one in the series. The minimum ecological flow established in the Basin Hydrological Plan to be supplied from the HP1 plant is 0.35 m^3/s, as a constant flow throughout the year [2]. Therefore, the minimum operating flow of HP1 is the sum of the minimum turbine flow and the minimum ecological flow established (Table 1).

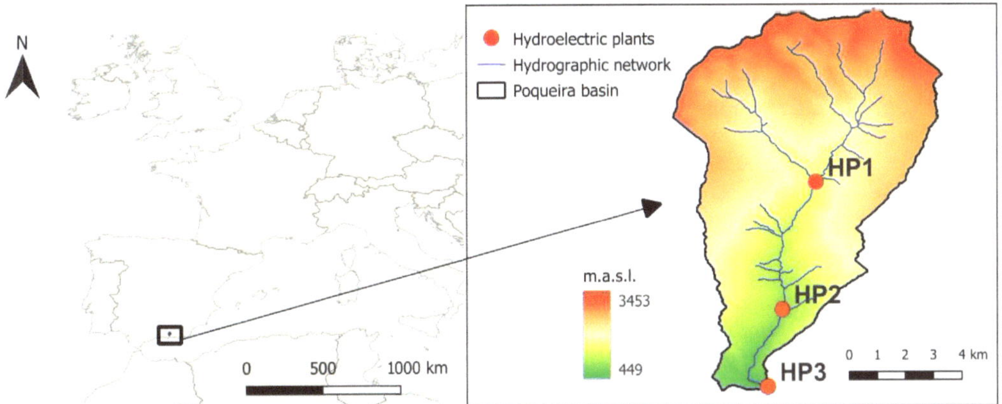

Figure 1. Location of the Poqueira River basin in southern Spain and the three RoR hydroelectric plants system in the study area.

Finally, an assessment of the operationality of the hydropower plant was performed with the analysis of several scenarios, including the hydropower plant operation rules and varying levels of water stored in the loading chamber.

2.2.1. Dependency Structure among Variables

First, the forcing agents and target variables were identified at the annual and monthly scales since these are the time scales of interest in the operation of hydroelectric power plants. Forcing agents are the variables that determine the increase in humidity in the basin that supplies the hydroelectric plants. Since this is an area influenced by a high mountain climate, in addition to rainfall, the presence of snow strongly determines the hydrological dynamics of the contributing basin [11,16]. Therefore, rainfall, snowfall, snow water equivalent, snowmelt, and evaposublimation were identified as possible forcing agents for the dependence analysis.

In terms of target variables, operationality has been defined in the scope of this study as the probability of being able to produce energy in the hydroelectric plant in 20 years. Operationality is, therefore, the complement of the probability of failure [13], with failure being defined as the day on which there is no energy production due to the flow being lower than the minimum operating flow. Therefore, the number of days on which the plant did not operate due to a circulating flow lower than the minimum operating flow was generated from the data series of daily flows available. The relationship between the two variables was then analyzed in order to quantify the probability of failure based on the mean daily flow. Thus, the final target variables are the mean daily flow (Q_{mean}) and the number of days of failure ($N_{failure}$) at the annual and monthly scales.

Based on the 20-year series of the variables considered, classical descriptive analysis techniques were applied by means of adjustments and analysis of the correlation coefficients between the series of forcing agents and target variables at different time scales—annual and monthly. The adjustments made between the variables were parametric relationships of the first-, second-, and third-degree polynomials, exponentials, and potentials. Due to space limitations in the Results section, only the best correlations obtained from all the parametric adjustments made for each variable are shown. However, all the parametric fits can be made available upon request to the authors.

Finally, the best parametric adjustments obtained for each time scale were selected.

2.2.2. Bayesian Dynamics Forecast of Water Inputs and Analysis of Operationality

In order to generate predictions of the target variables, the mean daily flow, and the number of days of failure, along with their associated uncertainty, the simulation of the forcing agent by Monte Carlo was combined with two approaches or methods—parametric and non-parametric. Figure 3 shows the calculation sequence with rainfall (R) as the forcing agent.

The parametric method (PM) is based on the parametric relationship between the forcing agent (e.g., rainfall or snowfall) and the target variable generated in the previous section. This method starts with the cumulative distribution function (cdf) of the forcing agent accumulated at the corresponding time scale and continues with the following calculation sequence, which shows rainfall as an example of the forcing agent (Figure 3):

1. A total of 250 sets of 20-year series of equally probable rainfall were obtained with Monte Carlo using the cdf of the available rainfall data series.
2. The mean daily flow was calculated from the best parametric fit with the accumulated rainfall; thus, the forecast was included in the calculation.
3. The number of days of failure was calculated from its best parametric relationship with the mean daily flow (Appendix A).

Regarding the non-parametric method (NPM), techniques already used in various areas of Europe were applied [13,22]. In this case, we started directly from the cumulative distribution functions of the rainfall, the mean daily flow, and the number of days of failure at the annual and monthly scales. Each of these functions was divided into 6 parts, considering the criterion that in each of these sections there should be at least 3 data points of the average flow or number of days of failure. Then, the cdf of each of the partitions of the different cumulative distribution functions was developed. Based on these developments, the calculation sequence of the non-parametric method was as follows (Figure 3):

1. A total of 250 sets of 20-year series of equally probable monthly rainfall were obtained with Monte Carlo using the distribution of the available monthly rainfall data series.
2. The obtained rainfall values were used to generate 250 repetitions of the 20-year mean daily flow data series, applying quantile mapping to the distribution functions generated in the partitions of the distribution functions of the measured data series, as this procedure is analogous to generating 250 repetitions of a series of number of days of failure.

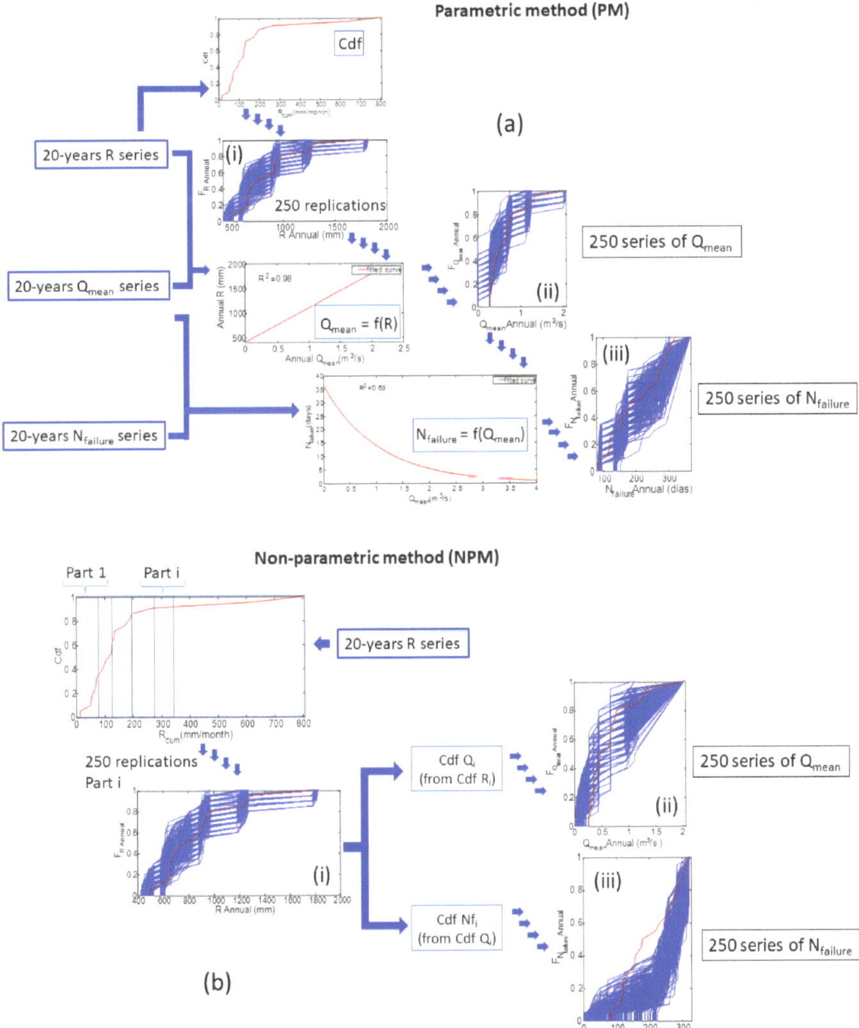

Figure 3. Calculation sequence of Bayesian dynamics forecast of water inputs: (**a**) parametric method (PM) and (**b**) non-parametric method (NPM). Blue lines are cdf of the 250 sets of 20-year series of equally probable rainfall (i), mean daily flow (ii), or number of days of failure (iii) obtained with Monte Carlo. In red, the empirical cdf obtained from the observed data.

Finally, the operationality of the hydropower plants was analyzed in terms of the complementary probability associated with the occurrence of failure based on the flow regime. To accomplish this analysis, the cdf of the target variables (the mean daily flow and the number of days of failure at the annual and monthly scales) was intercepted at the y-axis at the corresponding quartiles of 1, 2, and 3. Thus, it is possible to know with a 25, 50, or 75% probability whether the number of days of failure per year or month does not exceed a certain value above which the production of hydroelectric energy becomes unfeasible.

2.2.3. Scenario Analysis

Once the forecast of the number of days of failure for a given month is known, it is useful to know the effect of possible storage in a small load chamber to maximize the use of the incoming natural fluvial contributions.

The variable that determines the failure in the operation of a plant is a mean daily flow lower than the minimum daily operation flow, assigning to the variable $N_{failure}$ the failure or not in the operation of the plant. Scenario analysis was performed with a variable storage volume, which was null in case it did not exist, in order to simulate the operation of the plant as a pure RoR plant.

The number of days of failure was calculated on a daily scale from the available flow data series. Thus, the variable $N_{failure}$ was assigned the value of failure or not, not only depending on the availability of the flow provided but also taking into account the volume stored in the load chamber, if any. Then, the number of days of failure as a measure of operationality was aggregated at the corresponding time scale. Three were analyzed: no storage in the load chamber, which is equivalent to pure RoR plants, and a small load chamber volume equivalent to 3 days, which is the capacity of the chambers in other RoR plants in the area with similar characteristics (i.e., power generation capacity). Also, for comparison with traditional hydropower plants with larger storage systems, a chamber volume equivalent to 30 days is shown. This last value, oversized for the characteristics of the plant under study, is shown only for visual effect in order to clearly elucidate the effect of storage.

Figure 4 shows the flow chart of the plant operation calculation:

1. It is checked whether the incoming flow is equal to or higher than the minimum turbined flow to operate the plant.

    - If it is equal to or higher, a new volume of available water in the water chamber is calculated, which is equal to the existing volume plus that which exceeds the incoming flow on the day after the release. The day is assigned as a non-failure in the operation of the plant.
    - If it is lower or null, it is checked to see whether the incoming flow can be supplemented with the stored volume available in the load chamber.
        - If it can be supplemented, the new volume of the load chamber is calculated after extracting the required volume. The day is assigned as a non-failure in the plant's operation.
        - If it cannot be supplemented, the day is assigned as a failure in the plant's operation.

2. The calculation continues until the 20-year series is complete.

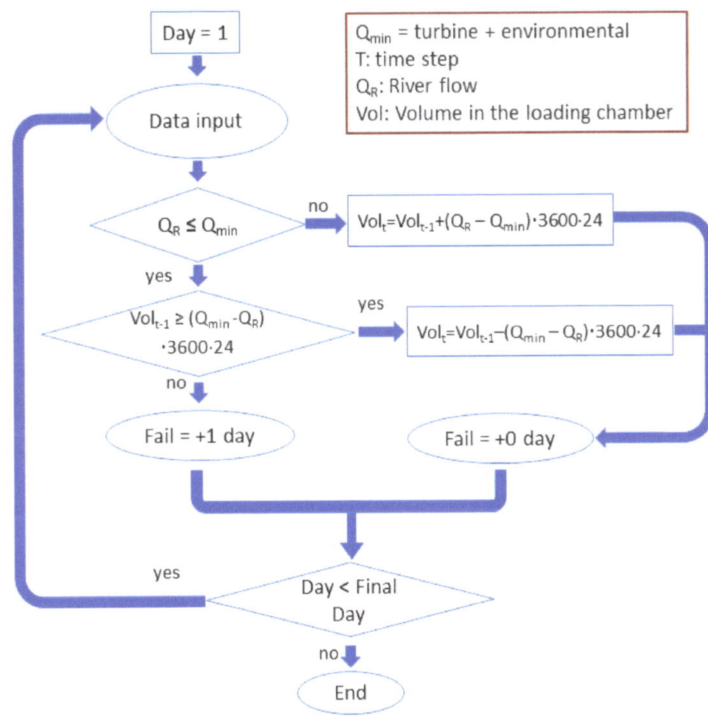

Figure 4. Flux diagram of the scenario analysis from the calculation of $N_{failure}$ for hydropower plant management.

3. Results and Discussion

3.1. Dependency Structure between Variables

3.1.1. Annual Scale

At an annual scale, a good linear relationship was observed between the maximum annual daily flow (Q_{max}) and the mean annual daily flow (Q_{mean}) (Table 3), with an R^2 value of 0.82. Regarding the relationship of the maximum annual daily flow with the average annual rainfall (R) and maximum annual daily rainfall (R_{max}), lower linear correlations were observed, with R^2 values of 0.8 and 0.52, respectively. The low correlation in this latter analysis can be explained by the snow influence in the area, which determined that the day with the maximum daily flow was not the same as the day with the maximum daily rainfall.

Table 3. R^2 of the linear correlations among the following variables: Q_{mean} (mean annual daily river flow), Q_{max} (maximum annual daily river flow), R (annual rainfall), ES (annual evaposublimation), S (annual snowfall), SWM (annual snowmelt), SWE (annual snow water equivalent), and $N_{failure}$ (annual number of days of failure). Values over 0.6 are in bold.

	$N_{failure}$	Q_{mean}	Q_{max}	R	R_{max}	S	R − ES	SWM	SWE
$N_{failure}$	1	**0.63**	0.33	**0.65**	0.19	0.36	**0.64**	0.38	0.41
Q_{mean}	**0.63**	1.00	**0.82**	**0.98**	0.41	**0.67**	**0.97**	**0.75**	**0.72**
Q_{max}	0.33	**0.82**	1.00	**0.80**	0.52	0.43	**0.82**	0.51	0.51
R	**0.65**	**0.98**	**0.80**	1.00	0.41	**0.67**	**0.99**	**0.74**	**0.72**
R_{max}	0.19	0.41	0.52	0.41	1.00	0.15	0.43	0.19	0.34
S	0.36	**0.67**	0.43	**0.67**	0.15	1.00	0.58	**0.98**	**0.81**
R − ES	**0.64**	**0.97**	**0.82**	**0.99**	0.43	0.58	1.00	**0.67**	**0.66**
SWM	0.38	**0.75**	0.51	**0.74**	0.19	**0.98**	**0.67**	1.00	**0.85**
SWE	0.41	**0.72**	0.51	**0.72**	0.34	**0.81**	**0.66**	**0.85**	1.00

The best correlations at the annual scale were found between the annual mean daily flow and the average annual rainfall (R), reaching an R^2 of 0.98, and the average rainfall minus the evaposublimation flux (R − ES) (Table 3), since it represents the net precipitation that directly contributes to runoff generation. Regarding the correlation between the annual mean daily flow and the snow-related variables, average snowfall (S), average snow water equivalent (SWE), and average snowmelt (SWM), slightly lower linear correlation coefficients were obtained. oscillating between 0.67 and 0.75, with the highest correlation being with snowmelt (0.75). These lower values were due to the highly variable snowpack with the presence of several accumulation and melting cycles during the snow season, as well as a non-negligible evaposublimation flux [11].

Regarding the annual number of days of failure ($N_{failure}$), the best correlations were found with both the mean annual daily flow and the annual rainfall, obtaining R^2 values of 0.63 and 0.65, respectively.

3.1.2. Monthly Scale

At a monthly scale, not only the linear correlation was considered but also the quadratic and exponential correlation, following previous studies that have flow-related variables [27,28] and use rainfall as the independent variable to better capture the non-linear effects of the hydrological water balance at shorter time scales. In addition, correlation with the value of some variables in the antecedent month was included in the analysis in order to analyze a possible monthly lag effect in the circulating flow.

Table 4 shows the correlation coefficients obtained between the monthly mean daily flow variable and the rest of the variables identified at a monthly scale for the best adjustments obtained for the sake of greater conciseness of the manuscript. However, the values obtained for all the adjustments per month carried out in this study can be made available upon request to the authors.

Table 4. Best correlation coefficients of the monthly mean daily flow with the variables analyzed. Values over 0.6 are in bold.

	September	October	November	December	January	February	March	April	May	June	July	August
$N_{failure}$	0.05	**0.9**	**0.88**	**0.89**	**0.86**	**0.91**	**0.91**	**0.87**	**0.91**	**0.94**	**0.93**	**0.77**
Q_{m_max}	**0.74**	**0.9**	**0.94**	**0.98**	**0.99**	**0.97**	**0.99**	**0.94**	**0.95**	**0.99**	**0.99**	**1**
R_m	0.35	**0.84**	0.36	0.54	**0.65**	**0.72**	**0.73**	0.02	0.46	0.25	**0.75**	0.06
R_{m_prev}	0.05	0.35	**0.72**	0.5	**0.92**	**0.77**	0.59	**0.74**	0.16	0.38	0.36	**0.71**
R_{m_max}	0.07	**0.76**	0.03	0.48	0.44	0.49	0.41	0.22	0.34	0.46	**0.73**	0.36
$R_{m\_max\_prev}$	0.01	**0.7**	0.55	0.26	**0.87**	0.51	0.38	0.43	0.34	0.35	0.57	**0.69**
$R_m - ES_m$	0.35	**0.84**	0.37	0.54	0.55	**0.72**	**0.68**	0.03	0.45	0.24	**0.75**	0.56
S_m	0.24	0.47	0.07	0.51	0.35	**0.76**	0.46	0.02	0.38	0.39	--	--
S_{m_prev}	--	0.29	**0.83**	0.05	**0.89**	0.49	**0.62**	**0.7**	0.39	0.39	0.5	--
SWM_m	0.27	0.28	0.4	0.49	**0.7**	**0.67**	0.51	**0.62**	**0.75**	**0.91**	**0.65**	--
SWE_m	0.02	0.49	0.16	0.45	**0.9**	**0.9**	**0.89**	**0.79**	**0.71**	**0.9**	**0.61**	--

Analogously to the annual scale, a good correlation was again observed between the maximum monthly daily flow and the mean monthly flow, with R^2 values between 0.74 and 1 (Table 4).

Regarding the monthly rainfall (R_m), good correlations were observed in the months of October, January, February, March, and July, with correlation coefficients between 0.65 and 0.84. When the same analysis was performed with the rainfall in the previous month (R_{m_prev}), the highest correlation coefficients were found in January, February, April, and November, with even higher R^2 values between 0.72 and 0.92.

As for the monthly snowfall (S_m), there was a good correlation only in the month of February (0.76, Table 4), reaching higher values in the months of November, January, March, and April with the snowfall of the previous month (S_{m_prev}). Finally, with respect to the monthly snow water equivalent (SWE_m) and the snowmelt (SWM_m), the correlation coefficient values were higher than 0.51 in the months from January to July, exceeding 0.62 in most months (Table 4).

At the monthly scale, the correlation between the monthly number of days of failure ($N_{failure}$) and the monthly mean river flow ($Q_{m\ mean}$) reached R^2 values of over 0.77 in most months (Table 4), exceeding 0.86 in most cases. Thus, the number of days of failure can be reproduced once the required forecasts of monthly river flows are available.

As already pointed out by previous studies at the study site [11], the peculiar snow dynamics result in the seasonality of the streamflow response in this headwater catchment. These results confirm this statement as the highest correlations with respect to the mean monthly flow occurred in the winter months, with monthly rainfall improving, in some cases, the correlation with respect to the rainfall of the previous month. In contrast, in the spring and summer months, the best correlations were obtained with the snowmelt and snow water equivalent. Despite being clear that the precipitation and snow depths impacted the monthly river flow, their effect was not instantaneous. Thus, this non-instantaneous time relationship needs to be considered, as already applied in previous studies [29], when looking for the best forcing variable at the monthly scale.

Considering that the main source of uncertainty in the flow regime in Mediterranean watersheds is due to variability in the occurrence of meteorological agents (mainly rainfall) [30], a more detailed analysis of lag times in the influence of rainfall accumulated in the antecedent months on the river flows was carried out. This analysis helped to solve the low correlation values obtained between the monthly mean flow and the monthly rainfall (Table 4) in certain months (e.g., December, May, and June). Thus, Table 5 shows the best correlations found for the monthly mean flow with one or several antecedent months of accumulated rainfall, as appropriate. The correlation coefficients in the table refer to the first-, second-, and third-degree polynomial adjustments, exponential adjustments, and potential adjustments in the case of December. The parametric expressions of these adjustments can be found in Table A1 in Appendix A. Correlation coefficients higher than 0.7 were reached in all months, and thus, these adjustments were used to reproduce the monthly river flow dynamics once the required forecasts of monthly rainfall were available. It can also be observed how, in most months, better correlations were obtained than those resulting from considering the current month or only a lag of one month (Table 4). The only exception is September, when no good correlation was found, neither with the rainfall in September nor extending the analysis to the rainfall of previous periods accumulated at various time scales, which is in accordance with previous studies in the basin encompassing the study area [13].

Table 5. Best correlation coefficients between accumulated rainfall and mean monthly river flow.

River Flow Forecast	Forecasting from Rainfall of	R^2
January	December	0.92
February	December and January	0.91
March	February and March	0.83
April	February to April	0.80
May	February to May	0.71
June	February to June	0.80
July	February to June	0.80
August	February to June	0.82
September	September	0.33
October	September and October	0.83
November	October and November	0.94
December	November and December	0.78

Two trends can be observed that group the forecast into three blocks. From October to December, the highest correlation was obtained between the rainfall of the previous month and that of the current month, i.e., the October flow was obtained with the accumulated rainfall of September and October, and in the same way, with November and December. In these months, the flow was mainly produced directly by the occurrence of rainfall–runoff events, mainly in liquid form, which explains this correlation between the previous month and the current month.

In the months of January and February, the occurrence of solid rainfall events or snowfall begins to be more frequent, so water accumulates in the snow layer and the proportion of direct runoff decreases, introducing a certain lag time in terms of river flows. Therefore, it was observed that the rainfall of the current month did not improve the average daily flow forecast for that month. Thus, the flow of January correlates to a greater extent with the rainfall of December, while the flow of February correlates with the accumulated rainfall of December and January.

The third block corresponds to the months from March to August, where snowmelt is the main variable describing the average daily flow in this period. In this case, starting with the February rainfall, the rainfall of the current month was accumulated until June. That is, the river flow in March was obtained from the accumulated rainfall of February and March; that of April was obtained from the accumulated rainfall of February, March, and April; and so on, until June. July and August were included in this block because their flow is related to the accumulated rainfall from February to June. Again, the role of accumulated water as snow may explain these correlations. Snowmelt begins in spring around the month of March and extends until the months of May-June, according to the hydrometeorological dynamics of the year [11,16], so the rainfall that occurs in these months, together with that of the previous months in which part of it would have accumulated as snow in Sierra Nevada, determines the runoff of the current month. In the case of July and August, there is little or no rainfall in these months, so the average flow in these months is due to the rainfall accumulated in the previous months.

In the case of the target variable number of days of failure ($N_{failure}$), parametric adjustments were also obtained with the mean river flow variable (Q_{mean}), given that, from the results shown in Table 4, there was a good correlation between these variables in most months. In this case, the effect of the lag time influence of the river flow on the number of days of failure was not observed, so the parametric adjustments were made with the mean river flow of the same month as the target forecast. The parametric expressions of these adjustments can be found in Table A2 in Appendix A.

3.2. Bayesian Dynamics Forecast of Inflows to the Plant

3.2.1. Mean Daily Flow and Number of Days of Failure

In Figure 5, the cdf of both the parametric and non-parametric methods is shown for the target variables Q_{mean} and $N_{failure}$. The empirical cdf of each variable obtained from observed data is represented in red. With both methods, the distribution of the mean river flow was reproduced using the 250 replications, with greater dispersion in the case of the non-parametric method, where, for instance, for 1 m^3/s of the mean river flow, the probability was between 0.63 and 0.95 using the parametric method and between 0.4 and 0.95 using the non-parametric method (Figure 5).

In the case of the target variable, the number of days of failure, the parametric method properly reproduced the shape of the empirical cdf. In contrast, the non-parametric method clearly overestimated the value of this variable; that is, this method gave lower probabilities for a higher number of days of failure than what actually occurred, and the stochastic replications did not encompass the empirical cdf distribution of the variable.

A comparison of the basic statistics of both the observed and simulated series is presented in Table 6. The quantiles of the simulated series are of the same order of magnitude as the observed series, notably for the parametric method (e.g., 0.48 m^3/s for the 50th percentile of the mean annual daily river flow). For the number of days of failure, the non-parametric method adequately reproduced the extremes of the distribution (e.g., 71 days for the fifth percentile and 310 days for the 90th percentile), although this method overestimated the value of the variable for the 50th percentile (267 days instead of 179 days in the observed series).

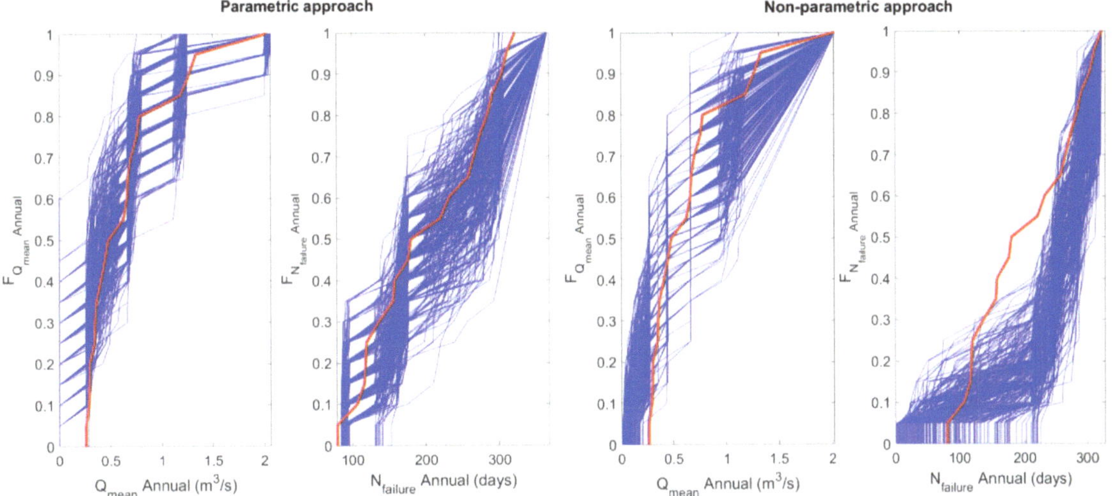

Figure 5. Cumulative distribution functions (cdf) of the mean annual daily river flow and the number of days of failure in the parametric and non-parametric methods. In red, the empirical cdf obtained from the observed data.

Table 6. The 5th, 50th, and 90th percentiles (F_5, F_{50}, F_{90}) of the observed and simulated mean annual river flow (Q_{mean}, Qp_{mean}, and Qnp_{mean} m^3/s) and number of days of failure ($N_{failure}$, $Np_{failure}$, and $Nnp_{failure}$ days), using the parametric and non-parametric methods. The statistics of the simulated series represent the average over the 250 replicates generated for each variable.

Percentile	Q_{mean}	Qp_{mean}	Qnp_{mean}	$N_{failure}$	$Np_{failure}$	$Nnp_{failure}$
F_5	0.26	0.004	0.056	79	88	71
F_{50}	0.48	0.48	0.51	179	210	267
F_{90}	1.26	1.15	1.48	303	325	310

Regarding the monthly scale, in most months, both methods properly reproduced the shape of the empirical distribution functions of the mean monthly daily flow (Figures 6 and 7), except in September, when the non-parametric method (Figure 7) encompassed the shape of the empirical distribution function. However, a greater dispersion was observed in the monthly daily flow predictions made using the non-parametric method in all the remaining months (Figure 7), being greater in September, October, December, and May.

Similarly, for the variable number of days of failure, the parametric method (Figure 8) best reproduced the shape of the empirical distribution function than the non-parametric method (Figure 9), with the worst fits being those for September and January. In all cases (Figures 8 and 9), the dispersion was greater for this variable, with a tendency to assign lower probabilities to higher failure values, i.e., to more days of failure in the month, in the same way as at the annual scale, especially in the months from November to June.

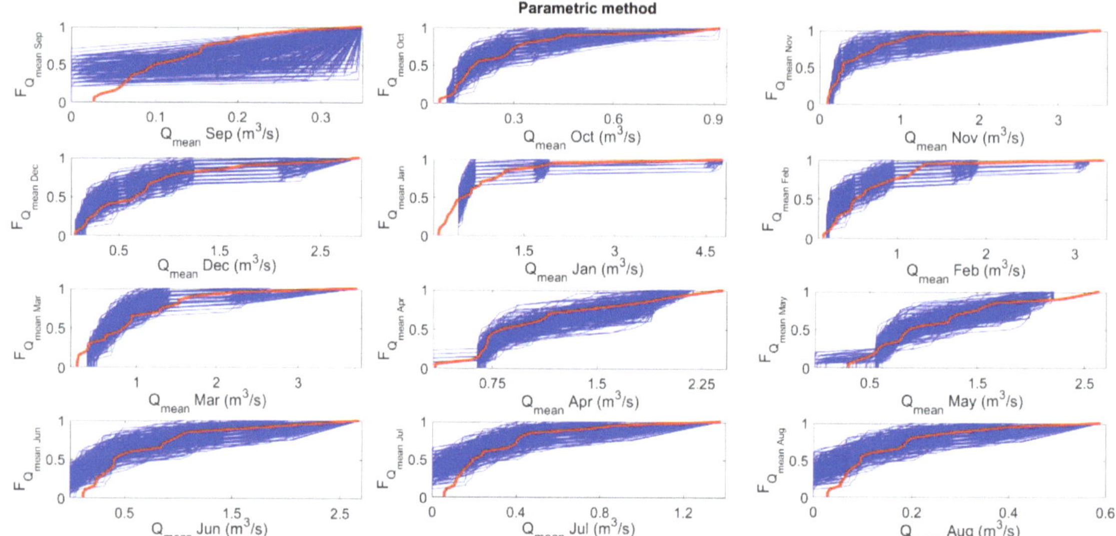

Figure 6. Cumulative distribution functions of monthly river flow using parametric method. In red, the empirical cdf obtained from the observed data.

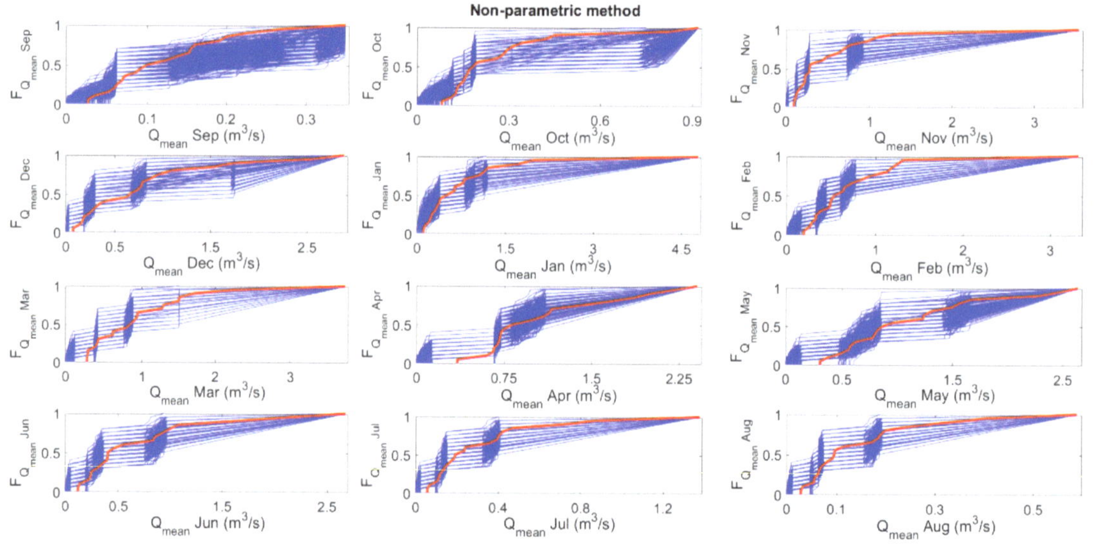

Figure 7. Cumulative distribution functions of monthly river flow using non-parametric method. In red, the empirical cdf obtained from the observed data.

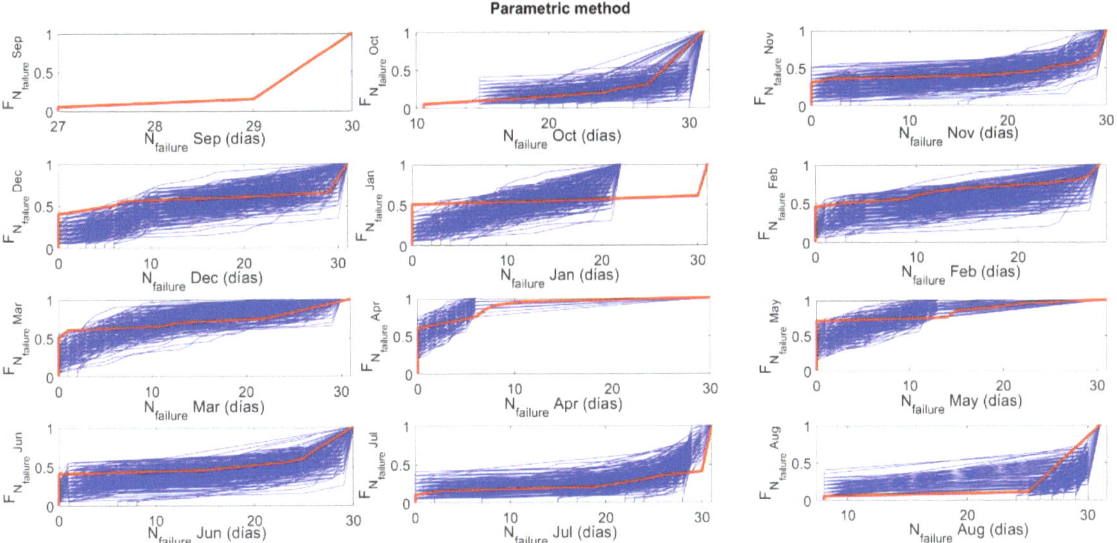

Figure 8. Cumulative distribution functions of monthly number of days of failure using parametric method. In red, the empirical cdf obtained from the observed data.

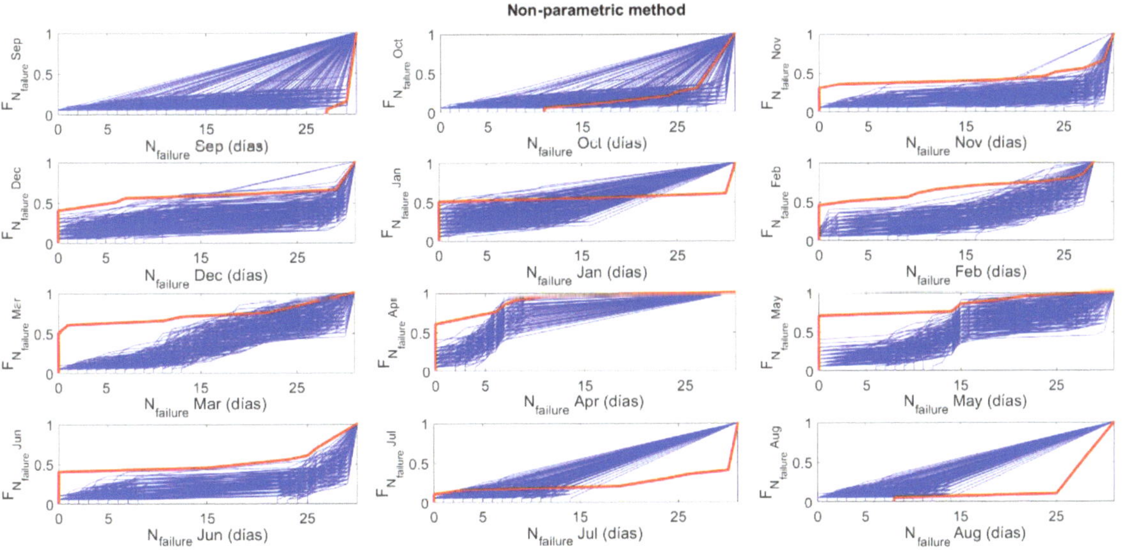

Figure 9. Cumulative distribution functions of monthly number of days of failure using non-parametric method. In red, the empirical cdf obtained from the observed data.

3.2.2. Operationality Assessment

In the decision-making support process, the distribution functions of quartiles 1, 2, and 3 were obtained for both the mean daily river flow and the number of days of failure at the annual and monthly scales. At the annual scale (Figure 10), with both methods, the probability of having an annual mean river flow of 0.45 m^3/s is 25%, whereas the probability of having an annual mean river flow of 0.73 m^3/s is 50%, with the greatest difference between the methods found in the case of quartile 3 (75% probability). Regarding the variable number of days of failure, as shown in Figure 10, with the non-parametric method, a tendency to assign a higher number of days of failure to the three quartiles was observed. Unlike with the mean daily river flow, great differences between the methods were found in the cases of quartiles 1 and 2, whereas the results obtained with both approaches are practically the same for a 75% probability.

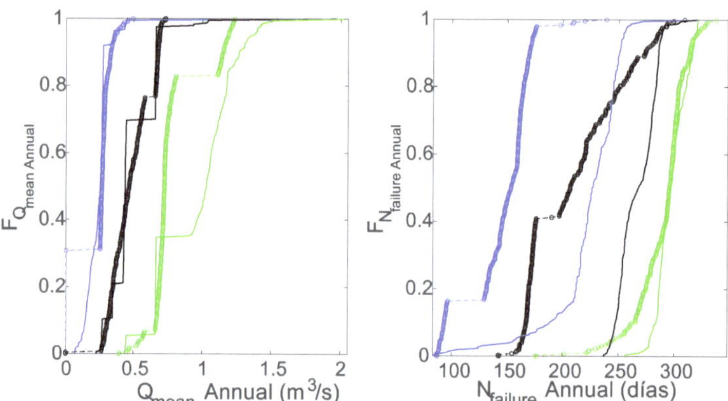

Figure 10. Cumulative distribution function of quartiles 1 (blue), 2 (black), and 3 (green) for annual river flow and number of days of failure using parametric method (circles) and non-parametric method (solid line).

At the monthly scale and with respect to the mean monthly daily river flow (Figure 11), the results reveal that both methods show similar tendencies in terms of the distribution of quartiles 1 and 2 in all months, with some greater differences in the 75th percentile, especially in the months of October and December and from January to April.

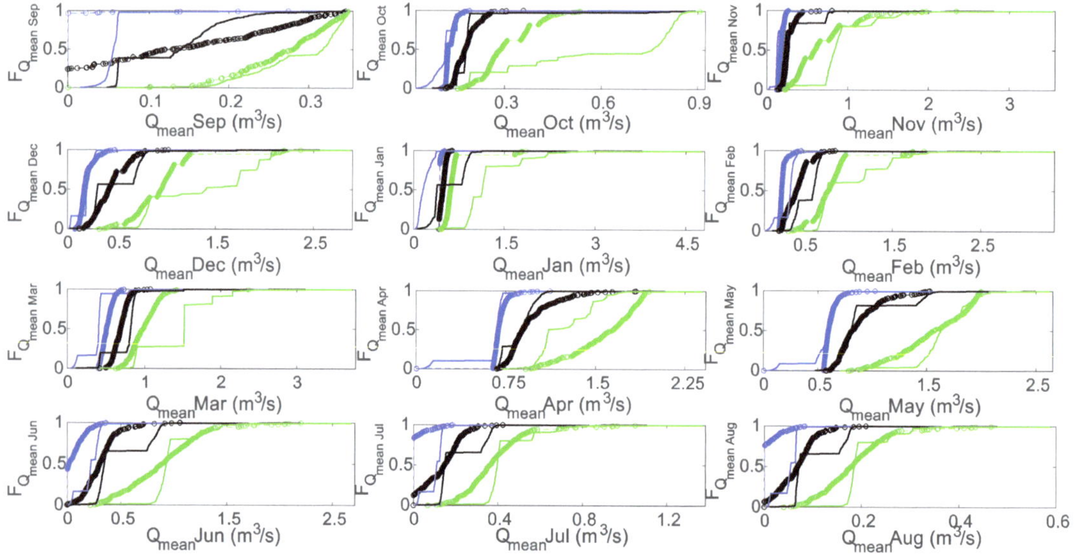

Figure 11. Cumulative distribution function of quartiles 1 (blue), 2 (black), and 3 (green) for monthly river flow using parametric method (circles) and non-parametric method (line).

Regarding the number of days of failure (Figure 12), the results show that higher operationality occurs in the snowmelt season between April and May, with a 25% probability of having fewer than 1 and 2 days of failure, respectively (Figure 12), according to the parametric method, which increases to 5–12 days with a 75% probability. In the winter months (December to March), there is a 25% probability of having fewer than 12 days of failure (Figure 12). These results again reveal the influence of snowmelt, along with its effect on the increase in flow rates in the spring months and, thus, fewer days of failure.

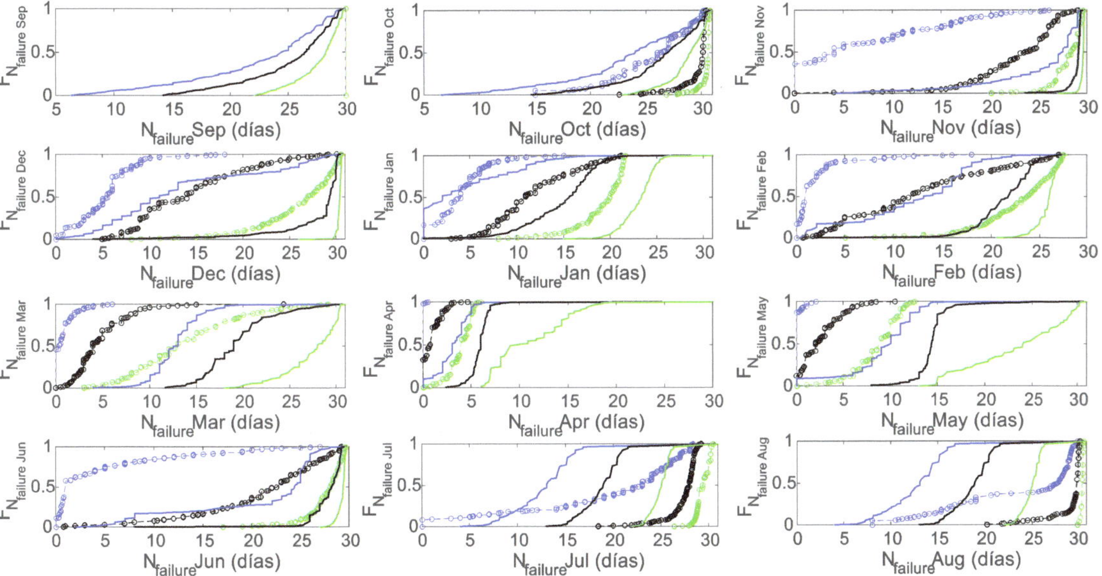

Figure 12. Cumulative distribution function of quartiles 1 (blue), 2 (black), and 3 (green) for monthly number of days of failure using parametric method (circles) and non-parametric method (line).

3.3. Scenario Analysis

Figure 13 shows the distribution functions of the number of days of failure at the annual scale for the different storage scenarios. Obviously, there is a shift in the cdf toward the left (lower number of days of failure) of the stochastic cdf with a higher volume stored, and thus, higher frequencies for a lower number of days of failure when the storage is included can be observed, being more pronounced in the case of 30 days of storage. Thus, with a 90% probability, the number of days of failure in the scenario without storage ranges between 239 and 354 days. However, in the scenario with 3 days of storage, 0 days of failure were reached in one of the simulations, although it can be considered that with a 90% probability, the range decreases to between 226 and 350 days of annual failure. Finally, for the scenario with 30 days of storage, the range of the number of days of failure changes from 151 to 350 days, i.e., the lower limit of this interval significantly decreased.

At the monthly time scale (Figure 14), the same trend was observed, with the effect of the water storage being greater in the months between June and November, when there are very few rainfall events and the snowmelt has almost finished. Thus, the effect of the decrease in the flow regime is compensated by the storage in the loading chamber. As an example, for this scale, in July, the range of probabilities for 15 days of failure is between 0.036 and 0.76 if there is no storage, between 0.1 and 0.76 in the scenario of 3 days of storage, and increasing to between 0.25 and 0.87 for the scenario of higher storage. This variation in percentages is greater in the month of November, where the probability of 15 days of plant operation failure is between 0.04 and 0.74 when there is no storage, between 0.1 and 0.87 when there are 3 days of storage, and increasing to between 0.25 and 0.9 for the case of 30 days of storage.

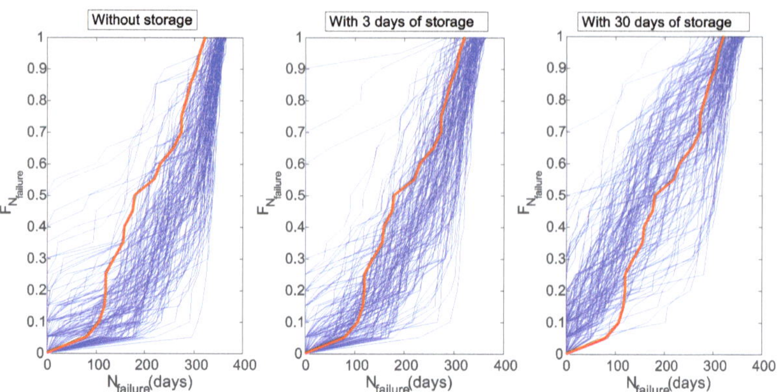

Figure 13. Cumulative distribution functions of number of days of failure in different storage scenarios: without storage, with 3 days of storage, and with 30 days of storage at annual scale. In red, the empirical cdf obtained from the observed data.

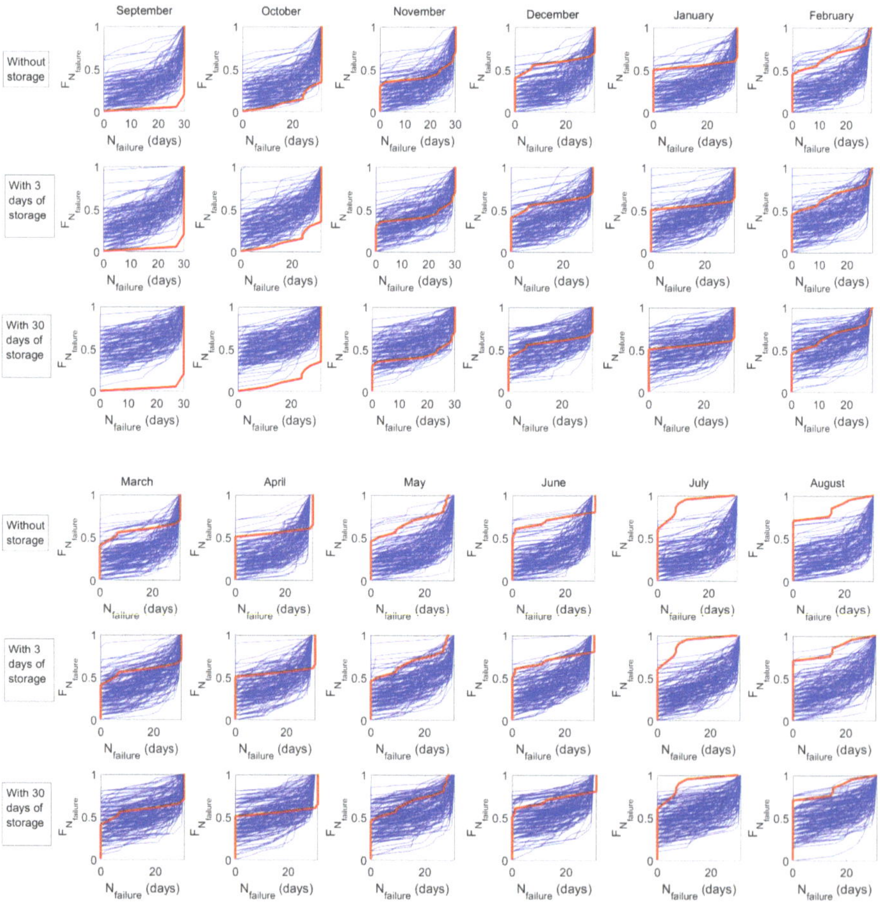

Figure 14. Cumulative distribution functions of number of days of failure in different storage scenarios: without storage, with 3 days of storage, and with 30 days of storage at monthly scale. In red, the empirical cdf obtained from the observed data.

4. Discussion

The operation of run-of-river plants in Mediterranean mountainous basins is highly sensitive to the natural inflows to the plants. The challenge of snow-dominated Mediterranean mountainous RoR plants is the strong seasonality and interannual variability of river flows, as well as the input data availability to implement other existing data-demanding forecasting tools. Thus, this study proposes a new simple seasonal forecasting system that integrates the dynamic effect of the precipitation regime as a valuable tool for decision-making and action plans for strongly heterogeneous Mediterranean watersheds, where the straightforward application of other existing tools is not feasible nowadays due to the limitations of the input data requirements.

A seasonal forecasting system with limited uncertainty and sufficient reliability for decision-making means an improvement in their seasonal and annual planning, as well as a reduction in opportunity costs due to the lack of such a forecast.

First, significant forcing agents that allow for anticipating the regime of river inflows to RoR hydroelectric power plants need to be identified at each temporal scale. The best correlations were found between the mean daily flow and the average rainfall at the annual scale. At the monthly scale, different lag times in the influence of rainfall accumulated in the antecedent months had to be considered to capture its non-instantaneous impact on the river flows. These variable monthly lag times corresponded to the three main types of monthly river flows that mainly depend on rainfall (September–December), snowfall (January–February) and snowmelt (March–August), respectively. As for the monthly number of days of failure, an effect of the lag time influence of rivers was not observed, so the mean river flow of the same month can be considered the target forecast.

The results of this study are conditioned by the quality of the input data and the goodness of the calibration obtained with the model used to generate the distributed maps of the hydrometeorological variables analyzed, as well as the flow data series. Furthermore, the parametric approach is based on the best fits obtained among the variables, which introduces additional uncertainty associated with them.

The best statistical forecasts of the target variables were obtained with the parametric method in this case study. However, the non-parametric approach constitutes a very interesting option as a first evaluation as well as when parametric adjustments among variables with a high correlation are not available for the study site.

Despite the limitations stated above, our results allow for implementing a simple methodology to evaluate the operationality of the plant with natural inflows. The study of the operation of hydroelectric plants at annual and monthly scales allows for the inclusion of forecast precipitation and flow data generated by the European Centre for Medium-Range Weather Forecasts (ECMWF), available through the Copernicus Climate Change Service (C3S) [31–33]. With these data, the simple forecasting system presented here allows for performing a monthly forecast of the operationality and the number of days of failure of run-of-river hydroelectric plants six months ahead. Thus, the potential impact on water resource management and renewable energy generation is straightforward. The application of this approach in any other RoR plant only requires the proper identification of the forcing variable or target forecasts at each temporal scale of interest and parametric adjustments if the parametric approach is the chosen option. Nevertheless, the uncertainty of the forecasts will depend on the length and quality of the available hydrometeorological data series.

Previous studies have assessed the impact of changes in hydrometeorological conditions on various aspects of electrical power and energy systems (e.g., electricity generation, electricity consumption, etc.) [34]. Making energy infrastructures resilient to climate change requires dedicated policies and sophisticated decision-making measures to build adaptive capacity [35]. In this context, the proposed methodology constitutes a useful tool to assess uncertainty in the operationality of RoR plants and can be supported by forecast information through climate services [36,37]. The development of decision support tools for water management in hydroelectric plants is determined to minimize the impact of the variability in the flow regime in the medium and long term in the case of RoR hydroelectric plants. In these plants, the availability of water storage is limited or non-existent, so this type of tool makes it possible to limit the uncertainty associated with the production of the hydroelectric plant on a monthly and seasonal scale, at least six months in advance. Thus, this methodology can assist hydropower systems as the managers can plan the production of the plant at the beginning of the calendar year until the end of the snowmelt, as well as maintenance shutdowns of the plants during months of lower productivity.

5. Conclusions

A Bayesian dynamics stochastic model was developed in this study based on the simulation of rainfall with Monte Carlo at the annual and monthly scales, combining two methods—parametric and non-parametric. The best statistical forecasts of the target variables, the mean daily river flow and the number of days of failure, were obtained with the parametric method based on the best adjustments at each temporal scale considered. The results show a greater dispersion in the variable number of days of failure, with a tendency to assign lower probabilities to higher failure values. The operationality assessment performed showed the influence of snowmelt and its effect on the increase in flow rates in the spring months, with a 25% probability of having 1 and 2 days of failure in April and May, respectively, while in the winter months, this probability of 25% corresponds to having fewer than 12 days of failure and close to 30 days of failure in the remaining months.

A scenario analysis was carried out with the inclusion of water storage in the load chamber of hydroelectric plants and assessing its effect on the variable number of days of failure as a measure of the plant's operationality. The results show the expected trends: the higher the load chamber considered, the higher the operationality level, and 239 to 151 annual days of failure without storage and with 30 days of storage, respectively, with a 90% probability. Regarding the monthly scale, the effect of water storage is greater in the months between June and November, with an operationality level of 0.04–0.74 for 15 days of failure without storage in November and of 0.25–0.9 with 30 days of storage.

Author Contributions: Conceptualization, R.G.-B. and C.A.; methodology, C.A. and R.G.-B.; validation, R.G.-B. and C.A.; formal analysis, C.A. and M.J.P.; investigation, R.G.-B., E.C. and C.A.; resources, C.A. and M.J.P.; writing—original draft preparation, R.G.-B.; writing—review and editing, R.G.-B. and C.A.; supervision, C.A. and M.J.P.; project administration, C.A. and M.J.P.; funding acquisition, C.A., E.C. and M.J.P. All authors have read and agreed to the published version of the manuscript.

Funding: This work has been funded by project 1381239-R "Herramienta de pronóstico estocástico de caudal para gestión de centrales hidroeléctricas en cuencas mediterráneas a distintas escalas temporales" with the economic collaboration of the European Funding for Rural Development (FEDER) and the Office for Economy, Knowledge, Enterprises and University of the Andalusian Regional Government; and by proyect PID2021-12323SNB-I00 "Incorporating hydrological uncertainty and risk analysis to the operation of hydropower facilities in Mediterranean mountain watersheds (HYPOMED)", funding by Spanish Ministry of Science and Innovation MCIN/AEI/10.13039/501100011033/FEDER, UE.

Data Availability Statement: The data are not publicly available due to the authors do not have permission to share data.

Acknowledgments: Authors are thankful for the support and technical knowledge provided by the Poqueira hydropower system managers and personnel and the hydrological data provided by all the weather and hydrological networks in the study site.

Conflicts of Interest: The authors declare no conflicts of interests. The funders had no role in the design of the study; in the collection, analyses, or interpretation of data; in the writing of the manuscript; or in the decision to publish the results.

Appendix A

Tables A1 and A2 show, respectively the fits and correlation coefficients (R^2) at annual and monthly scale, between the mean river flow (Q_{mean}) and rainfall (R), and between the number of days of failure ($N_{failure}$) and the mean river flow (Q_{mean}).

Table A1. Fits and correlation coefficients (R^2) between mean river flow (Q_{mean}) at annual and monthly scale and rainfall (R).

River Flow Forecast	Forecasting from Rainfall of	Relationship	R^2
Annual	Annual	$Q_{mean} = 0.0014{\cdot}R - 0.57$	0.99
January	December	$Q_{mean} = 7.70e^{-6}{\cdot}R^2 + 1.07e^{-3}{\cdot}R + 0.40$	0.92
February	December and January	$Q_{mean} = 3.14e^{-3}{\cdot}R + 0.13$	0.91
March	February and March	$Q_{mean} = 0.353{\cdot}e^{-0.0042{\cdot}R}$	0.83
April	February to April	$Q_{mean} = -7.40e^{-8}{\cdot}R^3 + 8.29e^{-5}{\cdot}R^2 - 0.024{\cdot}R + 2.74$	0.80
May	February to May	$Q_{mean} = -6.56e^{-8}{\cdot}R^3 + 8.05e^{-5}{\cdot}R^2 - 0.026{\cdot}R + 3.11$	0.71
June	February to June	$Q_{mean} = 7.97e^{-6}{\cdot}R^2 - 0.0025{\cdot}R + 0.40$	0.80
July	February to June	$Q_{mean} = 2.39e^{-8}{\cdot}R^3 - 2.68e^{-5}{\cdot}R^2 + 0.011{\cdot}R - 1.31$	0.80
August	February to June	$Q_{mean} = 7.25e^{-9}{\cdot}R^3 - 7.66e^{-6}{\cdot}R^2 + 0.0032{\cdot}R - 0.37$	0.82
September	September	$Q_{mean} = -1.64e^{-6}{\cdot}R^3 + 2.37e^{-4}{\cdot}R^2 - 0.0075{\cdot}R + 0.14$	0.33
October	September and October	$Q_{mean} = 0.092e^{0.00626{\cdot}R} + 7.52e^{-17}{\cdot}e^{0.1112{\cdot}R}$	0.83
November	October and November	$Q_{mean} = 0.107e^{0.00567{\cdot}R} + 7.29e^{-6}{\cdot}e^{0.0277{\cdot}R}$	0.94
December	November and December	$Q_{mean} = 0.0015{\cdot}R^{1.142}$	0.78

The variable number of days of failure acts as a discrete variable in its extreme values, given by the limits of turbine operation in the hydroelectric plant. Therefore, to prepare the parametric adjustments for this variable, the same limits have been respected as for the turbine. That is, when the flow variable on the corresponding scale gives a value lower than the minimum turbine speed, the maximum value of the corresponding scale is assigned to the variable number of days of failure. Similarly, the flow value in the river flow above which there is no operational failure in the hydroelectric plant at the scale considered has been identified. Between these values, a parametric adjustment has been made for each month.

Table A2. Fits and correlation coefficients (R2) between number of days of failure ($N_{failure}$) and mean river flow (Q_{mean}) at annual or monthly scale. The forecasting river flow month is the same than the forecast target month for $N_{failure}$.

$N_{failure}$ Forecast	Forecasting from River Flow of	Relationship	R^2
Annual mean	Annual mean	$N_{failure} = 162.82{\cdot}Q_{mean}^2 - 467.64{\cdot}Q_{mean} + 414.07$	0.89
January	January	$N_{failure} = 700.1{\cdot}e^{-8.201{\cdot}Q_{mean}}$	0.95
February	February	$N_{failure} = 211.5{\cdot}e^{-6.123{\cdot}Q_{mean}}$	0.93
March	March	$N_{failure} = 326.3{\cdot}e^{-5.923{\cdot}Q_{mean}}$	0.91
April	April	$N_{failure} = -0.81{\cdot}e^{-0.87{\cdot}Q_{mean}} + 170.8{\cdot}e^{-5{\cdot}Q_{mean}}$	0.88
May	May	$N_{failure} = 148.6{\cdot}e^{-4.49{\cdot}Q}$	0.86
June	June	$N_{failure} = -5.52{\cdot}Q_{mean}^3 + 36.48{\cdot}Q_{mean}^2 - 74.47{\cdot}Q_{mean} + 44.99$	0.93
July	July	$N_{failure} = 86.52{\cdot}Q_{mean}^3 - 164.9{\cdot}Q_{mean}^2 + 42.47{\cdot}Q_{mean} + 28.53$	0.98
August	August	$N_{failure} = -220.9{\cdot}Q_{mean}^3 + 71.35{\cdot}Q_{mean}^2 - 4.79{\cdot}Q_{mean} + 31.04$	0.99
September	September	--	--
October	October	$N_{failure} = 178.5{\cdot}Q_{mean}^3 - 251.7{\cdot}Q_{mean}^2 + 69.02{\cdot}Q_{mean} + 26.04$	0.97
November	November	$N_{failure} = -5.35{\cdot}Q_{mean}^3 + 35.83{\cdot}Q_{mean}^2 - 71.45{\cdot}Q_{mean} + 41.14$	0.89
December	December	$N_{failure} = -11.34{\cdot}Q_{mean}^3 + 61.55{\cdot}Q_{mean}^2 - 97.95{\cdot}Q_{mean} + 47.86$	0.84

References

1. Yildiz, V.; Vrugt, J.A. A toolbox for the optimal design of run-of-river hydropower plants. *Environ. Model. Softw.* **2019**, *111*, 34–152. [CrossRef]
2. Contreras, E.; Herrero, J.; Crochemore, L.; Pechlivanidis, I.; Photiadou, C.; Aguilar, C.; Polo, M.J. Advances in the Definition of Needs and Specifications for a Climate Service Tool Aimed at Small Hydropower Plants' Operation and Management. *Energies* **2020**, *13*, 1827. [CrossRef]
3. Manzano-Agugliaro, F.; Tahera, M.; Zapata-Sierra, A.; Del Juaidi, A.; Montoya, F.G. An overview of research and energy evolution for small hydropower in Europe. *Renew. Sustain. Energy Rev.* **2017**, *75*, 476–489. [CrossRef]
4. EnAppSys. European Electricity Fuel Mix Summary. Quarterly EU Market Summary, January to March. 2020. Available online: https://b74bc22f-390f-4347-ba45-b13ad13072ee.filesusr.com/ugd/9b26cb_80a91d80583e4c0e8c71ba3211517e3c.pdf (accessed on 27 March 2024).
5. Saeed, A.; Shahzad, E.; Khan, A.U.; Waseem, A.; Iqbal, M.; Ullah, K.; Aslam, S. Three-Pond Model with Fuzzy Inference System-Based Water Level Regulation Scheme for Run-of-River Hydropower Plant. *Energies* **2023**, *16*, 2678. [CrossRef]
6. Tsuanyo, D.; Amougou, B.; Aziz, A.; Nka Nnomo, B.; Fioriti, D.; Kenfack, J. Design models for small run-of-river hydropower plants: A review. *Sustain. Energy Res.* **2023**, *10*, 3. [CrossRef]

7. Sojka, M. Directions and Extent of Flows Changes in Warta River Basin (Poland) in the Context of the Efficiency of Run-of-River Hydropower Plants and the Perspectives for Their Future Development. *Energies* **2022**, *15*, 439. [CrossRef]
8. Pimentel, R.; Photiadou, C.; Little, L.; Huber, A.; Lemoine, A.; Leidinger, D.; Lira-Loarca, A.; Lückenkötter, J.; Pasten-Zapata, E. Improving the usability of climate services for the water sector: The AQUACLEW experience. *Clim. Serv.* **2022**, *28*, 100329. [CrossRef]
9. Tobias, W.; Manfred, S.; Klaus, J.; Massimiliano, Z.; Bettina, S. The future of Alpine Run-of-River hydropower production: Climate change, environmental flow requirements, and technical production potential. *Sci Total Environ.* **2023**, *890*, 163934. [CrossRef] [PubMed]
10. Viviroli, D.; Dürr, H.; Messerli, B.; Meybeck, M.; Weingartner, R. Mountains of the world, water tower for humanity: Typology, mapping, and global significance. *Water Resour. Res.* **2007**, *43*, W07447. [CrossRef]
11. Pimentel, R.; Herrero, J.; Polo, M.J. Subgrid Parameterization of Snow Distribution at a Mediterranean Site Using Terrestrial Photography. *Hydrol. Earth Syst. Sci.* **2017**, *21*, 805–820. Available online: https://hess.copernicus.org/articles/21/805/2017/ (accessed on 27 March 2024). [CrossRef]
12. Sakki, G.K.; Tsoukalas, I.; Kossieris, P.; Makropoulos, C.; Efstratiadis, A. Stochastic simulation-optimization framework for the design and assessment of renewable energy systems under uncertainty. *Renew. Sust. Energy Rev.* **2022**, *168*, 112886. [CrossRef]
13. Aguilar, C.; Polo, M.J. Assessing minimum environmental flows in nonpermanent rivers: The choice of thresholds. *Environ. Model. Soft.* **2016**, *79*, 102–134. [CrossRef]
14. Kottegoda, N.T.; Rosso, R. *Applied Statistics for Civil and Environmental Engineers*, 2nd ed.; Blackwell Publishing Ltd.: Oxford, UK, 2008; ISBN 978-1-4051-7917-1.
15. Kottegoda, N.T.; Natale, L.; Raiteri, E. Monte Carlo Simulation of rainfall hyetographs for analysis and design. *J. Hydrol.* **2014**, *519*, 1–11. [CrossRef]
16. Polo, M.J.; Herrero, J.; Pimentel, R.; Pérez-Palazón, M.J. The Guadalfeo Monitoring Network (Sierra Nevada, Spain): 14 Years of Measurements to Understand the Complexity of Snow Dynamics in Semiarid Regions. *Earth Syst. Sci. Data.* **2019**, *11*, 393–407. Available online: https://essd.copernicus.org/articles/11/393/2019/ (accessed on 27 March 2024). [CrossRef]
17. Authomatic Hydrological Information System (SAIH). Andalusian Government. Available online: http://www.redhidrosurmedioambiente.es/saih/ (accessed on 27 March 2024).
18. AEMET: AEMET Open Data, AEMET [Data Set]. Available online: https://www.aemet.es/es/datos_abiertos, (accessed on 27 March 2024).
19. Gan, K.C.; McMahon, T.A.; Finlayson, B.L. Analysis of periodicity in streamflow and rainfall data by Colwell's indices. *J. Hydrol.* **1991**, *123*, 105–118. [CrossRef]
20. Brunner, M.I.; Björnsen Gurung, A.; Zappa, M.; Zekollari, H.; Farinotti, D.; Stähli, M. Present and future water scarcity in Switzerland: Potential for alleviation through reservoirs and lakes. *Sci. Total Environ.* **2019**, *666*, 1033–1047. [CrossRef] [PubMed]
21. Polo, M.J.; Herrero, J.; Aguilar, C.; Millares, A.; Moñino, A.; Nieto, S.; Losada, M.A. WiMMed, a Distributed Physically-Based Watershed Model (I): Description and Validation. In *Theorical, Experimental and Computational Solutions, Proceedings of the International Workshop on Environmental Hydraulics 09, Valencia, Spain, 28–29 October 2009*; Taylor & Francis: Valencia, Spain, 2009; pp. 225–228.
22. Egüen, M.; Aguilar, C.; Solari, S.; Losada, M.A. Non-stationary rainfall and natural flows modeling at the watershed scale. *J. Hydrol.* **2016**, *538*, 767–782. [CrossRef]
23. Pastén-Zapata, E.; Pimentel, R.; Royer-Gaspard, P.; Sonnenborg, T.O.; Aparicio-Ibañez, J.; Lemoine, A.; Pérez-Palazón, M.J.; Schneider, R.; Photiadou, C.; Thirel, G.; et al. The effect of weighting hydrological projections based on the robustness of hydrological models under a changing climate. *J. Hydrol. Reg. Stud.* **2022**, *41*, 101113. [CrossRef]
24. Michaud, J.; Sorooshian, S. Effects of rainfall-sampling errors on simulations of desert flash floods. *Water Resour. Res.* **1994**, *30*, 2765–2775. [CrossRef]
25. Zhao, F.; Zhang, L.; Chiew, F.H.S.; Vaze, J.; Cheng, L. The effect of spatial rainfall variability on water balance modelling for south-eastern Australian catchments. *J. Hydrol.* **2013**, *493*, 16–29. [CrossRef]
26. Godinho, F.; Costa, S.; Pinheiro, P.; Reis, F.; Pinheiro, A. Integrated procedure for environmental flow assessment in rivers. *Environ. Process.* **2014**, *1*, 137–147. [CrossRef]
27. Fill, H.D.; Steiner, A.A. Estimating instantaneous peak flow from mean daily flow data. *J. Hydrol. Eng.* **2003**, *8*, 365–369. [CrossRef]
28. Taguas, E.V.; Ayuso, J.L.; Peña, A.; Yuan, Y.; Sanchez, M.C.; Giraldez, J.V.; Perez, R. Testing the relationship between instantaneous peak flow and mean daily flow in a Mediterranean Area Southeast Spain. *Catena.* **2008**, *75*, 129–137. [CrossRef]
29. Ho, L.T.T.; Dubus, L.; De Felice, M.; Troccoli, A. Reconstruction of Multidecadal Country-Aggregated Hydro Power Generation in Europe Based on a Random Forest Model. *Energies* **2020**, *13*, 1786. [CrossRef]
30. Polo, M.J.; Aguilar, C.; Millares, A.; Herrero, J.; Gómez-Beas, R.; Contreras, E.; Losada, M.A. Assessing risks for integrated water resource management: Coping with uncertainty and the human factor. *Proc. IAHS* **2014**, *364*, 285e291. Available online: https://piahs.copernicus.org/articles/364/285/2014/ (accessed on 14 February 2024). [CrossRef]
31. Crochemore, L.; Ramos, M.-H.; Pechlivanidis, I.G. Can Continental Models Convey Useful Seasonal Hydrologic Information at the Catchment Scale? *Water Resour. Res.* **2020**, *56*, e2019WR025700. [CrossRef]
32. Pechlivanidis, I.; Crochemore, L.; Bosshard, T. Seasonal streamflow forecasting—Which are the drivers controlling the forecast quality? In Proceedings of the EGU General Assembly 2020, Vienna, Austria, 4–8 May 2020.

33. Climate Change Copernicus. Available online: https://www.copernicus.eu/en/services/climate-change (accessed on 27 March 2024).
34. Mohammadi, Y.; Palstev, A.; Polajžer, B.; Miraftabzadeh, S.M.; Khodadad, D. Investigating Winter Temperatures in Sweden and Norway: Potential Relationships with Climatic Indices and Effects on Electrical Power and Energy Systems. *Energies* **2023**, *16*, 5575. [CrossRef]
35. Adaptation Challenges and Opportunities for the European Energy System. Building a Climate-Resilient Low-Carbon Energy System. Available online: https://www.eea.europa.eu/publications/adaptation-in-energy-system (accessed on 27 March 2024).
36. Buizer, J.; Jacobs, K.; Cash, D. Making short-term climate forecasts useful: Linking science and action. *Proc. Natl. Acad. Sci. USA* **2016**, *113*, 4597–4602. [CrossRef]
37. Dutton, J.A.; James, R.P.; Ross, J.D. Probabilistic Forecasts for Energy: Weeks to a Century or More. In *Weather & Climate Services for the Energy Industry*; Troccoli, A., Ed.; Springer: Cham, Switzerland, 2018; pp. 161–177; ISBN 978-3-319-68418-5.

 *energies*

Article

Machine Learning and Weather Model Combination for PV Production Forecasting

Amedeo Buonanno, Giampaolo Caputo *, Irena Balog, Salvatore Fabozzi, Giovanna Adinolfi, Francesco Pascarella, Gianni Leanza, Giorgio Graditi and Maria Valenti

Italian National Agency for New Technologies, Energy and Sustainable Economic Development (ENEA), 00196 Rome, Italy; amedeo.buonanno@enea.it (A.B.); irena.balog@enea.it (I.B.); salvatore.fabozzi@enea.it (S.F.); giovanna.adinolfi@enea.it (G.A.); francesco.pascarella@enea.it (F.P.); gianni.leanza@enea.it (G.L.); giorgio.graditi@enea.it (G.G.); maria.valenti@enea.it (M.V.)
* Correspondence: giampaolo.caputo@enea.it

Abstract: Accurate predictions of photovoltaic generation are essential for effectively managing power system resources, particularly in the face of high variability in solar radiation. This is especially crucial in microgrids and grids, where the proper operation of generation, load, and storage resources is necessary to avoid grid imbalance conditions. Therefore, the availability of reliable prediction models is of utmost importance. Authors address this issue investigating the potential benefits of a machine learning approach in combination with photovoltaic power forecasts generated using weather models. Several machine learning methods have been tested for the combined approach (linear model, Long Short-Term Memory, eXtreme Gradient Boosting, and the Light Gradient Boosting Machine). Among them, the linear models were demonstrated to be the most effective with at least an *RMSE* improvement of 3.7% in photovoltaic production forecasting, with respect to two numerical weather prediction based baseline methods. The conducted analysis shows how machine learning models can be used to refine the prediction of an already established PV generation forecast model and highlights the efficacy of linear models, even in a low-data regime as in the case of recently established plants.

Keywords: machine learning; PV forecasting; microgrids; signal processing

Citation: Buonanno, A.; Caputo, G.; Balog, I.; Fabozzi, S.; Adinolfi, G.; Pascarella, F.; Leanza, G.; Graditi, G.; Valenti, M. Machine Learning and Weather Model Combination for PV Production Forecasting. *Energies* **2024**, *17*, 2203. https://doi.org/10.3390/en17092203

Academic Editor: Younes Mohammadi

Received: 22 March 2024
Revised: 24 April 2024
Accepted: 25 April 2024
Published: 3 May 2024

1. Introduction

Climate change is deeply related to anthropic activities on earth, especially those of the industrial, agriculture, energy production, and transport sectors. Decarbonization is a challenging task in such contexts. It requires novel methods, technologies, strategies, and systems. In the last few decades, considerable efforts have been made by scientists and researchers to facilitate the decarbonization pathway. This is particularly true in the energy sector with numerous implemented initiatives to reduce its Greenhouse Gas (GHG) emissions [1]. These initiatives increasingly also have an economic assessment of the problem [2] underlining how the effects of global warming have negative repercussions on local or global economies, and also how risk mitigation interventions are much less expensive than the consequences of possible extreme events due to ongoing climate change [3]. From a political point of view, both national and international energy transition plans set ambitious decarbonization objectives. Initiatives such as the European Green Deal, Fit for 55%, and the National Ecological Transition Plans emphasize the central role of electrification in different sectors (mobility, heating, etc.), and the substantial contribution of renewable energy sources to electricity generation. In this context, microgrids certainly represent one of the most promising models of transformation of the electricity system. The development of microgrids has emerged as a catalyst for the seamless integration of renewable energy sources into the broader energy ecosystem. Microgrids can function as cohesive entities, capable of islanded operating (grid-off mode) or are able to function in concert with the main AC grid (grid-on mode). Microgrids represent a solution favoring the integration of renewable energy sources into both Alternating Current (AC) and Direct Current (DC) grids by power electronics converters.

The operation of microgrids depends on characteristics, functioning modes, and the reliability of their internal resources and the power systems they are interfaced with [4]. They can be managed through diverse controller configurations, such as hierarchical, decentralized, or centralized ones. Hierarchically controlled microgrids are of particular significance; each of them is characterized by a Master controller endowed with the capability to oversee and optimize internal resources in alignment with specific energy or economic strategies. Moreover, local Slave controllers are assigned to individual microgrid resources or groups of resources, ranging from renewable power plants and loads to energy storage systems. These controllers are in charge of measures and data transmission, the execution of commands, and alarm activation in the case of critical operation. Data collected and acquired are typically transmitted and processed by Energy Management Systems (EMSs) and specialized platforms, facilitating the planning and optimization of power flows [5].

Within this transformative landscape, Italy has made significant strides, with its photovoltaic (PV) plants boasting an installed capacity of 25 GW as of 2022. Notably, approximately 33% of these installations fall within the 200–1000 kW range [6]. Such sources offer a sustainable and eco-friendly energy solution, but their intermittent and variable production patterns can pose operational challenges that can be effectively addressed through proper modeling [7] and forecasting methodologies [8]. Accurate forecasts play a pivotal role in enabling meticulous planning, strategic commitment, the efficient management of available resources in the renewable energy sector [9], and in ancillary service provision [10]. Errors in energy forecasts can result in significant economic losses for microgrids, including unnecessary energy purchases from the main grid or excessive energy storage. Precise forecasts help us to mitigate these inefficiencies, reduce energy waste, and optimize investments in energy infrastructure, ultimately resulting in substantial economic savings.

For the optimal operation of microgrids, the availability of accurate forecasts is non-negotiable. These forecasts extend to the prediction of generation from renewable sources and from projected microgrid demand.

1.1. Literature Review

While ground-based observations from solar metric stations provide valuable data, they are often burdened by high costs and may not offer continuous, long-term coverage. As an alternative, meteorological satellites provide the means for the indirect determination of solar parameters, including cloudiness, albedo, and solar radiation reaching the earth's surface. Contemporary meteorological forecast models, particularly numerical weather prediction (NWP) ones, have become essential for estimating renewable energy resources, notably solar radiation. Meteorological forecast models, nowadays, are able to predict renewable resources well [11], and their valuable data can be used in further forecasts of solar power plant production (PV [12] and Concentrated Solar Power systems). The NWP models could offer great spatial and temporal coverage against observation data, which are spatially sparse and lack long temporal resolution. Those models can simulate the future and past atmospheric conditions, and they have become very valuable tools in the management of renewable energy plants [13]. Significantly, meteorological forecast models have demonstrated their proficiency in predicting renewable resources and have evolved tools in forecasting and managing renewable energy systems. Recent years have witnessed a surge in the adoption of Machine Learning (ML) techniques predicting PV output [14]. They are deeply analyzed in the literature sector [15]. Several ML models, including Multiple Linear Regression (MLR), Support Vector Machine (SVM), Random Forest (RF), Gradient Boosting (GB), Fully Connected Neural Network (FCNN), Bayesian Neural Network (BNN), Support Vector Regression (SVR), and Regression Tree (RT), have been harnessed and subjected to comparative analyses across diverse studies. For example, in [15], several approaches, grouped into direct (methods that output directly the PV power) and indirect (methods that require the solar irradiance forecast, the estimation of the plane of array irradiance, and the PV performance model) forecasting techniques, are evaluated.

A comparative analysis of MLR, SVR, RF, GB, and FCNN methodologies is reported in [16] for a day-ahead solar power forecast. It is carried out taking into account historic power production and regional weather prediction related to 152 PV systems. The conducted comparison underscores the fact that GB and RF techniques excel in predicting production for both individual and aggregated PV systems when compared with other considered methods. The BNN method is compared to SVR and RT techniques in [17]. Another study is reported in [18], where the PV production for the University of Manchester plant is forecasted considering different-sized datasets and diverse time horizons.

In [19], the authors employed MLR and Artificial Neural Network (ANN) methods for predicting PV production within a solar microgrid. The paper concluded that MLR offers simplicity and computational efficiency but struggles with capturing complex non-linear relationships, whereas ANN excels in capturing non-linear patterns albeit demanding larger datasets and computational resources with reduced interpretability. Furthermore, the authors in [20] examined enduring techniques for predicting energy consumption, PV, and wind power production.

Furthermore, in [21], cloud cover, humidity, and temperature impacts on PV generation predictions were evaluated by 175 time series. They were obtained measuring the production of an actual rooftop-mounted PV system installed in Utrecht (Netherlands). Moreover, the authors offered a concise overview of prediction methodologies employed in microgrids, particularly for short-term forecasting.

In the context of a microgrid, PV production forecasting was also discussed in [22], and the Tunicate Swarm Algorithm (TSA)-based Least-Square Support Vector Machine (LSSVM) was applied. It offers the advantages of robustness to noisy data and the ability to capture complex patterns, while suffering from the need for hyperparameter tuning and is computationally demanding. The TSA-based Multilayer Perceptron Neural Network (TSA-MLPNN) provides flexibility in capturing non-linear relationships and uncertainty estimation but demands significant computational resources and may be sensitive to overfitting. On the other hand, the Whales Optimization Algorithm (WOA)-based LSSVM offers advantages in convergence speed and robustness to local optima, but it may require careful hyperparameter tuning and lacks interpretability.

Lastly, in [23], a blended Fuzzy-PSO smart forecasting method is deployed, and its precision is documented and contrasted with Fuzzy and Fuzzy-GA prediction models. The authors underline the advantages of the Fuzzy-PSO smart forecasting method, which include the ability to combine the strengths of both fuzzy logic and Particle Swarm Optimization (PSO), potentially leading to improved accuracy and robustness in prediction. However, this method may require more computational resources due to the optimization process involved in PSO, and its interpretability could be reduced compared to standalone Fuzzy logic models.

Recently, the hybridization of data-driven approaches based on the history of observed production and of physical approaches that employ both weather forecast and a mathematical model of the PV system are emerging as effective solutions [24]. For example, in [25], an ANN employs clear sky solar radiation and weather forecast data to obtain a PV production prediction, while, in [26], an ANN that has inputs and also the output of a five-parameter equivalent model of a PV module (using datasheet data or using an optimization method to identify the parameters) is considered. The studies observe that PV production forecasting is improved by the combined (hybridized) methods.

1.2. Beyond the State of the Art

In this work, we propose a combined approach of an already established NWP-based PV production model with an ML model that leverages the past observations of production to improve the final performance. Differently from works present in the literature, our approach assesses different types of ML models and two baseline models (a physical and a data-driven one). The ML model does not have direct access to weather forecast variables. The novelty of our approach is hence in the combination of these different elements, where some of them can be assumed to be already available on a considered site, to improve the final forecasting result.

By improving the accuracy of PV production forecasts, our research plays a crucial role in ensuring that solar energy is harnessed to its full potential, furthering the microgrid's support to minimize carbon footprints and advance the cause of the sustainable energy transition.

The analyzed case study refers to the ENEA—Italian National Agency for new technologies, energy and economic sustainable development—microgrid realized at the Research Centre of Portici (Italy). It is a demonstrator of "Multivector Integrated Smart Systems and Intelligent microgrids for accelerating the energy transition" (MISSION) project furnished with renewables (solar and wind), a Combined Heat Power (CHP) plant, storage devices, and a data center critical load.

The objective of achieving carbon-neutral microgrids is actively pursued by optimizing and managing the available resources in alignment with energy and environmental objectives [27]. The proposed models are implemented in the MISSION demonstrator, the functionality of which has strong connections with precise PV output predictions, ensuring appropriate resource management.

A promising aspect of the proposed combined method consists of its significant performance improvement, also in a low-data regime, as seen in the case of recently installed plants. This research is poised to make a substantive contribution to the overarching mission of cultivating sustainable, climate-resilient energy systems and driving the realization of carbon-neutral microgrids. In our previous article [28], we described some preliminary results that were extended with more models for comparison, a more detailed discussion of results, and a more extensive literature review.

The manuscript is organized as follows. Section 2 is dedicated to prediction models and the considered combined approach description. In Section 3, details about the model training process are provided. The conducted tests and obtained results are reported and commented on in Section 4.

2. Materials and Methods

2.1. Weather Research and Forecasting Model

The Weather Research and Forecasting model (WRF) is one member of the NWP model family designed for both atmospheric research and operational forecasting applications [29]. The model is used in different research areas in a wide range of meteorological applications where different time and horizontal resolutions, from tens of meters to thousands of kilometers, could be applied [30].

In this work, the computational spatial resolution of the WRF model is set to $10 \, \text{km} \times 10 \, \text{km}$ with 151×151 simulation grid points. The computational domain covers the entire region of the Italian peninsula with a center at $41.25°$ latitude and $13.5°$ longitude. The model uses 30 sigma atmospheric vertical levels and 4 soil levels. Input boundary conditions are given with four-time-daily runs of the Global Forecast System [31]. The WRF model starts its forecast at 00:00 UTC for the next 48 h where the first hours are considered as a model spin-up. The described WRF model provides different atmospheric outputs. In the assessed application, the authors consider only atmospheric temperature (T_A) and Cloud Cover (CC). The outputs of the described atmospheric forecasting model are used as data inputs for the PV plant production model that will be detailed below.

2.2. BaselineP Model

In PV plant installation sites, where observations of solar irradiance, in particular Global Horizontal Irradiance (GHI) are available, the maximum potential of global irradiance at the horizontal surface can be estimated. The specific maximum potential radiation is estimated when ground horizontal irradiance is considered with no cloud condition.

This information, together with the astronomical sun position and extraterrestrial global horizontal irradiance, provide an input for the "clear-sky" model.

Observations of GHI are expensive, and, with the lack of these data, satellite data and NWP models are used to estimate atmospheric temperature, cloud coverage, relative humidity, and radiation balance outcomes.

First, Cloud Cover (CC) is used from a weather model to decrease the intensity of the GHI clear sky value ($GHI_{clear\ sky}$) and to obtain the GHI forecast for a specified location (GHI_{pred}). The mathematical expression of the previous statement is written in Equation (1):

$$GHI_{pred} = GHI_{clear\ sky} \cdot (offset + (1 - offset) \cdot (1 - CC)) \tag{1}$$

where the *offset* is 0.35 and CC is 0 for a no clouds condition, and 1 for completely covered cloudy conditions [32]. When GHI_{pred} for the desired location is calculated, the decomposition model of solar radiation can be applied.

Each PV module has its own orientation, azimuth angle (α) with respect to the south direction, and tilt (β) of the panel with respect to the horizontal plane. The decomposition model allows us to compute Global Irradiance on any oriented PV surface (GI) as reported in Equation (2). The incident radiation GI on the panel surface is composed of three components: the direct radiation I_{dir}, the diffuse radiation I_{diff}, and the radiation reflected from the ground I_{ref}:

$$GI = I_{dir} + I_{diff} + I_{ref} \tag{2}$$

Now, the temperature of PV module (T_M) can be computed as in Equation (3) [33]:

$$T_M = T_A + (NOCT - 20)\frac{GI_{pred}}{G} \tag{3}$$

where T_A stands for ambient temperature [°C] (output from WRF model), $NOCT$ is the Nominal Operating Cell temperature [°C] calculated for a wind speed at a PV module height of 1 m/s, an ambient temperature of 20 °C, and an irradiance value of $G = 800$ W/m^2.

Finally, it is possible to calculate the PV production forecast for a specific plant (PV_{pred}) in the following 24 and 48 h by using the PVWatts model represented in Equation (4):

$$PV_{pred} = \eta\, P_n\, \frac{GI_{pred}}{GI_0}(1 + K(T_M - T_0))(1 - A) \tag{4}$$

where η is the inverter efficiency, P_n is the nominal power of the PV plant [W], GI_0 is the global solar radiation at standard test condition (=1000 W/m^2), K is the temperature coefficient of PV modules [%°C^{-1}], T_M is the temperature of the PV module, T_0 is the reference cell temperature at the standard test conditions (25 °C), and A represents the system losses [34].

Given the great generality of the WRF model, the BaselineP model can be readily applied to any location.

In this work, we have considered $NOCT = 47$ °C, $P_n = 9000$ W, $K = -0.45\%$°C^{-1}, $A = 14\%$, $\eta = 0.95$ in Equations (3) and (4).

2.3. BaselineD Model

When there is constant monitoring of PV production and ground solar radiation available, an uncomplicated model to forecast PV production could be used, here called BaselineD model. This model represents the relationship between measurements of PV production and GHI observations, as described in Equation (5):

$$PV_{obs} = m \cdot GHI_{obs} + q \tag{5}$$

The parameters m and q are fitted based on the available monitored PV production and GHI values. This mathematical function is then applied on the outputs of Equation (1) (GHI_{pred}) to calculate the forecast of PV production (PV_{pred}). This model could be used when observations are available. The found linear correlations are valid only for the exact PV plant and cannot be used for different PV locations.

In Figure 1, the relationship between PV plant production (PV_{obs}) and the observation of GHI (GHI_{obs}), and their corresponding correlation coefficients for 3 January 2021, is depicted. The example of a cloudy day is shown to present that an unpredictable radiation condition of PV production also has linear dependence with GHI. In this study, 365 relationships are obtained, one for each day of the year, and the relationship depicted by Equation (5) differs for each day. The correlation of the previous day can be used for the forecast of the next day because the astronomical sun–earth position has small variations. Daily correlations are valid only for the considered PV plant and the same approach could be used where there are available daily measurements of PV production and GHI.

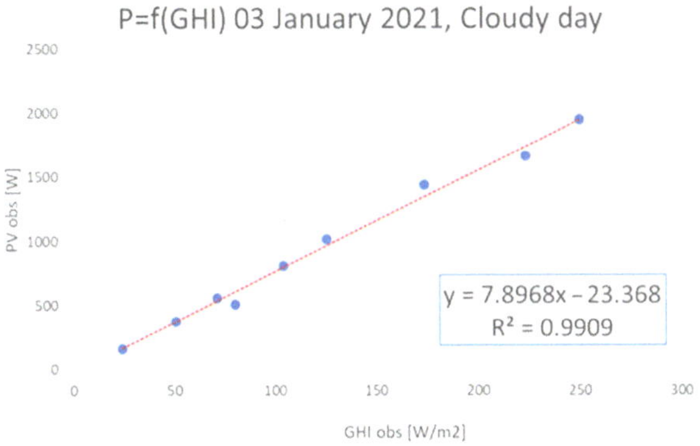

Figure 1. Correlation between the PV production and GHI observations. The blue dots are the observations of GHI and PV production, and red dashed line is the computed regression line.

A conceptual scheme of both BaselineP and BaselineD models is presented in Figure 2.

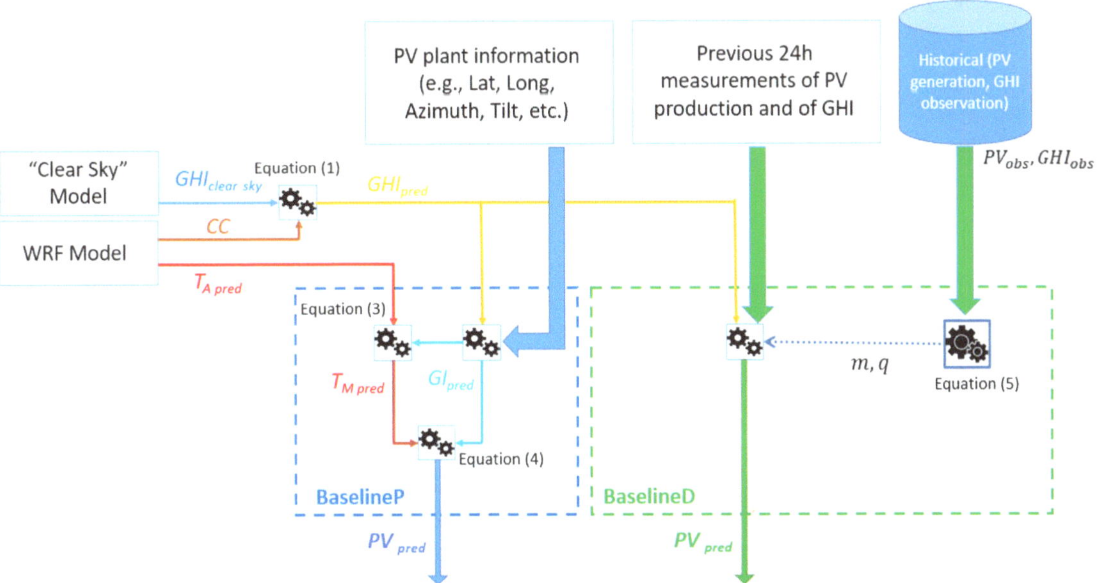

Figure 2. Conceptual scheme of BaselineP (blue) and BaselineD (green).

2.4. Proposed Approach

This section focuses on forecasting a target value (y) for a specific time (t) and for the following $H - 1$ time steps where H is the forecasting horizon. There are many possible approaches, but a general scheme involves the following:

- Target values;
- Past covariates: variables influencing the target value, observed in the previous W time steps;
- Future covariates: variables impacting the target value, related to the time t and to the subsequent $H - 1$ time steps, and that are known at the prediction time.

This concept is outlined in Equation (6), where the term y_{t-i} denotes the target value in the past, at the time step t-i; x^p_{t-j} (with $p \in \{1, \dots, P\}$) is the value of the p-th past covariate at the time step t-j; and z^f_{t+k} (with $f \in \{1, \dots, F\}$) indicates the value of the f-th future covariate at the time step $t + k$:

$$y_{t:t+H-1} = f\left(y_{t-1}, \dots, y_{t-W}, x^1_{t-1}, \dots, x^1_{t-W}, \dots, \dots, x^P_{t-W}, z^1_t, \dots, z^1_{t+H-1}, \dots, z^F_t, \dots, z^F_{t+H-1}\right) \qquad (6)$$

A schematic representation of Equation (6) is presented in Figure 3, where the past target values together with past and future covariates are employed to predict the next H target values.

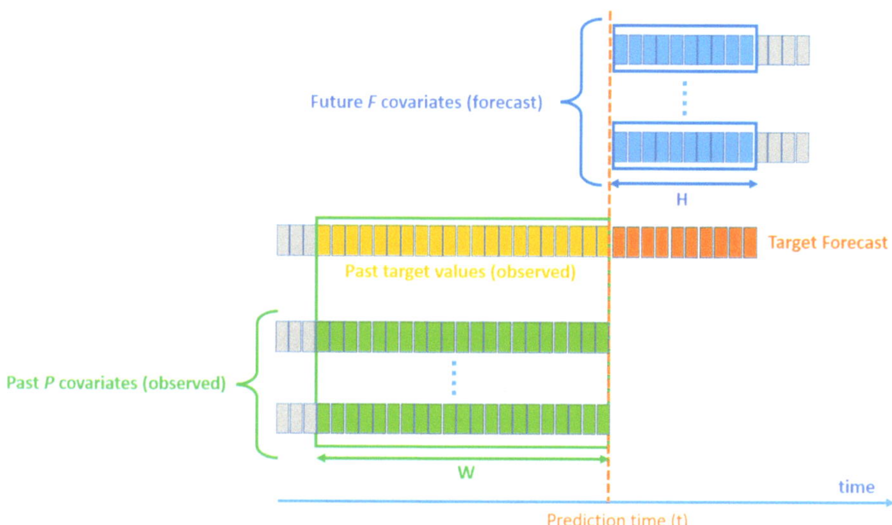

Figure 3. Generic forecasting problem's conceptual layout using future and past covariates.

In a scenario involving the production forecast of PV systems, the past covariates can be represented by various measurements such as solar irradiance and temperature. Conversely, the future covariates can be acquired either through weather forecasts or by utilizing PV production forecasts generated by baseline models, as performed in this work and illustrated in Figure 4.

The weather forecast module generates predictions for various weather variables for the next H hours. Subsequently, the baseline model utilizes these forecasts to provide a corresponding PV production forecast that can be employed by the ML model as future covariates, together with observed production, to enhance and refine the result.

Several ML regression models can be considered. In this work, we have focused on a linear model, Long Short-Term Memory (LSTM), eXtreme Gradient Boosting (XGB), and the Light Gradient Boosting Machine (LGBM), which are briefly described below. Even though these ML models are often employed in forecasting tasks, the originality of our approach is using their combination with baseline models in order to have a more precise energy prediction model that leverages both historical data and a PV production forecast obtained from BaselineP or BaselineD.

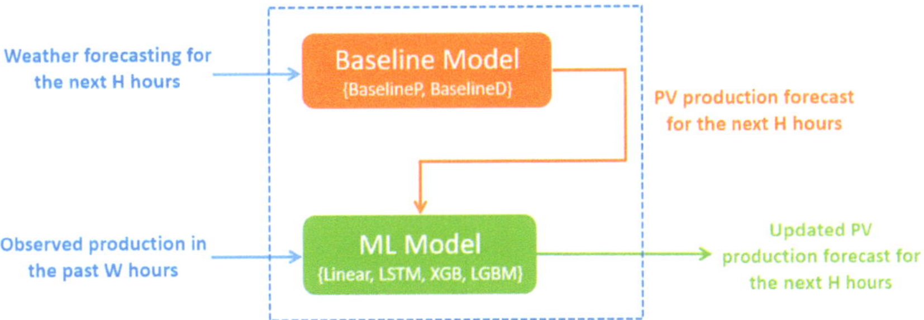

Figure 4. Proposed approach's conceptual scheme.

2.4.1. Linear Model

In the context of linear models [35], a mathematical representation capturing the relationship between the target variable to be predicted at time t (denoted as y_t), the historical target values $\{y_k\}_{k=t-1}^{t-W}$, and the future covariates $\{z_k\}_{k=t}^{t+H-1}$ is reported in Equation (7):

$$y_t = \delta_0 + \left(\sum_{k=t-1}^{t-W} \alpha_k y_k\right) + \left(\sum_{k=t}^{t+H-1} \gamma_k z_k\right) \tag{7}$$

Here, δ_0 represents the intercept, while α_k are the weights associated with the past target values and γ_k are the weights associated with the future covariates.

A crucial prerequisite to the predictions involves obtaining estimations for the future covariates (z_k) and the past target values (y_k). In the present study, the authors adopt a multi-model approach wherein H distinct models are trained. Each model is designed to predict a single value of the target within the interval $[t, t + H - 1]$.

2.4.2. Long Short-Term Memory (LSTM)

Differently by Feed Forward Neural Networks (FFNNs), the Recurrent Neural Networks (RNNs) are specialized in handling sequential data, and the computed output is not only dependent on the input but also on the hidden state of the system that is updated as the sequence is processed [35]. Utilizing the input (x_t) and the state from the preceding step (h_{t-1}), it is possible to update the current state (h_t) and compute the output (o_t) through the following equations:

$$h_t = \sigma_h(W_x \cdot x_t + W_h \cdot h_{t-1} + b_h) \tag{8}$$

$$o_t = \sigma_o(V_h \cdot h_t + b_o) \tag{9}$$

Here, σ_h and σ_o denote activation functions for the state and output, respectively; W_x, W_h, and V_h are the weight matrix for the input–state connection, the recurrent connection between states, and for the state–output connection, respectively; and b_h and b_o serve as the bias vectors for the state and output, respectively. The Long Short-Term Memory (LSTM) [36] is a specific RNN designed to mitigate issues like vanishing and exploding gradients by incorporating a memory element (c_t).

2.4.3. Gradient Boosting Methods

XGB [37] and LGBM [38] are two powerful ML algorithms belonging to the family of Gradient Boosting methods. The former employs a regularizer on the tree complexity to avoid overfitting and the latter makes the tree expand, leaf-wise. They are widely adopted due their speed, scalability, and robustness.

2.5. Datasets

To train and evaluate ML models, we need the PV production forecast obtained from BaselineP and BaselineD, the weather forecast, and the observed PV production. All data cover the year 2021 (from 1 of January to 31 of December).

2.5.1. PV Production

The observed PV production data are obtained by a PV installation above a motorcycle parking lot at the ENEA Research Centre located in Portici (NA), Italy. It consists of 36 glass/mono-crystalline panels, producing a nominal power of 9 kWp with a total covering area of 60 m^2. The installation has a tilt angle of 7° and is 28° south-east oriented (with 0° representing the south). The generated electrical energy is utilized for internal purposes after being delivered to the centre.

2.5.2. The Weather Forecast and the Baselined PV Production Forecast

The weather forecast values have been considered for the entire year of 2021. They represent the GHI and the predicted temperature for each day throughout the year on hourly basis.

These values are employed to predict the PV generation employing the BaselineP and BaselineD models described in Section 2.2 and Section 2.3, respectively.

In Figures 5 and 6, the actual and the predicted PV generation are shown for a sunny day and cloudy day (when the value of the daily clearness index (the ratio between the global irradiance and extraterrestrial irradiance on a horizontal surface) is greater than 0.65, the day is considered sunny otherwise cloudy) for the year 2021, respectively.

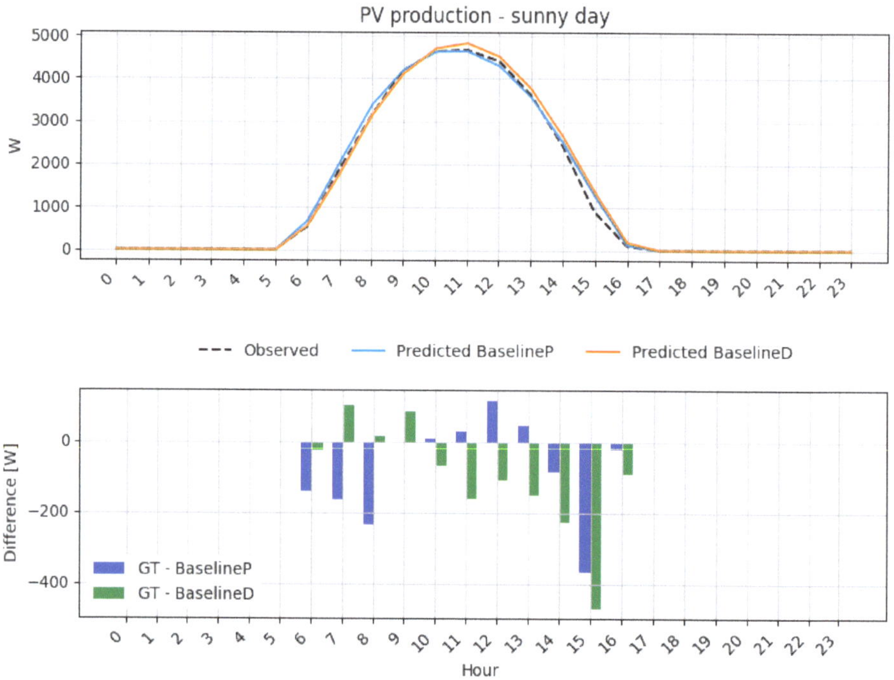

Figure 5. (**Top**) the actual and predicted PV production, using BaselineP and BaselineD, for a sunny day in the test set; (**bottom**) difference between the ground truth (observed) and the prediction results obtained using BaselineP and BaselineD.

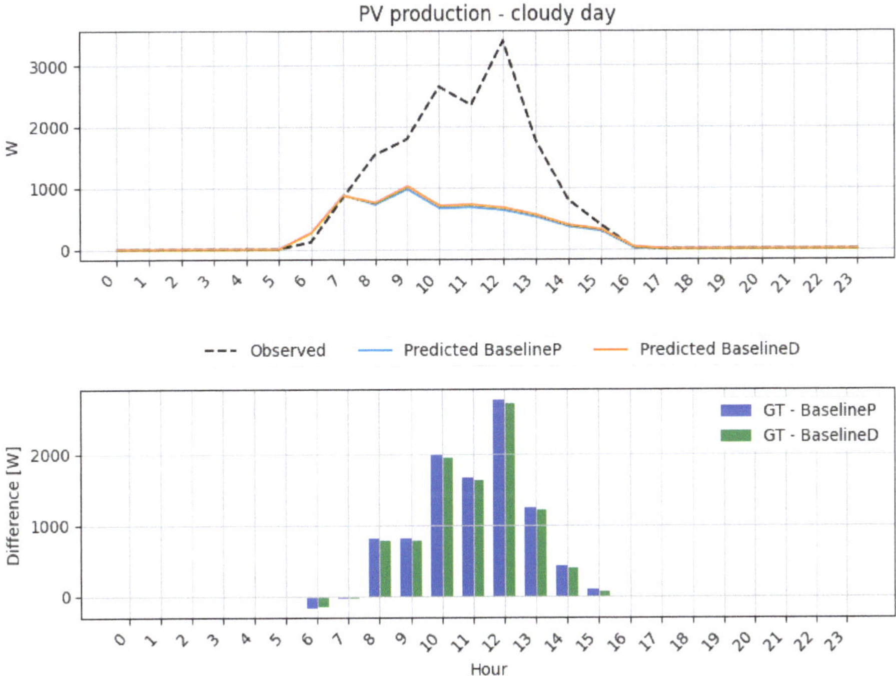

Figure 6. (**Top**) the actual and predicted PV production, using BaselineP and BaselineD, for a cloudy day in the test set; (**bottom**) difference between the ground truth (observed) and the prediction results obtained using BaselineP and BaselineD.

2.6. Metrics

The Root-Mean-Square-Error (*RMSE*) and the Coefficient of Variation (*CV*) are the metrics employed to evaluate the performance of the implemented models. The *RMSE* quantifies how closely a model's predictions $\hat{y}(t)$ align with the actual target values $y(t)$, often referred to as the ground truth. In this case, N represents the number of values taken into account. Differently, *CV* offers insights into the dispersion of errors relative to the average observed value $\bar{y}$. The *RMSE*, expressed in Watts, is computed using Equation (10), while the *CV* is dimensionless and is computed using Equation (11):

$$RMSE = \sqrt{\frac{1}{N}\sum_{t=1}^{N}(y(t) - \hat{y}(t))^2} \tag{10}$$

$$CV = \frac{RMSE}{\bar{y}} * 100 \tag{11}$$

3. Model Training

The considered dataset is divided into three sets: the training set, validation set, and test set. Namely, the training set goes from 1 January 2021 0:00 to 30 September 2021 23:00, the validation set is internal to the training set and goes from 1 August 2021 0:00 to 30 September 2021 23:00, and the test set goes from 1 October 2021 0:00 to 31 December 2021 23:00.

The authors utilized the pre-known information of the sunrise and sunset time (obtained from Sunrise Sunset API (https://sunrise-sunset.org/api accessed on 26 February 2024)) to refine the predictions.

The implementation of the models was carried out utilizing Python, and some libraries such as Darts [39] version 0.27.1, Numpy version 1.24.3, pandas version 1.5.3, and Scipy version 1.10.1.

The experiments were conducted on a Personal Computer (PC) equipped with an Intel Core i7-9700 CPU running at 3.00 GHz with eight cores, 16 GB of RAM, and a NVIDIA GeForce GTX 1050Ti GPU. The operating system used was Windows 10 Pro.

4. Discussion

Figures 7 and 8 depict the forecasting for two representative days in the test set, a sunny and a cloudy day, respectively, together with the difference between the ground truth and the prediction.

Table 1 presents the outcomes of the PV generation forecasting, utilizing only the two baseline models, BaselineP and BaselineD, and the combined models LinearD, LinearP, LSTMD, LSTMP, XGBD, XGBP, LGBMD, and LGBMP, where the final P and D indicate which is the employed baseline model, BaselineP or BaselineD, respectively. All models (except baselines) employ a 48 h window for past target values, with a forecasting horizon of 24 h.

For each metric, considering the days in the test set, the authors calculated the average and standard deviation (in parentheses). The Wilcoxon signed rank test was used to compare the metrics obtained from the combined approach and the related baseline. The null hypothesis under consideration was that the paired samples obtained by the baseline and the model that utilizes it originate from the same distribution. In Tables 1 and 2, * represents the rejection of the null hypothesis considering a p-value < 0.05, associated with a statistically significant difference between the baseline and related combined model. Table 2 presents the average and standard deviation of *RMSE* computed over the days of the test set for all considered models, differentiating between sunny and cloudy days, as defined previously.

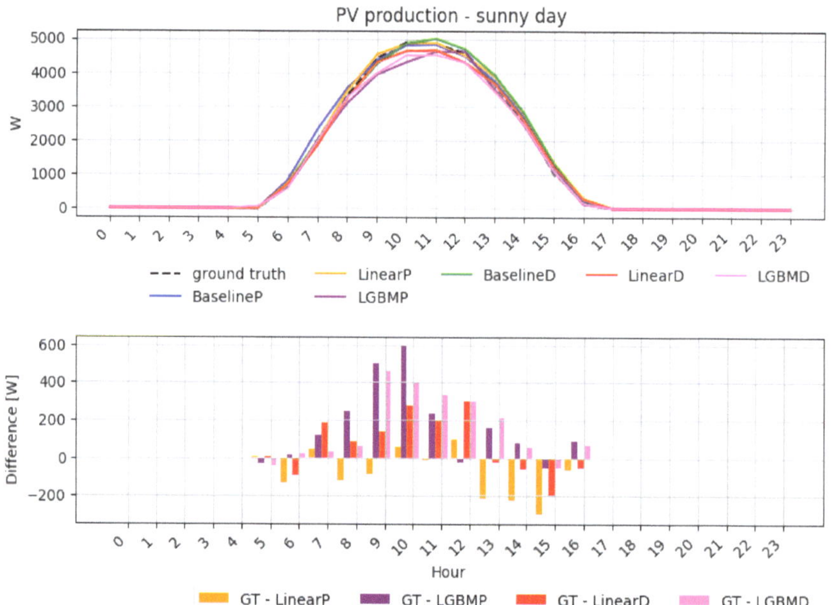

Figure 7. (**Top**) ground truth compared to the forecasting results of PV production obtained using only the baselines (BaselineP and BaselineD) and four combined models (LinearP, LinearD, LGBMP, LGBMD) for a cloudy day; (**bottom**) prediction error.

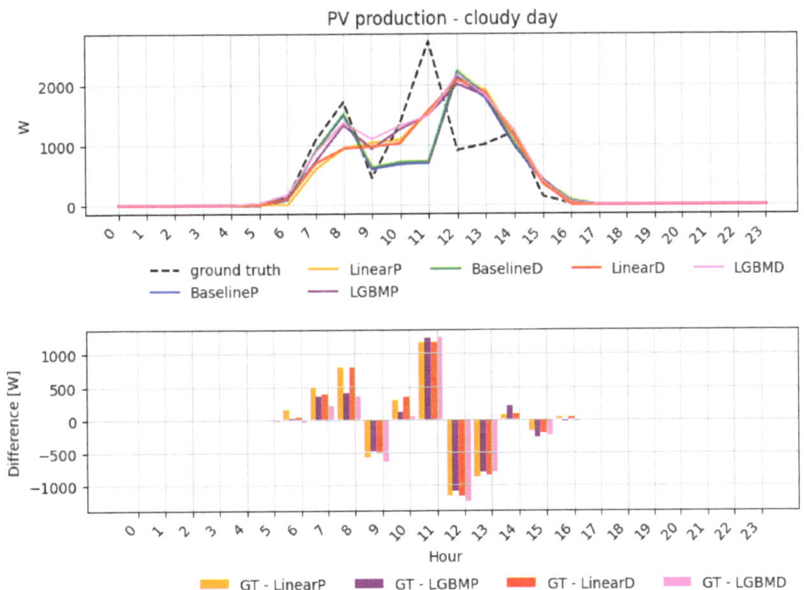

Figure 8. (**Top**) ground truth compared to the forecasting results of PV production obtained using only the baselines (BaselineP and BaselineD) and four combined models (LinearP, LinearD, LGBMP, LGBMD) for a cloudy day; (**bottom**) prediction error.

Table 1. PV generation forecasting results. For the days in the test set, the average (standard deviation) of *RMSE* and *CV* is calculated.

Type	Model	*RMSE*	*CV*
D	BaselineD	436.87 (253.95)	100.19 (91.62)
	LinearD	**420.72 (221.13)** *	**99.30 (95.83)** *
	LSTMD	435.18 (245.29)	100.15 (95.84)
	XGBD	502.75 (259.69) *	115.01 (106.85) *
	LGBMD	445.17 (245.45)	99.33 (85.65)
P	BaselineP	470.73 (248.73)	107.50 (96.66)
	LinearP	**438.47 (231.66)** *	103.26 (96.53) *
	LSTMP	454.46 (260.33) *	102.53 (91.77) *
	XGBP	487.92 (255.42)	112.51 (106.10)
	LGBMP	443.08 (250.66) *	**99.91 (90.17)** *

Table 2. PV generation forecasting results divided for sunny and cloudy days. For the days in the test set, the average (standard deviation) of *RMSE* is calculated.

Type	Model	Sunny Days	Cloudy Days
D	BaselineD	**223.57 (159.13)**	485.62 (246.31)
	LinearD	267.35 (170.66) *	**455.77 (216.37)** *
	LSTMD	260.65 (176.97) *	475.07 (228.16)
	XGBD	358.11 (188.82) *	535.81 (262.36) *
	LGBMD	287.85 (184.33) *	481.12 (243.511)
P	BaselineP	281.16 (159.12)	514.06 (245.22)
	LinearP	**254.81 (149.02)** *	480.44 (226.68) *
	LSTMP	257.04 (195.29)	499.59 (252.20)
	XGBP	335.36 (193.64) *	522.80 (255.04)
	LGBMP	290.70 (213.31)	**477.92 (245.50)** *

From Table 1, we can observe that BaselineD already produces promising results, having considered aspects not modeled by BaselineP and having discovered the observation of real data. Among the evaluated models, the linear ones outperform the other ones in terms of *RMSE* and consistently outperform the baselines, with a statistically significant improvement of 6.9% in the P case and of 3.7% in D case. Moreover, in the D case, compared to baseline, the linear model returns to a lower *RMSE* in 58.1% of all test set days (and in 67.1% of cloudy days), and in 69.8% of all test set days (and in 68.6% of cloudy days) in the P case.

In the D case, when specifically addressing only sunny days within the test set (Table 2), the baseline is optimal, but, considering only the cloudy days in the test set (83% of the total), the most effective solution becomes the linear model.

In P case, instead, linear model is the best model for sunny days and is the second-best approach after LGBM for cloudy days.

In Figure 9 the difference between the ground truth and the prediction is estimated on an hourly basis (from 5:00 to 17:00) and only for baselines and linear models. For the P case, the median error of the linear model is nearer to zero than that of the baseline. For the D case, instead, the median error of the baseline is lower, but the errors are usually more dispersed, presenting a higher range than the linear model.

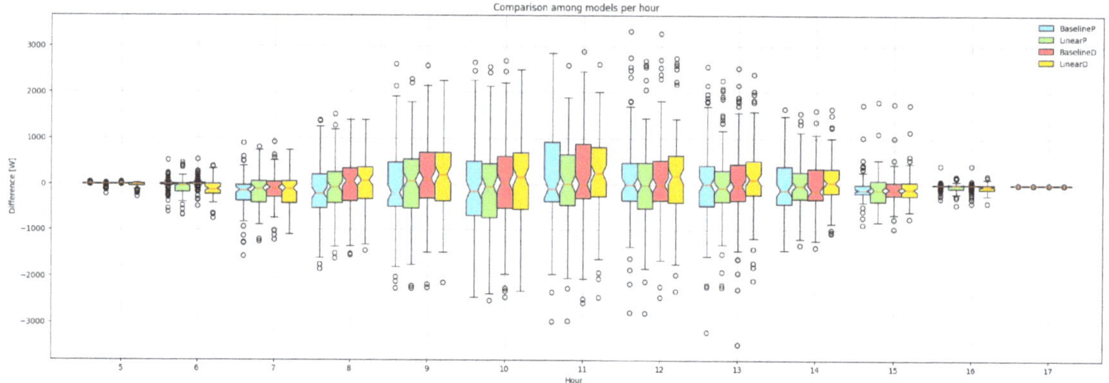

Figure 9. Box-and-whisker plot of the difference between ground truth and the prediction for the BaselineP, BaselineD, LinearP, and LinearD models. The white circles are the outliers.

Although the usage of one year of data is a limitation of this study, it allows authors to demonstrate the efficacy of the proposed methodology and of linear models even in a low-data regime, which is frequently the case in scenarios within a newly established plant. Integrating additional years into the dataset has the potential to help the ML models and to enable the exploration of more complex models that may have benefits from an extensive dataset. Moreover, the authors would like to point out that the proposed approach is general, and it is suitable to be applied in emerging photovoltaic systems where a weather forecast module and historical energy production dataset are available.

5. Conclusions

The forecasting of PV power holds essential significance in accurately strategizing and managing resources for grids and microgrids. It plays a pivotal role in aligning supply and demand, thereby aiding in preventing grid imbalances.

In this study, the authors explored the application of an approach leveraging information derived from both historically observed PV power generation and PV power forecasts generated by numerical weather prediction models. The conducted analysis highlights how machine learning models can be utilized to enhance the prediction of an already established PV generation forecast model. Several machine learning methods, including the linear model, Long Short-Term Memory, eXtreme Gradient Boosting, and the Light Gradient Boosting Machine, were tested for the combined approach. However, the linear models proved to be the most effective, showing at least a 3.7% improvement in *RMSE* in PV production forecasting compared to two numerical weather prediction-based baseline methods. Among the employed machine learning models, the linear models demonstrated their validity, surpassing baseline performances and showcasing their effectiveness with only one year of data.

Author Contributions: Conceptualization, A.B. and G.C.; methodology, A.B., G.C., I.B. and G.A.; software, A.B., G.C. and I.B.; data curation, G.C., I.B., F.P. and G.L.; writing—original draft preparation, A.B., G.C., I.B., S.F. and G.A.; writing—review and editing, A.B., G.C., I.B., S.F., G.A., G.G. and M.V.; visualization, A.B., G.C. and I.B.; supervision, M.V. and G.G. All authors have read and agreed to the published version of the manuscript.

Funding: This research was funded by the Research Fund for the Italian Electrical System under the contract agreement "Accordo di Programma Mission Innovation 2021–2024—Project MISSION (POA Smart Grid)" between ENEA and the Ministry of the Environment and Energetic Safety (MASE).

Data Availability Statement: The dataset related to PV generation is available on request from the authors.

Conflicts of Interest: The authors declare no conflicts of interest.

References

1. Kumi, E.N.; Mahama, M. Greenhouse gas (GHG) emissions reduction in the electricity sector: Implications of increasing renewable energy penetration in Ghana's electricity generation mix. *Sci. Afr.* **2023**, *21*, e01843. [CrossRef]
2. Tol, R.S.J. A meta-analysis of the total economic impact of climate change. *Energy Policy* **2024**, *185*, 113922. [CrossRef]
3. Rezai, A.; Taylor, L.; Foley, D. Economic Growth, Income Distribution, and Climate Change. *Ecol. Econ.* **2018**, *146*, 164–172. [CrossRef]
4. Adinolfi, G.; Ciavarella, R.; Graditi, G.; Ricca, A.; Valenti, M. A Planning Tool for Reliability Assessment of Overhead Distribution Lines in Hybrid AC/DC Grids. *Sustainability* **2021**, *13*, 6099. [CrossRef]
5. Vinothine, S.; Arachchige, L.N.W.; Rajapakse, A.D.; Kaluthanthrige, R. Microgrid Energy Management and Methods for Managing Forecast Uncertainties. *Energies* **2022**, *15*, 8525. [CrossRef]
6. Gestore dei Servizi Energetici. Rapporto Statistico Solare Fotovoltaico 2022. Available online: https://www.gse.it/documenti_site/Documenti%20GSE/Rapporti%20statistici/GSE%20-%20Solare%20Fotovoltaico%20-%20Rapporto%20Statistico%202022.pdf (accessed on 26 February 2024).
7. Buonanno, A.; Caliano, M.; Di Somma, M.; Graditi, G.; Valenti, M. A Comprehensive Tool for Scenario Generation of Solar Irradiance Profiles. *Energies* **2022**, *15*, 8830. [CrossRef]
8. Yang, D.; Wang, W.; Gueymard, C.A.; Hong, T.; Kleissl, J.; Huang, J.; Perez, M.J.; Perez, R.; Bright, J.M.; Xia, X.; et al. A review of solar forecasting, its dependence on atmospheric sciences and implications for grid integration: Towards carbon neutrality. *Renew. Sustain. Energy Rev.* **2022**, *161*, 112348. [CrossRef]
9. Graditi, G.; Buonanno, A.; Caliano, M.; Di Somma, M.; Valenti, M. *Machine Learning Applications for Renewable-Based Energy Systems*; EAI/Springer Innovations in Communication and Computing; Springer: Cham, Switzerland, 2023; Volume Part F665, pp. 177–198. [CrossRef]
10. Bekhit, R.; Bianco, G.; Delfino, F.; Ferro, G.; Noce, C.; Orrù, L.; Parodi, L.; Robba, M.; Rossi, M.; Valtorta, G. A platform for demand response and intentional islanding in distribution grids: The LIVING GRID demonstration project. *Results Control Optim.* **2023**, *12*, 100294. [CrossRef]
11. Climate Models | NOAA Climate.gov. Available online: https://www.climate.gov/maps-data/climate-data-primer/predicting-climate/climate-models (accessed on 26 February 2024).
12. Fuoco, D.; Mendicino, G.; Senatore, A.; Balog, I.; Caputo, G.; Spinelli, F.; Lepore, M.; Franconiero, D.; Mautone, P.; Oliviero, M. Modelli Previsionali di Producibilità: Ambiti Applicativi. Rapporto Tecnico di Ricerca Industriale D5.3a. Available online: http://www.comesto.eu/wp-content/uploads/2020/11/D5.3a_Modelli-previsionali-di-producibilit%C3%A0_ambiti-applicativi.pdf (accessed on 26 February 2024).
13. Best Practices Handbook for the Collection and Use of Solar Resource Data for Solar Energy Applications: Third Edition—IEA-PVPS. Available online: https://iea-pvps.org/key-topics/best-practices-handbook-for-the-collection-and-use-of-solar-resource-data-for-solar-energy-applications-third-edition/ (accessed on 26 February 2024).
14. Ledmaoui, Y.; El Maghraoui, A.; El Aroussi, M.; Saadane, R.; Chebak, A.; Chehri, A. Forecasting solar energy production: A comparative study of machine learning algorithms. *Energy Rep.* **2023**, *10*, 1004–1012. [CrossRef]
15. Gupta, P.; Singh, R. PV power forecasting based on data-driven models: A review. *Int. J. Sustain. Eng.* **2021**, *14*, 1733–1755. [CrossRef]

16. Visser, L.; AlSkaif, T.; van Sark, W. Benchmark analysis of day-ahead solar power forecasting techniques using weather predictions. In Proceedings of the 2019 IEEE 46th Photovoltaic Specialists Conference (PVSC), Chicago, IL, USA, 16–21 June 2019; pp. 2111–2116.
17. Theocharides, S.; Theristis, M.; Makrides, G.; Kynigos, M.; Spanias, C.; Georghiou, G.E. Comparative Analysis of Machine Learning Models for Day-Ahead Photovoltaic Power Production Forecasting. *Energies* **2021**, *14*, 1081. [CrossRef]
18. Scott, C.; Ahsan, M.; Albarbar, A. Machine learning for forecasting a photovoltaic (PV) generation system. *Energy* **2023**, *278*, 127807. [CrossRef]
19. Kallio, S.; Siroux, M. Photovoltaic power prediction for solar micro-grid optimal control. *Energy Rep.* **2023**, *9*, 594–601. [CrossRef]
20. Dutta, S.; Li, Y.; Venkataraman, A.; Costa, L.M.; Jiang, T.; Plana, R.; Tordjman, P.; Choo, F.H.; Foo, C.F.; Puttgen, H.B. Load and Renewable Energy Forecasting for a Microgrid using Persistence Technique. *Energy Procedia* **2017**, *143*, 617–622. [CrossRef]
21. Gaboitaolelwe, J.; Zungeru, A.M.; Yahya, A.; Lebekwe, C.K.; Vinod, D.N.; Salau, A.O. Machine Learning Based Solar Photovoltaic Power Forecasting: A Review and Comparison. *IEEE Access* **2023**, *11*, 40820–40845. [CrossRef]
22. Tayab, U.B.; Yang, F.; Metwally, A.S.M.; Lu, J. Solar photovoltaic power forecasting for microgrid energy management system using an ensemble forecasting strategy. *Energy Sources Part A Recover. Util. Environ. Eff.* **2022**, *44*, 10045–10070. [CrossRef]
23. Teferra, D.M.; Ngoo, L.M.; Nyakoe, G.N. Fuzzy-based prediction of solar PV and wind power generation for microgrid modeling using particle swarm optimization. *Heliyon* **2023**, *9*, e12802. [CrossRef] [PubMed]
24. Mayer, M.J. Benefits of physical and machine learning hybridization for photovoltaic power forecasting. *Renew. Sustain. Energy Rev.* **2022**, *168*, 112772. [CrossRef]
25. Ogliari, E.; Dolara, A.; Manzolini, G.; Leva, S. Physical and hybrid methods comparison for the day ahead PV output power forecast. *Renew. Energy* **2017**, *113*, 11–21. [CrossRef]
26. Niccolai, A.; Dolara, A.; Ogliari, E. Hybrid PV Power Forecasting Methods: A Comparison of Different Approaches. *Energies* **2021**, *14*, 451. [CrossRef]
27. Fabozzi, S.; Graditi, G.; Valenti, M. Techno-economic design of a smart multienergy microgrid. In Proceedings of the 2022 AEIT International Annual Conference (AEIT), Rome, Italy, 3–5 October 2022.
28. Buonanno, A.; Caputo, G.; Balog, I.; Adinolfi, G.; Pascarella, F.; Leanza, G.; Fabozzi, S.; Graditi, G.; Valenti, M. Combined Machine Learning and weather models for photovoltaic production forecasting in microgrid systems. In Proceedings of the 2023 International Conference on Clean Electrical Power (ICCEP), Santa Margherita Ligure, Italy, 27–29 June 2017; pp. 216–222.
29. WRF Model Users Site. Available online: https://www2.mmm.ucar.edu/wrf/users/ (accessed on 26 February 2024).
30. WRF Community. Weather Research and Forecasting (WRF) Model, UCAR/NCAR. 2000. Available online: https://www2.mmm.ucar.edu/wrf/users/ (accessed on 26 February 2024).
31. Global Forecast System (GFS) | National Centers for Environmental Information (NCEI). Available online: https://www.ncei.noaa.gov/products/weather-climate-models/global-forecast (accessed on 26 February 2024).
32. Larson, D.P.; Nonnenmacher, L.; Coimbra, C.F. Day-ahead forecasting of solar power output from photovoltaic plants in the American Southwest. *Renew. Energy* **2016**, *91*, 11–20. [CrossRef]
33. CEI 82-25: 2008 Guide for Design and Installation of Photovoltaic. Available online: https://www.intertekinform.com/en-au/standards/cei-82-25-2008-319110_saig_cei_cei_735215/ (accessed on 26 February 2024).
34. Dobos, A.P. PVWatts Version 5 Manual. 2014. Available online: www.nrel.gov/publications (accessed on 26 February 2024).
35. Murphy, K.P. *Probabilistic Machine Learning: An Introduction*; Massachusetts Institute of Technology: Cambridge, MA, USA, 2022.
36. Hochreiter, S.; Schmidhuber, J. Long short-term memory. *Neural Comput.* **1997**, *9*, 1735–1780. [CrossRef] [PubMed]
37. Chen, T.; Guestrin, C. XGBoost: A Scalable Tree Boosting System. In Proceedings of the KDD '16: 22nd ACM SIGKDD International Conference on Knowledge Discovery and Data Mining, San Francisco, CA, USA, 13–17 August 2016; Association for Computing Machinery: New York, NY, USA, 2016; pp. 785–794. [CrossRef]
38. Ke, G.; Meng, Q.; Finley, T.; Wang, T.; Chen, W.; Ma, W.; Ye, Q.; Liu, T.Y. LightGBM: A Highly Efficient Gradient Boosting Decision Tree. In *Proceedings of the 31st International Conference on Neural Information Processing Systems (NIPS'17)*; Curran Associates Inc.: Red Hook, NY, USA, 2017; pp. 3149–3157. Available online: https://proceedings.neurips.cc/paper_files/paper/2017/file/6449f4 4a102fde848669bdd9eb6b76fa-Paper.pdf (accessed on 26 February 2024).
39. Herzen, J.; Lässig, F.; Piazzetta, S.G.; Neuer, T.; Tafti, L.; Raille, G.; Van Pottelbergh, T.; Pasieka, M.; Skrodzki, A.; Huguenin, N.; et al. Darts: User-Friendly Modern Machine Learning for Time Series. *J. Mach. Learn. Res.* **2022**, *23*, 1–6. Available online: http://jmlr.org/papers/v23/21-1177.html (accessed on 26 February 2024).

energies

Article

Assessing the Impact of Climate Changes, Building Characteristics, and HVAC Control on Energy Requirements under a Mediterranean Climate

António M. Raimundo [1,*] and A. Virgílio M. Oliveira [2]

1 University of Coimbra, Department of Mechanical Engineering, Pólo II, Rua Luís Reis Santos, 3030-788 Coimbra, Portugal

2 Polytechnic Institute of Coimbra, Coimbra Institute of Engineering, Rua Pedro Nunes–Quinta da Nora, 3030-199 Coimbra, Portugal; avfmo@isec.pt

* Correspondence: antonio.raimundo@dem.uc.pt

Citation: Raimundo, A.M.; Oliveira, A.V.M. Assessing the Impact of Climate Changes, Building Characteristics, and HVAC Control on Energy Requirements under a Mediterranean Climate. *Energies* **2024**, *17*, 2362. https://doi.org/10.3390/en17102362

Academic Editors: Boštjan Polajžer, Davood Khodadad, Younes Mohammadi and Aleksey Paltsev

Received: 16 April 2024
Revised: 10 May 2024
Accepted: 12 May 2024
Published: 14 May 2024

Abstract: Despite efforts to mitigate climate change, annual greenhouse gas emissions continue to rise, which may lead to the global warming of our planet. Buildings' thermal energy needs are inherently linked to climate conditions. Consequently, it is crucial to evaluate how climate change affects these energy demands. Despite extensive analysis, a comprehensive assessment involving a diverse range of building types has not been consistently conducted. The primary objective of this research is to perform a coherent evaluation of the influence of climate changes, construction element properties, and the Heating, Ventilation, and Air Conditioning (HVAC) system type of control on the energy requirements of six buildings (residential, services, and commercial). The buildings are considered to be located in a temperate Mediterranean climate. Our focus is on the year 2070, considering three distinct climatic scenarios: (i) maintaining the current climate without further changes, (ii) moderate climate changes, and (iii) extreme climate changes. The buildings are distributed across three different locations, each characterized by unique climatic conditions. Buildings' envelope features a traditional External Thermal Insulation Composite System (ETICS) and expanded polystyrene (EPS) serves as thermal insulation material. Two critical design factors are explored: EPS thickness ranging from 0 (no insulation) to 12 cm; and horizontal external fixed shading elements varying lengths from 0 (absence) to 150 cm. Six alternative setpoint ranges are assessed for the HVAC system control: three based on the Predicted Mean Vote (PMV) and three based on indoor air temperature (T_{air}). Results were obtained with a validated in-home software tool. They show that, even under extreme climate conditions, the application of thermal insulation remains energetically favorable; however, its relative importance diminishes as climate severity increases. Then, proper insulation design remains important for energy efficiency. The use of external shading elements for glazing (e.g., overhangs, louvers) proves beneficial in specific cases. As climate changes intensify, the significance of shading elements grows. Thus, strategic placement and design are necessary for good results. The HVAC system's energy consumption depends on the level of thermal comfort requirements, on the climate characteristics, and on the building's type of use. As climate change severity intensifies, energy demands for cooling increase, whereas energy needs for heating decrease. However, it is essential to recognize that the impact of climate changes on HVAC system energy consumption significantly depends on the type of building.

Keywords: climate change; buildings' energy requirements; HVAC control; buildings' thermal insulation; external solar shadings; buildings' type of use; Mediterranean climate; buildings climatization

1. Introduction

The Intergovernmental Panel on Climate Change (IPCC) report on climate change mitigation in 2022 [1] highlights significant trends in the global emissions of radiatively active substances (e.g., greenhouse gases (GHGs) and aerosols). Despite climate change mitigation efforts, annual greenhouse emissions grew on average by 2.2% per year from 2000 to 2019, compared with 1.3% per year from 1970 to 2000. Slightly different values for these emissions are reported on the Emissions Gap Report 2022 of the United Nations Environment Programme [2], where an average annual growth rate of 2.6% per year from 2000 to 2009 and 1.1% per year from 2010 to 2019 is reported. According to both reports, a peak was reached in 2019, followed by a decrease in 2020 due to COVID-19-related restrictions; it is also suggested that, in 2021, the level of total global emissions of GHGs and aerosols will be like, or even surpass, the 2019 level. According to the IPCC report [1], the building sector was responsible for 32% of the final energy consumption and 19% of the global equivalent of CO_2 emissions. These facts underscore the urgent need for sustainable practices and targeted policies to mitigate climate change and reduce emissions in the building sector.

1.1. Overview

Based on coherent and consistent assumptions about driving forces, such as demographic and socioeconomic development, technological change, energy consumption, and land use, the Intergovernmental Panel on Climate Change (IPCC) regularly presents plausible alternative forecasts for the future evolution of global emissions of radiatively active substances (e.g., greenhouse gases (GHGs) and aerosols) [3–5]. The likelihood of each emission scenario depends on the level of sustainability occurring in the global economy. Using these alternative emission forecasts, the IPCC has developed a series of "climate projections", which are commonly referred to as "climate scenarios".

In the second Assessment Report of the IPCC, published in 1996 [3], a set of alternative climate projections known as the "IS92 scenarios" was presented. Later, the IPCC Special Report on Emissions Scenarios [4] introduced the "SRES scenarios", comprising 40 distinct scenarios grouped into four families: A1, A2, B1, and B2. These scenarios vary in terms of their accumulated emissions and global warming potential. Specifically: SRES scenario families B1 and B2 can be considered to have a moderate impact; SRES scenario families A1 and A2 are associated with a high impact. Globally, these scenarios can be ordered from lowest to highest impact as follows: B1, B2, A1, A2.

In the fifth Assessment Report of the Intergovernmental Panel on Climate Change [5], four alternative scenarios for climate change are presented. These scenarios are known as Representative Concentration Pathways (RCP) and serve as critical tools for understanding and planning different future climates. Each RCP represents a different trajectory of GHGs emissions, shaped by various factors such as population size, economic activity, lifestyle, energy use, land use patterns, technology, and climate policy. They include a stringent mitigation scenario (RCP 2.6), two intermediate scenarios (RCP 4.5 and RCP 6.0), and one scenario with very high global emission of substances radiatively active (RCP 8.5).

- RCP 2.6 (stringent mitigation scenario): it assumes substantial and sustained reductions in GHGs emissions, representing a world where global efforts effectively limit climate change.
- RCP 4.5 (intermediate scenario): moderately reduced GHGs emissions reveal a future with some mitigation measures but are not as stringent as RCP 2.6.
- RCP 6.0 (intermediate scenario): it involves intermediate emission reductions and considers a world where climate action is taken, but not to the same extent as RCP 4.5.
- RCP 8.5 (high emissions scenario): it represents a future with very limited climate policies and very high global emissions of radiatively active substances, promoting a substantial environmental impact.

The land scenarios within the RCP framework offer a diverse range of potential futures, ranging from a net reforestation (RCP 2.6), some net reforestation (RCP 4.5), forestation similar to actual reality (RCP 6.0) and further deforestation (RCP 8.5). In terms of global emission of substances radiatively active and comparatively to present, scenario RCP 2.6 represents a future characterized by a substantial net reduction, scenario RCP 4.5 represents a future with some reduction, scenario RCP 6.0 represents a future with similar emissions, and scenario RCP 8.5 represents a future with a strong increase.

Relative to 1850–1900, global warming at the end of the 21st century (2081–2100) is projected to likely exceed 1.5 °C for RCP 4.5, RCP 6.0, and RCP 8.5 (high confidence), likely to exceed 2 °C for RCP 6.0 and RCP 8.5 (high confidence), more likely than not to exceed 2 °C for RCP 4.5 (medium confidence), but unlikely to exceed 2 °C for RCP 2.6 (medium confidence) [5].

The RCP scenarios cover a wider range of projections than SRES scenarios, as they also considered forecasts for land use and for climate policy. Globally, RCP 8.5 is broadly comparable to the SRES A2 scenario, RCP 6.0 to B2, RCP 4.5 to B1, and there is no equivalent scenario in SRES projections for RCP 2.6 [5].

1.2. State of the Art

Achieving good indoor environmental quality is crucial for promoting a pleasant sense of well-being and ensuring work efficiency [6,7]. Among the various factors that contribute to indoor environmental quality, thermal comfort stands out as particularly significant, even more so than visual and acoustic comfort or indoor air quality [8]. Furthermore, a substantial portion of a building's environmental impact results from energy consumption by the Heating, Ventilation, and Air Conditioning (HVAC) system [9,10]. Therefore, to minimize our ecological footprint, it is essential to maintain conditions of thermal comfort with low energy consumption.

The energy consumption of a building's air conditioning system—whether residential, commercial, or service-oriented—depends on several critical factors. These include the desired level of thermal comfort, the efficiency and type of control of the Heating, Ventilation, and Air Conditioning (HVAC) system, the building's architectural design and solar orientation, the characteristics of its passive construction elements, the thermal gains produced by the internal energy systems, the type of building occupancy, and the climatic conditions [7]. Moreover, given the extended lifespan of buildings (typically spanning 50–100 years), the likelihood of climate change occurring during their operational lifetime is substantial. Consequently, construction and refurbishment projects must account for sustainable operation in both the present and future climates [11–14].

Thermal comfort is influenced by both environmental conditions (such as air temperature, humidity, air velocity, and mean radiant temperature) and individual factors (including activity level and clothing characteristics) [15,16]. To maintain optimal thermal comfort indoors, HVAC systems adjust one or more parameters related to the thermal environment. The effectiveness of these systems hinges on two critical factors: equipment energy efficiency and the proficiency of the control system in ensuring thermal comfort and indoor air quality [7]. A wide variety of possibilities exists for HVAC control systems. The most common involves constraining environmental parameters within a specified range, without considering individual occupant factors [7,9]. Unfortunately, these procedures do not guarantee the desired thermal comfort quality and often result in higher energy consumption compared to occupant-based control methodologies [6,7,17,18].

The building characteristics that lead to the lowest value of energy demand for climatization strongly depends on the climate of the building location [10]. This holds true not only for extreme cold and hot climates but also for temperate regions, including the Mediterranean, where marked seasonal variations occur, with both cold and hot seasons [19], both necessitating HVAC systems to achieve indoor thermal comfort [9,10,20,21]. Consequently, the selection of the best constructive solutions for buildings located in these climates remains challenging.

The production of energy—whether thermal, mechanical, or electrical—from fuels, particularly fossil fuels, results in a significant emission of greenhouse gases (GHGs), which have a major impact on global warming [1,2]. This drawback can be mitigated by two primary approaches: producing energy from renewable sources and reducing overall energy consumption. Consequently, buildings must be designed to operate sustainably.

Achieving this goal involves minimizing energy usage while relying on renewable energy sources [11–14].

Energy consumption for air conditioning depends on the climate characteristics, the building's type of use, the quality of its passive and active constructive elements, the level of thermal comfort assured, and the HVAC system energy efficiency and the proficiency of its operation control [7,10,12], and represents a very significant portion of the building's energy consumption [22,23].

Buildings, whether new or existing, are significant energy consumers. Then, they must have an active role in mitigating climate change, namely by ensuring thermal comfort conditions with reduced energy consumption. Given their long lifespan—often exceeding 50 years [20]—it becomes imperative to identify solutions that reduce energy consumption by HVAC systems in buildings. This holds true for both current climatic conditions and possible alternatives (scenarios) arising from ongoing climate change.

1.3. Objectives and Scope

It is widely acknowledged that climate change will result in global warming [1,5,13]. Furthermore, a connection is predicted between the current climate characteristics and those anticipated due to climate change. Consequently, future climate scenarios for specific locations are typically derived from the present climate conditions at those sites [24–26], among others. Therefore, in studies like the one at hand, the current climate of the building's location holds relevance and must be taken into account.

It has been well-established that due to a warmer climate, the energy requirements for heating buildings will decrease, whereas the energy demands for cooling will rise [12,13,26,27], among others. The extent of the reduction in heating energy needs and the magnitude of the increase in cooling energy requirements depend on several factors, including the building's use, the characteristics of its passive and active construction elements, and the specific climate conditions. Consequently, this dynamic can lead to either an increase or a decrease in energy consumption for air conditioning. So, the main objective of this research is to conduct a comprehensive assessment of how climate changes, properties of construction elements, and the type of HVAC system control impact the energy requirements for climatization in a wide range of buildings (including residential, service, and commercial structures) placed in a Mediterranean climate.

The building stock comprises six types of structures: residential, including apartments within multifamily buildings and detached houses; service buildings with permanent occupancy, such as clinics; and service buildings with intermittent use, including schools and bank branches. Additionally, there is a commercial building, and a supermarket, which also has intermittent utilization.

All buildings share the same type of passive construction solutions, both opaque and glazed. As is often recommended for this type of construction, the opaque elements of the building envelope are equipped with a traditional External Thermal Insulation Composite System (ETICS) based on expanded polystyrene (EPS) material [10,20,28–31]. EPS thicknesses ranging from 0 (no insulating material) to 12 cm were tested, along with horizontal external fixed shading elements varying in length from 0 (absence) to 150 cm.

HVAC System and Setpoint Ranges: The HVAC system in all the buildings relies on a chiller/heat-pump with consistent performance coefficients. For the HVAC control system, six alternative setpoint ranges were assessed: three based on the Predicted Mean Vote (PMV), and three based on the indoor air temperature (T_{air}).

To accurately represent the temperate Mediterranean climate, the buildings were hypothetically situated in three distinct locations, each characterized by a different climate intensity: mild, moderate, and intense. This study considered the year 2070, and three climatic scenarios were assumed: (i) NCC—no further climatic changes (maintenance of the current climate); (ii) MRS—mid-range scenario (RCP 4.5), representative of medium-intensity climate changes; and (iii) HRS—high-range scenario (RCP 8.5), representing strong climate changes.

2. Research Objects

Six buildings, each with varying acclimatized areas, occupancy levels, internal thermal gains, and distinct types of use, were selected to represent the building stock: (i) an apartment at midlevel of a multi-story building; (ii) a detached house; (iii) a clinic with hospitalization; (iv) a high school; (v) a bank branch; and (vi) a medium-sized supermarket.

To enable meaningful comparisons between the various buildings, we assumed that they were all constructed using identical passive construction solutions (including opaque, glazed, and shading elements), and each one is equipped with a Heating, Ventilation, and Air Conditioning (HVAC) system that exhibits consistent seasonal energy performance.

2.1. Buildings' Main Characteristics and Occupancy

Table 1 provides a summary of the key characteristics of these buildings. The net and gross areas exclude non-acclimatized spaces. For further details about the layout and main features of these buildings can be found in the work by Raimundo et al. [20].

Table 1. Summary of the characteristics of the 6 buildings considered: Np—maximum number of occupants, Nf—number of floors, A_{cl}—acclimatized floor area, A_{gf}—gross floor area, Ch—ceiling height, Vol—acclimatized volume, A_{opc}—opaque area of external envelope, A_{glz}—glazed area of external envelope, AR—aspect ratio = $(A_{opc} + A_{glz})/Vol$, EA—envelope area ratio = $(A_{opc} + A_{glz})/A_{cl})$, GA—glazed area ratio = A_{glz}/A_{cl}.

	Apartment	Detached House	Clinic	High School	Bank Branch	Supermarket
Np [persons]	4	4	151	1100	12	194
Nf [--]	1	3	2	4	1	1
A_{cl} [m²]	109.4	167.1	926.7	11,246.0	111.4	1035.3
A_{gf} [m²]	141.6	212.6	1161.2	14,147.5	134.7	1176.1
Ch [m]	2.62	2.96	3.72	3.84	2.60	3.60
Vol [m³]	286.6	494.6	3447.3	43,184.6	316.2	3727.1
A_{opc} [m²]	58.6	343.4	743.4	22,703.8	181.0	2830.6
A_{glz} [m²]	21.3	49.7	192.8	2975.3	37.2	96.6
AR [m^{-1}]	0.28	0.79	0.27	0.59	0.69	0.79
EA [--]	0.73	2.35	1.01	2.28	1.96	2.83
GA [--]	0.19	0.30	0.21	0.26	0.33	0.09

In general terms, occupancy and operating profiles exhibit the following characteristics:

- Across all buildings, occupancy and operating profiles vary based on the time of day, the day of the week, and the week of the year;
- When a building is unoccupied, the Heating, Ventilation, and Air Conditioning (HVAC) system remains off, and the lighting systems are either turned off or operate at very low power;
- Residential buildings are assumed to be unoccupied during the first fifteen days of August and permanently occupied during the remaining days of the year, by four people on Saturdays and Sundays, and between 6 P.M. and 8 A.M. on weekdays (Mondays to Fridays) and by one person the rest of the time;
- The clinic operates continuously throughout the year, with higher occupancy intensity between 8 A.M. and 8 P.M. on weekdays and on Saturdays;
- The school is only occupied between 8 A.M. and 6 P.M. on weekdays, it remains closed on Saturdays and Sundays and its operation follows the Portuguese academic calendar, so it operates at 100% during regular school periods; at 50% during the 1st examination phase (15–30 June); at 25% during the 2nd examination phase (1–15 July); at 25% during admission phase (16–31 July); and is closed during school holidays (the first 15 days of April, 1 to 31 August, and the last 15 days of December);

- The bank branch operates every weekday of the year and is occupied between 8 A.M. and 6 P.M., and it remains closed on Saturdays and Sundays;
- The supermarket operates every day of the year and it is occupied between 8 A.M. and 10 P.M., but with more intense activity on Saturdays and Sundays.

2.2. Opaque Elements of Buildings' Envelope

Each type of opaque construction element relies on a common base structure, consistent across all buildings and climates. The base structure most used in Portugal was assumed [20,32], which leads to buildings with substantial thermal inertia, a strategic choice for an effective mitigation of both overheating and cooling load peaks [10,13]. Table 2 outlines the base structure details for the opaque elements in contact with the exterior, including their thickness, useful thermal mass (Mt), and thermal transmission coefficient (U).

Table 2. Base structure of some opaque elements of the external envelope.

Element	Description (from Outside to Inside)	Values
Wall	Traditional plaster with 2 cm, bored brick of 22 cm, not-ventilated air space with 1 cm, bored brick of 11 cm, traditional plaster with 2 cm	*Thickness* = 38 cm Mt = 150 kg/m^2 U = 0.88 W/(m^2 K)
Pillar/Beam	Traditional plaster with 2 cm, inert reinforced concrete (iron volume less than 1%) with 22 cm, not-ventilated air space of 1 cm, bored brick of 11 cm, traditional plaster with 2 cm	*Thickness* = 38 cm Mt = 150 kg/m^2 U = 1.36 W/(m^2 K)
Floor above outside	Traditional plaster with 2 cm, lightened slab of 38 cm, light-sand concrete of 7.5 cm, screed (mortar) of 5.5 cm, oak wood with 2 cm	*Thickness* = 55 cm Mt = 150 kg/m^2 U = 1.17 W/(m^2 K)
Ground floor	Waterproofing layer, lightened slab of 38 cm, light-sand concrete of 7.5 cm, screed (mortar) of 5.5 cm, oak wood with 2 cm	*Thickness* = 54 cm Mt = 150 kg/m^2 U = 1.23 W/(m^2 K)
Accessible roof	Mosaic tile with 1 cm, screed (mortar) of 5.5 cm, waterproofing of 3 mm, light-sand concrete of 7.5 cm, lightened slab of 38 cm, traditional plaster with 2 cm	*Thickness* = 55 cm Mt = 150 kg/m^2 U = 1.39 W/(m^2 K)
Not accessible roof	Sandstone (inert) with 4 cm (or ceramic tile), waterproofing of 3 mm, screed (mortar) of 4 cm, lightened slab of 23 cm, traditional plaster with 2 cm	*Thickness* = 33 cm Mt = 150 kg/m^2 U = 2.40 W/(m^2 K)

The basic structure of each opaque construction element is enhanced by the application of expanded polystyrene (EPS) on the outer surface through an External Thermal Insulation Composite System (ETICS), often recognized as an efficient solution in terms of energy demands [10,20,28–31]. An additional advantage is its versatility, as it can be employed in both new constructions and building refurbishments. EPS thermal insulation material was chosen due to its economic and environmental benefits, its integrability into nearly all opaque elements, and its durability of at least 50 years [10,20,29].

EPS thicknesses ranging from 0 cm (without insulation) to 12 cm were tested, representing the economically viable range for buildings situated in temperate Mediterranean climates [10,20]. As an example, Table 3 displays the thermal transmission coefficient (U) values for the more relevant opaque elements of the external envelope, corresponding to different EPS thicknesses. Like the buildings, these values have already been considered in previous works [7,10,20]. Notably, the impact on the U value diminishes as the thickness of thermal insulation increases.

Table 3. Thermal transmission coefficient [W/(m^2 K)] of some opaque elements of the external envelope as function of EPS thicknesses.

EPS Thickness [cm]	Thermal Transmission Coefficient—U [W/(m^2 K)]					
	Wall	Pillar/Beam	Floor above Outside	Ground Floor	Accessible Roof	Non-Accessible Roof
0	0.88	1.36	1.17	1.23	1.39	2.40
1	0.72	1.01	0.90	0.94	1.03	1.49
2	0.62	0.83	0.75	0.78	0.84	1.12
3	0.54	0.69	0.64	0.65	0.70	0.88
4	0.48	0.59	0.56	0.57	0.60	0.73
5	0.43	0.52	0.49	0.50	0.52	0.62
6	0.39	0.46	0.44	0.45	0.47	0.54
7	0.36	0.42	0.40	0.41	0.42	0.49
8	0.33	0.38	0.36	0.37	0.38	0.43
9	0.30	0.34	0.33	0.33	0.34	0.38
10	0.28	0.32	0.30	0.31	0.32	0.35
11	0.26	0.29	0.28	0.28	0.29	0.32
12	0.25	0.28	0.27	0.27	0.28	0.30

2.3. Glazing Elements

The glazing system identified by Raimundo et al. [33] as the most economically advantageous for buildings located in Portugal was selected. The windows incorporate an aluminum frame with thermal barrier and double glazing (colorless of 6 mm + 11 mm air-layer + colorless of 4 mm), and they are externally protected by blinds made of horizontal plastic strips. This glazing system has a thermal transmission coefficient (U) and a solar factor ($g_\perp$) of $U_\mathrm{w} = 3.05$ W m^{-2} K^{-1} and $g_{\perp\mathrm{w}} = 0.79$ when the blind element is not active and of $U_\mathrm{wp} = 1.56$ W m^{-2} K^{-1} and $g_{\perp\mathrm{wp}} = 0.05$ when it is active.

2.4. External Fixed Shading Elements

Likely, climate change will lead to an increase in both the outside air temperature and the intensity of solar radiation [1,4,5], and, consequently, buildings will experience reduced energy requirements for heating and increased energy demands for cooling [12,26,27], among others. To reduce cooling needs without compromising natural interior lighting, an effective strategy is the implementation of external horizontal glazing shading systems [34–37].

Despite the existing glazing areas in the current architecture (referred to as the base architecture) being partially shaded by building elements such as balconies and facade cutouts, an assessment was conducted to evaluate the impact of installing horizontal external fixed shading elements on air conditioning energy consumption. However, the application of additional shades was only considered for glazing areas not already shaded by elements of the base architecture or when such shading had minimal relevance. Additionally, given the buildings' location in the northern hemisphere, no additional shading elements were considered for glazing oriented toward east-northeast, north, or west-northwest.

If present, all additional external fixed horizontal shading elements are positioned at the top of the respective window and have the same length, and shade lengths ranging from 0 cm (no shade) to 150 cm, in increments of 10 cm, have been tested. It is important to recognize that whereas external fixed glazing shades have the potential to reduce cooling energy demands, they may also increase heating energy requirements. Consequently, the energy impact of installing fixed glazing shades depends on the building's use type and the prevailing climate conditions.

2.5. Heating, Ventilation, and Air Conditioning System

In temperate Mediterranean climates, buildings rely on both heating and cooling systems to maintain thermal comfort. Among the available options, electric air-source heat pumps demonstrate reasonable performance in heating mode. Consequently, systems based on air-source chiller/heat pumps are commonly chosen [7,20,32]. Therefore, these are the Heating, Ventilation, and Air Conditioning (HVAC) systems considered. The indoor air renewal is ensured by Air Handling Units (AHUs) and/or air-extraction fans, both operating at an efficiency of 70% [7,20].

The HVAC systems are assumed to be equipped with a chiller/heat-pump classified as European class A+ [38], as it aligns with the equipment commonly installed in practice. The chiller has a seasonal energy efficiency ratio $SEER = 5.85$ in cooling mode and the heat-pump has a seasonal coefficient of performance $SCOP = 4.30$ in heating mode [7,10,38].

3. Methods and Conditions

The present study relies on a numerical assessment of the relationship between energy demand and consumption for air conditioning with the level of thermal comfort indoors, the building's type of use, the building's passive and active construction elements, and the climate specificities, considering alternative scenarios of climate change.

3.1. Calculation Tool

The version 5.07 of SEnergEd software [7,10,20,33], a validated in-home tool developed for research purposes, was employed in this study. This user-friendly software integrates algorithms for dynamically simulating the thermal and energy behavior of various building types (residential, commercial, and service). Its capabilities include assessing thermal comfort, analyzing environmental impact, and evaluating the economic aspects of a building's life cycle.

This software predicts the thermal behavior of buildings using a reformulated version of the dynamic hourly model known as 5R1C (which stands for five thermal resistances and one thermal capacitance) described in ISO 13790 [39]. The thermal behavior and energy needs are conditioned by the maximum useful capacity of the HVAC system installed in the building. Energy demands from other equipment and systems (such as domestic hot water, lighting, and appliances) are calculated dynamically based on their hourly operating profiles and installed power. By considering the energy performance of the equipment and systems, the energy demands are then converted into actual consumption.

The operation of the HVAC system can be controlled using either indoor air temperature (T_{air}) setpoints or predicted mean vote (PMV) setpoints [15,16]. Additionally, both control strategies incorporate an additional setpoint for air relative humidity (RH). This procedure is carried out following a predictive control algorithm model. In addition to the control by setpoints and with the ability to override them, hourly operating profiles of HVAC systems can be defined.

Further details about the SEnergEd software can be found elsewhere [7,10,20,33].

3.2. Control of the Climatization System Operation

According to standards ASHRAE 55:2004 [15] and ISO 7730:2005 [16], the predicted mean vote (PMV) is determined based on the overall thermal balance of the human body. Its absolute value correlates with the percentage of people who experience thermal discomfort, more specifically, $PMV = 0$ indicates thermal comfort, $PMV < 0$ means discomfort due to cold, and $PMV > 0$ is discomfort due to heat. The calculation of PMV value requires the knowledge of four environmental parameters (air temperature, air humidity, air velocity, and mean radiant temperature) and of three individual factors (clothing intrinsic insulation, metabolic rate, and external work).

In each hourly time-step, the SEnergEd software computes the following parameters within the thermal zone: air temperature, humidity, and mean radiant temperature. Then, the clothing's intrinsic thermal insulation, the person's physical activity and external work, and the air's velocity must be provided as input parameters. The values considered in this study for these parameters, typical of habits in Mediterranean temperate climates, are shown in Table 4. These values vary based on the building's type of use, the season of the year, and whether it is daytime or nighttime.

Table 4. Intrinsic clothing insulation, activity level, and indoor air velocity.

	During the:	During the:	Intrinsic Clothing Insulation [clo]	Activity Level [met]	Air Velocity [m/s]
Apartment and Dwelling	Winter	Day/Night	1.3/2.6		
	Spring and Autumn	Day/Night	1.0/2.0	1.2/0.8	0.2
	Summer	Day/Night	0.7/1.4		
Clinic	Winter	Day/Night	1.3/2.0		
	Spring and Autumn	Day/Night	1.0/2.0	1.4/0.8	0.2
	Summer	Day/Night	0.7/1.4		
School	Winter	Day	1.3		
	Spring and Autumn	Day	1.0	1.4	0.3
	Summer	Day	0.7		
Bank branch	Winter	Day	1.4		
	Spring and Autumn	Day	1.2	1.2	0.2
	Summer	Day	1.0		
Supermarket	Winter	Day/Night	1.5/1.5		
	Spring and Autumn	Day/Night	1.2/1.2	1.5/1.5	0.3
	Summer	Day/Night	0.7/0.7		

Six possibilities for the operation of the HVAC system were considered. In three of them, the control was performed by setpoints of the predicted mean vote ($PMV_{min} \leq PMV \leq PMV_{max}$) and in the other three, this control is performed by air temperature setpoints ($T_{min} \leq T_{air} \leq T_{max}$). A control with air relative humidity setpoints ($RH_{min} \leq RH \leq RH_{max}$) was associated with both controls (in the present study RH was maintained between 50 and 70%). In addition to the control by setpoints, and with the ability to override them, hourly operating profiles were defined.

Table 5 outlines the six possibilities considered for HVAC system control, along with the hypothesis of the non-existence of an HVAC system (NHS). A, B, and C represent PMV setpoints, separated by increments of 0.25. The three T_{air} setpoints are labeled as DT1, DT3, and DT5, where DT1 represents a temperature difference between the upper and lower limits of 1 °C, DT3 of 3 °C, and DT5 of 5 °C, respectively. In the case of the bank branch and the supermarket, the setpoint values of T_{air} are slightly lower than the corresponding ones for the other buildings, since it was considered that the occupants of those buildings usually wear clothing with higher thermal insulation.

The setpoint limits considered for both PMV and T_{air}, as shown in Table 5, are based on the endorsements outlined in the standard EN 16798-1 [40], which provides specific conditions that must be met in buildings to achieve defined levels of indoor environmental quality. Controls A and DT1 guarantee the highest quality of thermal comfort and align with the Category I level of this standard, recommended for spaces occupied by fragile individuals or those with special requirements. Controls B and DT3 counterpart Category II, endorsed for buildings to be used by people without special requirements, but with high expectations. Controls C and DT5 fall under Category III, suggested for spaces with moderate expectations. NHS concerns the situation where the building lacks an HVAC system.

Table 5. Types of control of operation of the buildings' climatization system.

Control	Control of HVAC System	
Type	**Apartment, Dwelling, Clinic, School, Bank Branch, and Supermarket**	
A	$-0.25 \leq PMV \leq +0.25$	$PPD \leq 6.3\%$
B	$-0.50 \leq PMV \leq +0.50$	$PPD \leq 10.2\%$
C	$-0.75 \leq PMV \leq +0.75$	$PPD \leq 16.8\%$
NHS	No HVAC system	

	Apartment, Dwelling, Clinic, and School	**Bank Branch**	**Supermarket**
DT5	$20 \leq T_{air} \leq 25\,°C$	$19 \leq T_{air} \leq 24\,°C$	$18 \leq T_{air} \leq 23\,°C$
DT3	$21 \leq T_{air} \leq 24\,°C$	$20 \leq T_{air} \leq 23\,°C$	$19 \leq T_{air} \leq 22\,°C$
DT1	$22 \leq T_{air} \leq 23\,°C$	$21 \leq T_{air} \leq 22\,°C$	$20 \leq T_{air} \leq 21\,°C$

3.3. Climate Scenarios

Research related to the thermal energy demand of buildings often relies on simulation tools, which necessitate a file containing a year's worth of hourly climate data specific to the building's location. Subsequently, to assess the impact of climate change on the thermal and energy behavior of buildings, appropriately prepared climate data files are essential. There are two primary approaches for creating these files: one involves predictions based on historical data, whereas the other relies on fundamental physical models [25]. In this study, we employ the historical model to generate the required hourly climate dataset files. For this, the "morphing procedure" proposed by Belcher and colleagues was used [24]. This approach involves generating future design weather data by adjusting present-day climate data using "correction coefficients" derived from "global climatic models" tailored to specific climate change scenarios. To derive the correction coefficients, the global climate model CGCM3.1/T47, developed by the "Canadian Center for Climate Modeling and Analysis" [41] was employed, which generates values for nearly all geographical locations on the planet, with a resolution of 3.75° × 3.75°.

The morphing of each individual weather parameter is accomplished using three alternative algorithms [24,25]: shifting, linear stretching (scaling factor), and a combination of both (shifting and stretching). The shifting method relies on an absolute change in the monthly mean value of the variable, and it is employed when a change in the mean is predicted for that specific weather parameter in that given climate change scenario. The linear stretching is used when a proportional change to either the mean or the variance of the individual weather parameter is predicted in that climate change scenario; for instance, this approach is suitable for variables like solar radiation, which becomes zero at night. A combination of shift and stretch is applied in cases where both the mean and variance of an individual weather parameter are expected to change (e.g., air temperatures), reflecting changes in average, minimum, and maximum daily values. Deeper details about the mathematical operations involved in the "morphing procedure" can be found in Belcher et al.'s paper [24].

To generate files for future climate scenarios based on the current climate, adjustments were made only to the values of dry bulb temperature, relative humidity, and direct, global, and diffuse solar radiation. The values of the components of solar radiation were obtained using the scale factor morphing procedure (linear stretching), ensuring that the solar radiation values align with the projected changes. The dry bulb temperature at each hour of each month was determined using a combination of the morphing procedures shifting and stretching, which led to changes in average, minimum, and maximum daily values. Unfortunately, the global climate model used (CGCM3.1/T47) does not provide predictions for the correction coefficient needed to obtain relative humidity. However, it does offer predictions for specific humidity [41]. Therefore, to derive the relative humidity of the considered future climate scenarios, it was necessary to first obtain the corresponding specific humidity values and then convert them into relative humidity using appropriate methods. The specific humidity value for each hour of each month was obtained using the linear stretching methodology.

The buildings (residential, services, and commercial) under consideration are hypothetically situated in a temperate Mediterranean climate. This climate type spans an extensive range of countries (including Greece, Italy, Portugal, Spain, Turkey), and several specific regions (such as parts of Albania, Australia, France, South Africa, and California). Temperate climates strike a balance: they are neither excessively hot in summer nor too cold in winter, and they avoid extreme dryness or excessive humidity. Despite this moderation, these climates exhibit substantial temperature differences between day and night, and present marked climatic variations across different seasons. The Köppen–Geiger climate classification designates these temperate Mediterranean climates as *Csa* or *Csb* [12,30].

Significant climatic disparities exist across regions within temperate Mediterranean climates (referred to here as MC) [7,19]. To accurately represent these climates, we hypothetically position buildings in three distinct locations, each characterized by different weather patterns. These locations correspond to the following MC types: (i) mild in winter and mild in summer (MC1); (ii) moderate in winter and moderate in summer (MC2); and (iii) intense in winter and intense in summer (MC3). The locals selected to represent these climate types are all located in Portugal and are Funchal (at an elevation above sea level $Z = 415$ m) for mild climate MC1; Ansião ($Z = 361$ m) for moderate climate MC2; and Mirandela ($Z = 600$ m) for intense climate MC3. These carefully selected localities provide a comprehensive snapshot of the diverse climatic variations within the temperate Mediterranean regions.

In this analysis, the year 2070 is considered and three distinct climate change scenarios are explored as follows: (i) no further climate changes (NCC); a mid-range scenario (MRS); and a high-range scenario (HRS). The NCC scenario assumes that the current climate remains unchanged, with no additional alterations beyond the existing climatic conditions. The MRS scenario represents medium-intensity climate changes, as projected by the IPCC Representative Concentration Pathway RCP 4.5 [5], representing some impact on climate, affecting ecosystems, weather patterns, and global temperatures. The HRS scenario emerges from extreme climate changes, as forecasted by the IPCC scenario RCP 8.5, and represents a severe impact on climate.

Various methodologies exist for classifying the different climate types. Among these, the approach based on heating degree days(HHD [°C·day/year]) and cooling degree days(CDD [°C·day/year]) provides a more direct link between outdoor weather conditions and energy requirements for heating and cooling, respectively [7,30,42].

In this study, the HDD and CDD values are defined with respect to reference temperatures of 20 °C and 25 °C, respectively, and are accordingly referred to as HDD_{20} and CDD_{25}. Their values for the three temperate Mediterranean climate types selected (MC1, MC2, MC3) and the three climate change scenarios considered (NCC, MRS, HRS) are summarized in Table 6. The corresponding annual average values of air temperature T_m (and its difference to the NCC scenario ΔT_m), of air relative humidity RH_m (and its difference to the NCC scenario ΔRH_m), and of horizontal global solar radiation $HGSR_m$ (and its difference to the NCC scenario $\Delta HGSR_m$) are also presented in Table 6.

Figure 1 displays boxplot graphs that provide a global overview of the climate predictions for the year 2070, for the three types of temperate Mediterranean climate selected (MC1, MC2, MC3) and the three climate change scenarios considered (NCC, MRS, HRS). As they are the most indicative, the values of air temperature, relative humidity, and global solar radiation incident on a horizontal plane are presented in this figure. In each case shown, the lower line indicates the minimum value, the bottom line of the box the first quartile (25th percentile), the line inside the box the median, the marker inside the box the mean, the top edge of the box the third quartile (75th percentile) and the upper line the maximum value.

Table 6. HDD_{20} and CDD_{25} values [°C·day/year] and annual average values of air temperature T_m (and its difference to the NCC scenario, ΔT_m) [°C], of air relative humidity RH_m (and its difference to the NCC scenario, ΔRH_m) [%], of horizontal global solar radiation $HGSR_m$ (and its difference to the NCC scenario, $\Delta HGSR_m$) [W/m^2], of maximum difference in air temperature during the year $\Delta T_{\max}$ (= $T_{\max} - T_{\min}$) [°C], and of average values of air temperature T_m (and its difference to the NCC scenario, ΔT_m) [°C] for the stations of the year, for the temperate Mediterranean climate types selected and the climate change scenarios considered.

Climate Type		NCC	MRS (RCP 4.5)	HRS (RCP 8.5)
MC1	HDD_{20}	1256	682	456
Mild	CDD_{25}	16	72	148
	$T_m\ (\Delta T_m)$	17.0 (--)	18.4 (+1.4)	19.4 (+2.4)
	$RH_m\ (\Delta RH_m)$	76 (--)	79 (+3)	74 (−2)
	$HGSR_m\ (\Delta HGSR_m)$	284 (--)	328 (+44)	329 (+45)
	$\Delta T_{\max}$	19.6	11.7	11.1
MC2	HDD_{20}	2111	1732	1357
Moderate	CDD_{25}	81	134	257
	$T_m\ (\Delta T_m)$	15.1 (--)	16.5 (+1.4)	18.3 (+3.2)
	$RH_m\ (\Delta RH_m)$	73 (--)	72 (−1)	69 (−4)
	$HGSR_m\ (\Delta HGSR_m)$	317 (--)	361 (+44)	362 (+45)
	$\Delta T_{\max}$	29.9	26.7	28.6
MC3	HDD_{20}	2762	2170	1739
Intense	CDD_{25}	144	152	276
	$T_m\ (\Delta T_m)$	13.6 (--)	15.3 (+1.7)	17.0 (+3.4)
	$RH_m\ (\Delta RH_m)$	69 (--)	72 (+3)	74 (+5)
	$HGSR_m\ (\Delta HGSR_m)$	305 (--)	323 (+18)	336 (+31)
	$\Delta T_{\max}$	35.9	28.4	30.0
MC1 + MC2 + MC3	Winter	10.6 (--)	12.6 (+2.0)	13.7 (+3.1)
$T_m\ (\Delta T_m)$	Spring	15.2 (--)	15.7 (+0.5)	17.0 (+1.8)
	Summer	20.7 (--)	22.2 (+1.5)	24.2 (+3.5)
	Autumn	14.3 (--)	16.3 (+2.0)	17.9 (+3.6)

As depicted in Table 6, both the values of HDD_{20} (heating degree days at 20 °C) and CDD_{25} (cooling degree days at 25 °C) increase as the severity of the present climate intensifies. However, their behavior diverges based on the impact of climate change. The value of HDD_{20} decreases as climate change becomes stronger. Conversely, the value of CDD_{25} rises with increasing of climate change intensity. These trends highlight the relationship between climate severity, ongoing climate changes, and temperature-related energy demands.

The global climate forecasts, as depicted in Table 6 and Figure 1, indicate an increase in the average air temperature (T_m) with the escalation of climate change intensity. Additionally, climate change alters temperature patterns throughout the year, affecting both the maximum temperature difference ($\Delta T_{\max} = T_{\max} - T_{\min}$), and the average air temperatures across seasons. Relatively to the NCC scenario, climate change leads to a decrease of $\Delta T_{\max}$, which is more pronounced in the MRS scenario than in the HRS one. The average values of air temperature T_m reveal that in the present climate (NCC scenario), the coldest season of the year is winter, followed by autumn, spring, and summer (the hottest). In the situation of further climate change, the order will be winter, spring, autumn, and summer, and all stations of the year will warm up with the increase in climate change intensity, but not in a uniform way. The station less affected by climate change will be the spring and the most affected will be the autumn. These predictions also highlight that there will be no substantial alteration in relative humidity values, and a definitive relationship between these values and climate change intensity remains elusive. Regarding horizontal global solar radiation, an elevation in the corresponding value is anticipated as climate change severity intensifies. Notably, the variation in the average value of horizontal global solar radiation ($HGSR_m$) is significant when transitioning from the NCC scenario (no further climatic changes) to the MRS (mid-range scenario, RCP 4.5), but less pronounced when moving from MRS to HRS (high-range scenario, RCP 8.5).

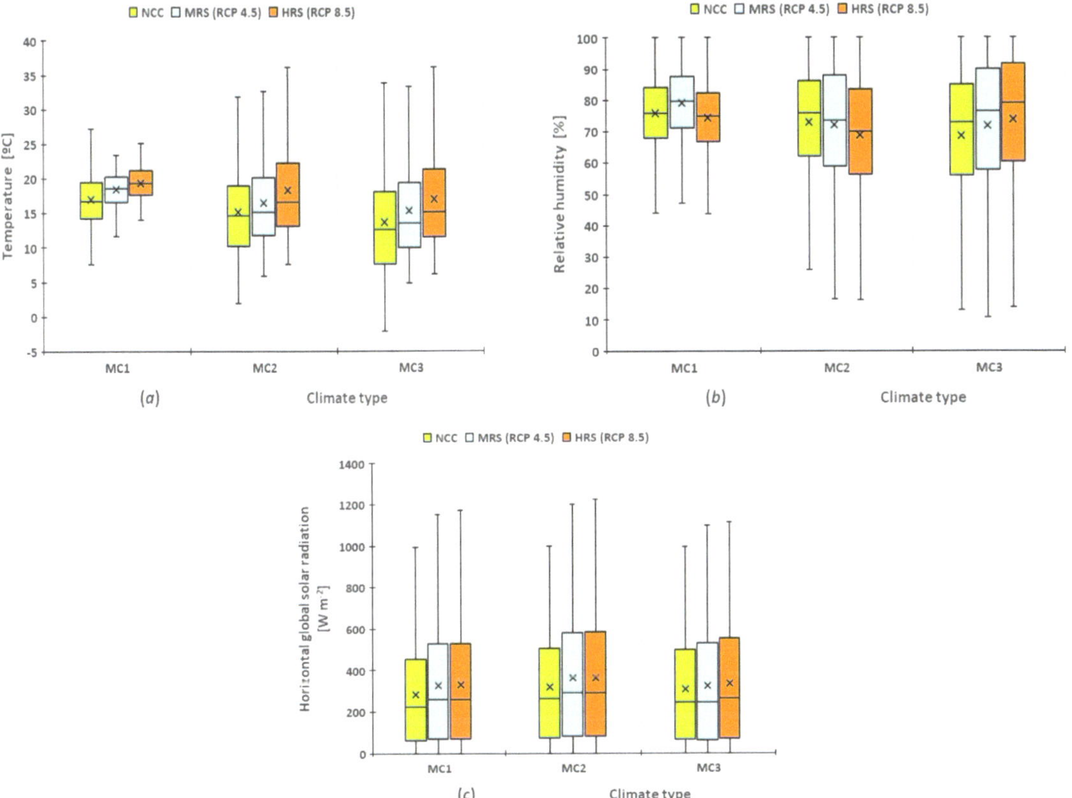

Figure 1. Boxplot representation of the climate in 2070, for the three Mediterranean climates selected (MC1, MC2, MC3) and the three climate change scenarios considered (NCC, MRS, HRS), by (**a**) air temperature, (**b**) relative humidity, and (**c**) global solar radiation on a horizontal plane.

In Table 6, considering the mild climate MC1, the average air temperature (T_m) exhibits the values of 17.0 °C for the NCC scenario, of 18.4 °C for the MRS scenario, and of 19.4 °C for the HRS scenario. The average horizontal global solar radiation ($HGSR_m$) values are 284 W/m^2 for NCC, 328 W/m^2 for MRS, and 329 W/m^2 for HRS. Taking the NCC scenario as reference, we observe an increase in average air temperature (ΔT_m) of +1.4 °C in the MRS scenario and of +2.4°C in the HRS one. Additionally, climate change leads to a decrease in the maximum difference in air temperature during the year (ΔT_{max}) of 7.9 °C for the MRS and of 8.5°C for the HRS. The change in $HGSR_m$ ($\Delta HGSR_m$) is of +44 W/m^2 for MRS and of +45 W/m^2 for HRS.

In the context of the moderate MC2 climate, Table 6 reveal average air temperatures (T_m) of 15.1 °C for the NCC scenario, of 16.5°C for the MRS scenario, and of 18.3 °C for the HRS scenario. The average horizontal global solar radiation ($HGSR_m$) exhibits the values of 317 W/m^2 for NCC, of 361 W/m^2 for MRS, and of 362 W/m^2 for HRS. Comparing these values to the NCC reference, we note an increase of ΔT_m = +1.4 °C in the MRS case and a more substantial rise of ΔT_m = +3.2 °C in the HRS case. Climate change leads to a decrease in the maximum difference in air temperature during the year (ΔT_{max}) of 3.2 °C for the

MRS and of 1.3 °C for the HRS. The change in $HGSR_m$ amounts to $\Delta HGSR_m$ = +44 W/m^2 in the MRS and $\Delta HGSR_m$ = +45 W/m^2 in the HRS.

For the intense climate MC3, Table 6 reveals the values of T_m = 13.6 °C for the NCC scenario, 15.3 °C for the MRS scenario, and 17.0 °C for the HRS scenario. In this type of climate, $HGSR_m$ = 305, 323, and 336 W/m^2 for the NCC, MRS, and HRS scenarios, respectively. Comparing previous values with the NCC reference, we observe an increase of ΔT_m = +1.7 °C in the MRS scenario and a more substantial rise of ΔT_m = +3.4 °C in the HRS one. Climate change leads to a decrease in the maximum difference in air temperature during the year ($\Delta T_{\max}$) of 7.5 °C for the MRS and of 5.9 °C for the HRS. The change in $HGSR_m$ amounts to $\Delta HGSR_m$ = +18 W/m^2 in the MRS and $\Delta HGSR_m$ = +31 W/m^2 in the HRS.

To assess whether the differences between scenarios are statistically significant, we employed a Student's t-test, considering a two-tailed distribution and two samples with unequal variance. Probabilities associated with this test were calculated for three key parameters: air temperature (T_{air}), relative humidity (RH), and horizontal global solar radiation ($HGSR$). Relative to the present climate (scenario NCC), the other two (mid-range (MRS) and high-range (HRS)), show a significant statistical difference ($p < 0.001$) for the three previous parameters (T_{air}, RH, and $HGSR$) in the three Mediterranean climates (mild (MC1), moderate (MC2), and intense (MC3)). The difference is also statistically significant ($p < 0.001$) between MRS and HRS scenarios for the parameters T_{air} and RH, but not for $HGSR$ ($p > 0.05$).

4. Results and Discussion

The energy perspective was employed to assess the relation between the thickness of thermal insulation and the length of horizontal external fixed glazing shades with the building type, the type of control of the Heating, Ventilation, and Air Conditioning (HVAC) system, and the severity of climate change. For this, three climate change scenarios projected for the year 2070 (NCC—no further climate changes, MRS—mid-range scenario, and HRS—high-range scenario), and six different buildings located in temperate Mediterranean climates (an apartment, a detached house, a clinic, a school, a bank branch, and a supermarket) were considered. The energy perspective considered includes only the operational energy, without accounting for embodied energy on materials or energy associated to buildings' end-of-life. Then, "energy demand" and "energy needs" refer to the "operational useful annual thermal energy" requirement for heating or for cooling, and "energy consumption" refers to "operational energy consumption by the HVAC system (electric energy) during an entire year".

The results presented in the subsequent sections are normalized per square meter (m^2) of the acclimatized spaces' floor area. Table 1 provides details on the net (A_{cl}) and gross (A_{gf}) floor areas of the buildings. As previously indicated in Table 5, six alternatives for HVAC system control were explored. These alternatives include three by predicted mean vote (PMV) setpoints (labeled as A, B, and C), and three by indoor air temperature (T_{air}) setpoints (labeled as DT1, DT3, and DT5). "A" corresponds to $-0.25 \leq PMV \leq +0.25$, "B" to $-0.50 \leq PMV \leq +0.50$, "C" to $(-0.75 \leq PMV \leq +0.75)$, "DT1" to a temperature difference between the upper and lower limits of 1 °C, "DT3" represents a difference of 3 °C, and "DT5" reflects a difference of 5 °C. Furthermore, the non-existence of an HVAC system (NHS) was accounted for.

This study involves a total of 67,392 cases (=16 shading lengths × 13 insulation thickness × 6 buildings × 3 climate change scenarios × 3 locations × 6 HVAC setpoint types). To handle this large number of cases efficiently, the following strategy was implemented: (1st) The in-home software was prepared to simulate the 3 climate change scenarios, the 3 locations and the 6 HVAC setpoint types in each run (1 run → simulation of 54 cases); (2nd) the simulations were conducted in two rounds (which reduces the cases considered to 9396): identification of the optimal thermal insulation thickness for buildings without additional shading (78 runs → 4212 cases); and 2nd round-identification of the optimal shading length only for buildings with the optimal thermal insulation thickness (96 runs → 5184 cases). With this strategy, it was necessary to prepare and make only 174 runs.

4.1. Optimal Thermal Insulation Thickness and External Shade Lengths

Achieving the right balance between insulation thickness, shade length, and HVAC control is crucial for maximizing energy efficiency while ensuring occupant comfort. Each building type, climate type, and climate change scenario requires tailored solutions.

The optimal thermal insulation thickness and glazing shade length depend on the specific perspective: energy efficiency, environmental impact, or economic cost. This issue has been already addressed by Raimundo's team [10,20], where the same buildings and locations were considered, but only for the present climate (NCC scenario). In the present work, only the energy perspective was considered.

In energy terms, the optimal values for thermal insulation thickness of the opaque elements of the building envelope and the length of external shades are those that result in the lowest energy consumption for climatization (both heating and cooling). The corresponding values were determined for each building type and each HVAC control through a two-step process: (1st) the optimal thickness of expanded polystyrene (EPS) insulation for the base architecture of the building (without any additional shading) was obtained; (2nd) the optimal length for external shades (considering the opaque elements insulated with the previously EPS thickness) was achieved. EPS thicknesses ranging from 0 cm (no insulation) up to 12 cm were checked, as this range aligns with economic viability for buildings in temperate Mediterranean climates [10,20]. Shade lengths within the range of 0 cm (no additional shading) to 150 cm were tested. For aesthetic reasons, the shade length was limited to 150 cm.

Due to the simplification adopted in the simulations (in steps of EPS thickness and of shade length), the optimal values of the thermal insulation thickness and of shade length can be slightly different from the values obtained. Also, they can even have a value greater than the maximum tested, respectively, 12 and 150 cm. Then, the values identified as "optimal" must be looked at as the "best solution" within the range of the values tested.

Based on a global analysis, it was discovered that the energy-optimal values for EPS thickness and external shade length differ depending on whether the HVAC system is controlled using PMV or T_{air} setpoints. However, these values are equal or nearly identical for control types A, B, and C, as well as for control types DT1, DT3, and DT5. Therefore, it suffices to specify whether the HVAC system control is based on PMV or T_{air} setpoints. The optimal values obtained for thermal insulation thickness and for shade length are summarized in Table 7, for the three types of Mediterranean climates (MC1, MC2, MC3) and the three climate change scenarios (NCC, MRS, HRS) considered, as a function of the type of HVAC system control.

The results presented in Table 7 indicate that the energy-optimal values for thermal insulation thickness and shade length tend to align with either the respective minimum values (0 cm, 0 cm) or the respective maximum values (12 cm, 150 cm, respectively) that were tested. Additionally, for assessment purposes, the buildings can be categorized as: (i) of permanent use (apartment, detached house, and clinic); (ii) of intermittent use and low internal thermal loads (school and bank branch); and (iii) of intermittent use and high internal thermal loads (supermarket).

In the case of buildings with permanent use, the energy-optimal thermal insulation thickness is consistently 12 cm across all types of HVAC system control, temperate Mediterranean climates, and climate change scenarios. This arises from the fact that, for these types of buildings (with low internal thermal loads) and passive constructive elements (with high thermal mass), the increase in thermal insulation thickness leads to a decrease in energy consumption for heating and an increase in energy consumption for cooling, as shown by the results of this study (figures not shown) and what is reported in the bibliography [10,20,30,42]. Additionally, with the increase in thermal insulation thickness, the rate of decrease in energy consumption for heating is greater than the rate of increase in consumption for cooling, which is reflected in a continuous decrease in energy consumption for air conditioning. In this type of buildings, the use of additional glazing shades is energetically advantageous when the HVAC system is controlled by PMV setpoints. On the other hand, these elements do not bring energy advantages when this control is performed by T_{air} setpoints.

Table 7. Energetically optimal values of thermal insulation (EPS) thickness (0 to 12 cm), and of external fixed horizontal shade length (Shd) (0 to 150 cm), as function of Mediterranean type of climate (MC1, MC2, MC3), type of control of HVAC system (PMV/T_{air}), and climate change scenario (NCC, MRS, HRS).

Apartment	NCC: No Further Climatic Changes						MRS: Mid-Range Scenario (RCP 4.5)						HRS: High-Range Scenario (RCP 8.5)					
Climate	MC1		MC2		MC3		MC1		MC2		MC3		MC1		MC2		MC3	
HVAC	EPS	Shd	EPS	Shd	EPS	Shd	EPS	Shd	EPS	Shd	EPS	Shd	EPS	Shd	EPS	Shd	EPS	Shd
Control	[cm]	[cm]	[cm]	[cm]	[cm]	[cm]	[cm]	[cm]	[cm]	[cm]	[cm]	[cm]	[cm]	[cm]	[cm]	[cm]	[cm]	[cm]
PMV	12	80	12	60	12	40	12	110	12	90	12	60	12	150	12	130	12	110
T_{air}	12	0	12	0	12	0	12	0	12	0	12	0	12	0	12	0	12	0
Dwelling	NCC: no further climatic changes						MRS: mid-range scenario (RCP 4.5)						HRS: high-range scenario (RCP 8.5)					
Climate	MC1		MC2		MC3		MC1		MC2		MC3		MC1		MC2		MC3	
HVAC	EPS	Shd	EPS	Shd	EPS	Shd	EPS	Shd	EPS	Shd	EPS	Shd	EPS	Shd	EPS	Shd	EPS	Shd
Control	[cm]	[cm]	[cm]	[cm]	[cm]	[cm]	[cm]	[cm]	[cm]	[cm]	[cm]	[cm]	[cm]	[cm]	[cm]	[cm]	[cm]	[cm]
PMV	12	100	12	70	12	50	12	140	12	110	12	90	12	150	12	130	12	110
T_{air}	12	0	12	0	12	0	12	0	12	0	12	0	12	0	12	0	12	0
Clinic	NCC: no further climatic changes						MRS: mid-range scenario (RCP 4.5)						HRS: high-range scenario (RCP 8.5)					
Climate	MC1		MC2		MC3		MC1		MC2		MC3		MC1		MC2		MC3	
HVAC	EPS	Shd	EPS	Shd	EPS	Shd	EPS	Shd	EPS	Shd	EPS	Shd	EPS	Shd	EPS	Shd	EPS	Shd
Control	[cm]	[cm]	[cm]	[cm]	[cm]	[cm]	[cm]	[cm]	[cm]	[cm]	[cm]	[cm]	[cm]	[cm]	[cm]	[cm]	[cm]	[cm]
PMV	12	150	12	130	12	100	12	150	12	130	12	110	12	150	12	150	12	150
T_{air}	12	0	12	0	12	0	12	0	12	0	12	0	12	0	12	0	12	0
School	NCC: no further climatic changes						MRS: mid-range scenario (RCP 4.5)						HRS: high-range scenario (RCP 8.5)					
Climate	MC1		MC2		MC3		MC1		MC2		MC3		MC1		MC2		MC3	
HVAC	EPS	Shd	EPS	Shd	EPS	Shd	EPS	Shd	EPS	Shd	EPS	Shd	EPS	Shd	EPS	Shd	EPS	Shd
Control	[cm]	[cm]	[cm]	[cm]	[cm]	[cm]	[cm]	[cm]	[cm]	[cm]	[cm]	[cm]	[cm]	[cm]	[cm]	[cm]	[cm]	[cm]
PMV	0	150	12	50	12	20	0	150	6	90	12	70	0	150	3	150	8	150
T_{air}	12	0	12	0	12	0	6	150	12	0	12	0	0	150	12	0	12	0
Bank	NCC: no further climatic changes						MRS: mid-range scenario (RCP 4.5)						HRS: high-range scenario (RCP 8.5)					
Climate	MC1		MC2		MC3		MC1		MC2		MC3		MC1		MC2		MC3	
HVAC	EPS	Shd	EPS	Shd	EPS	Shd	EPS	Shd	EPS	Shd	EPS	Shd	EPS	Shd	EPS	Shd	EPS	Shd
Control	[cm]	[cm]	[cm]	[cm]	[cm]	[cm]	[cm]	[cm]	[cm]	[cm]	[cm]	[cm]	[cm]	[cm]	[cm]	[cm]	[cm]	[cm]
PMV	0	150	12	70	12	40	0	150	12	110	12	80	0	150	0	150	12	150
T_{air}	12	0	12	0	12	0	12	0	12	0	12	0	12	0	12	0	12	0
Super	NCC: no further climatic changes						MRS: mid-range scenario (RCP 4.5)						HRS: high-range scenario (RCP 8.5)					
Climate	MC1		MC2		MC3		MC1		MC2		MC3		MC1		MC2		MC3	
HVAC	EPS	Shd	EPS	Shd	EPS	Shd	EPS	Shd	EPS	Shd	EPS	Shd	EPS	Shd	EPS	Shd	EPS	Shd
Control	[cm]	[cm]	[cm]	[cm]	[cm]	[cm]	[cm]	[cm]	[cm]	[cm]	[cm]	[cm]	[cm]	[cm]	[cm]	[cm]	[cm]	[cm]
PMV	0	150	0	150	0	150	0	150	0	150	0	150	0	150	0	150	0	150
T_{air}	0	150	12	150	12	150	0	150	12	150	12	150	0	150	12	150	12	150

In the case of the school and the bank branch, the values presented in Table 7 reveal that the energy-optimal thermal insulation thickness is greater when the HVAC system is controlled using T_{air} setpoints compared to PMV setpoints. As climate intensity increases (from MC1 to MC2 to MC3), the optimal insulation thickness tends to rise. Conversely, with more severe climate change scenarios (from NCC to MRS to HRS), the optimal thickness tends to decrease. When the Mediterranean climate is mild (MC1) and the HVAC system is controlled by PMV setpoints, the optimal solution is always the absence of thermal insulation (0 cm). Regardless of the climate type (MC1, MC2, or MC3), when T_{air} setpoints are used, the optimal EPS thickness is consistently 12 cm under the current climate scenario (NCC). Irrespective of the climate change scenario (NCC, MRS, or HRS), when the HVAC system is controlled by T_{air} setpoints, the optimal EPS thickness remains 12 cm in the case of moderate (MC2) and intense (MC3) Mediterranean climates. For these building types, additional glazing shades offer energy advantages when the HVAC system is controlled by *PMV* setpoints. However, these elements do not confer similar energy benefits when the HVAC system is controlled by T_{air} setpoints.

For the supermarket, the energy-optimal EPS thickness is 0 cm (i.e., no thermal insulation) when the control of the HVAC system is carried out by *PMV* setpoints and when the building is in a location with a mild Mediterranean climate (MC1). For all other situations, the optimal EPS thickness equals the maximum tested value of 12 cm. In this building, regardless of climate type or climate change scenario, the use of additional glazing shades is energetically advantageous when the HVAC system is controlled by both *PMV* and T_{air} setpoints.

Understanding the interplay between thermal insulation, glazing shading, climate characteristics, and HVAC control strategies is essential for energy-efficient design in buildings. Buildings' energy consumption is significantly influenced by the climatic conditions they face [7,10,12]. Thus, the severity of climate change will have a major influence on energy use for air conditioning. Consequently, optimizing building design in energy terms necessitates precise knowledge of the future climatic conditions [11–14]. Although achieving this precision is challenging, certain good practices can guide the process.

Within the tested climate change scenarios (NCC, MRS, and HRS) and the considered temperate Mediterranean climates (MC1, MC2, and MC3), as climate change severity increases, there is a tendency for reduced energy-optimal EPS thickness and for an increase in energy advantage of using glazing shades, but these trends are not highly significant. Then, buildings designed for good energy performance in the current climate (NCC) will maintain good performance in the future. Even if global warming reaches levels equivalent to the HRS scenario, well-designed buildings will remain energy-efficient. It should be recognized that good energy performance does not necessarily equate to optimal performance. Also, as local specific climate conditions play a significant role in buildings' energy demand, the extension of the previous statement to other climate types must be performed with caution, especially in hot and/or humid climates.

4.2. Energy Demands for Heating and for Cooling

Figure 2 illustrates the impact of climate change scenarios on annual energy demand for heating (left column) and for cooling (right column) across the six different building types incorporating the energy-optimal values for thermal insulation thickness and glazing shading length. Each case shown includes all types of setpoints (A, B, C, DT1, DT3, DT5) and temperate Mediterranean climates (MC1, MC2, MC3). As can be observed, the annual energy needs for heating and for cooling clearly depend on the building type of use, on the climate intensity (MC1, MC2, MC3) and on the climate change scenario (NCC, MRS, HRS).

In the published literature, it is commonly asserted that even in temperate climates, the demand for thermal energy for heating typically exceeds that for cooling. However, Figure 2 reveals that this statement does not universally hold true. Contrary to the general trend, the supermarket exhibits significantly higher energy needs for cooling than for heating in all climate change scenarios and climate types considered. Also, for all building types, cooling demands surpass heating demands in the HRS scenario (high-intensity climate changes), which emphasize the importance of addressing cooling requirements in future climate conditions. In the NCC (no further climatic changes) and MRS (medium-intensity climate changes) scenarios, heating demands exceed cooling demands for all buildings except the supermarket.

The supermarket experiences varying energy demands based on the season. Although winter requires minimal heating, summer necessitates substantial cooling efforts to maintain a comfortable environment for shoppers and staff. The low energy needs for heating are due to the high internal head loads and the occupants' high clothing insulation and activity level, whereas the substantial energy needs for cooling are due to the high internal head loads and the noticeable occupants' activity level. Otherwise, the thermal energy needs of the bank branch are both significant, for heating (in winter) due to occupants' sedentary activities and clothing thermal insulation below appropriate, and for cooling (in summer) mainly due to occupants' clothing thermal insulation above the recommended.

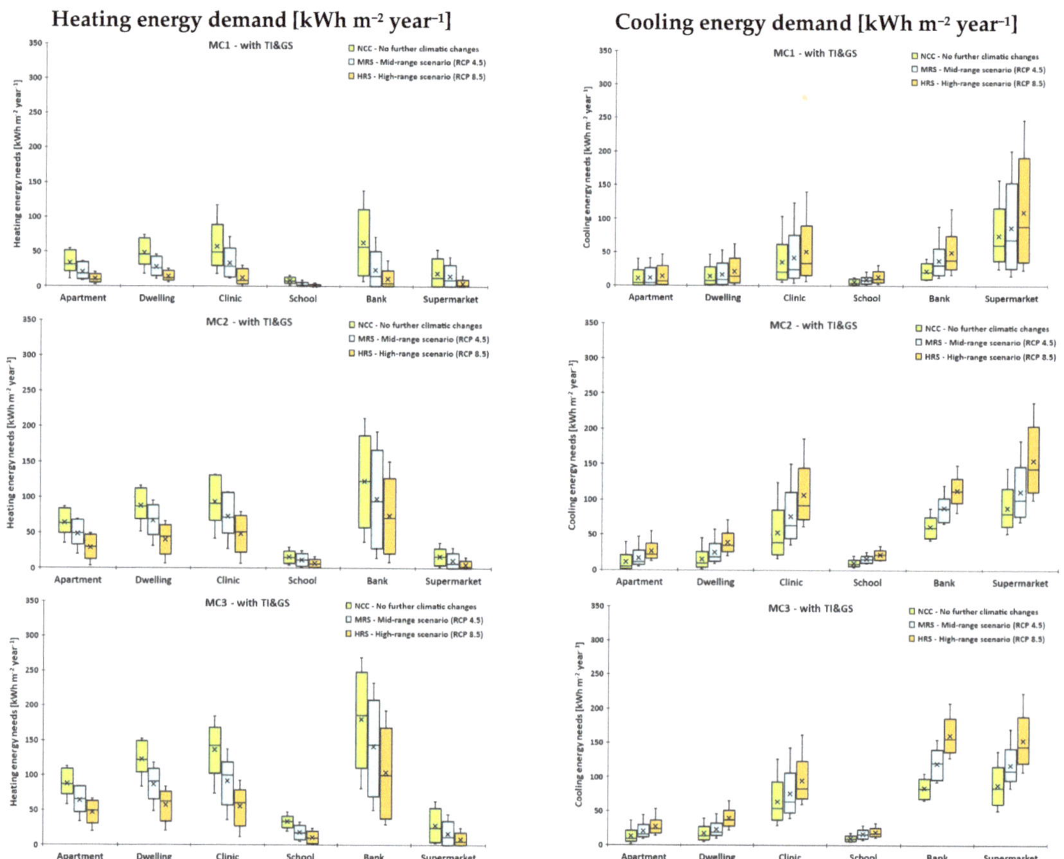

Figure 2. Buildings' annual energy demand for heating and for cooling for the three climate change scenarios and the three Mediterranean climate types (MC1, MC2, MC3) of buildings with the optimal values of thermal insulation thickness and glazing shading length (with TI&GS). Each case shown includes all setpoints (A, B, C, DT1, DT3, DT5).

Regardless of the climate change scenario, the buildings with the lowest energy demand for heating are the school (in the case of mild climate MC1) and the supermarket (in the cases of moderate MC2 and intense MC3 climates). The buildings with the highest energy demand for heating are the clinic (MC1 and MRS), the detached house (MC1 and HRS), and the bank branch (in all the other situations).

Irrespective of the climate type and across all climate change scenarios, the building with the lowest energy demand for cooling is the school. In climate types MC1 and MC2 and under all climate change scenarios, the supermarket experiences the highest energy requirements for cooling. In climate type MC3 and across all climate change scenarios, the building with the highest energy demands for cooling is the bank branch.

Regardless of the building type or the climate change scenario, thermal energy demands for climatization increase with the intensity of the climate (MC1 → MC2 → MC3). Irrespective of the building type or the climate intensity, the energy needs for heating decrease as the severity of climate change grows (NCC → MRS → HRS). Conversely, the energy needs for cooling rise as the severity of climate change increases. These two last statements align with extensive reports in the bibliography [12,26,27].

To assess the impact of climate change scenarios on energy demands for heating and cooling, the probability associated with a Student's *t*-test, considering a two-tailed distribution and two samples of unequal variance, was calculated using the NCC (present climate) scenario as the reference. The results obtained for the level of statistical difference, for scenarios mid-range (MRS) and high-range (HRS) and Mediterranean climates mild (MC1), moderate (MC2), and intense (MC3), are shown in Table 8.

Table 8. Student's *t*-test statistical significance of the difference relative to the NCC scenario of energy demands for heating and for cooling, for scenarios MRS and HRS and climate types MC1, MC2, and MC3. Legend: — $\rightarrow$ no statistical difference ($p > 0.05$), * $\rightarrow$ significant difference with $p < 0.05$, ** $\rightarrow$ significant difference with $p < 0.01$.

Building	Climate Type	Heating MRS	Demand HRS	Cooling MRS	Demand HRS
Apartment	MC1	—	*	—	—
	MC2	—	**	—	—
	MC3	—	**	—	—
Detached house	MC1	—	*	—	—
	MC2	—	**	—	*
	MC3	—	**	—	*
Clinic	MC1	—	*	—	—
	MC2	—	*	—	—
	MC3	—	**	—	—
School	MC1	—	*	—	—
	MC2	—	*	—	*
	MC3	—	**	—	*
Bank branch	MC1	—	—	—	—
	MC2	—	—	—	*
	MC3	—	—	—	**
Supermarket	MC1	—	—	—	—
	MC2	—	—	—	*
	MC3	—	—	—	*

The results shown in Table 8 reveal that, regardless of building and climate types, the difference relative to the present climate (NCC scenario) in energy demand for heating and for cooling is not statistically significant in the case of the mid-range scenario (MRS). Regardless of the type of climate, in the case of the climate change high-range scenario (HRS), the decrease in energy demand for heating is statistically significant for buildings with permanent use (apartment, detached house, and clinic) and the school, and not statistically significant for the bank branch and the supermarket. For the HRS scenario, the augmentation of energy demand for cooling is not statistically significant in the case of the mild climate (MC1), regardless of the type of building, and for the apartment and the clinic, regardless of the type of climate.

Figures 3 and 4 illustrate the annual energy demand for heating (left column) and cooling (right column) for buildings with permanent use (Figure 3) and for buildings with intermittent use (Figure 4), when placed in the moderate Mediterranean climate (MC2). Due to space limitations, only results related to climate MC2 were present, which fall between the mild (MC1) and intense (MC3) climates. The results shown in these figures demonstrate that the energy demands for heating and cooling indoor spaces depend on the building type of use, on the climate change scenario, on the type of operation of the climatization system, and on the existence of thermal insulation of opaque elements and glazing shading.

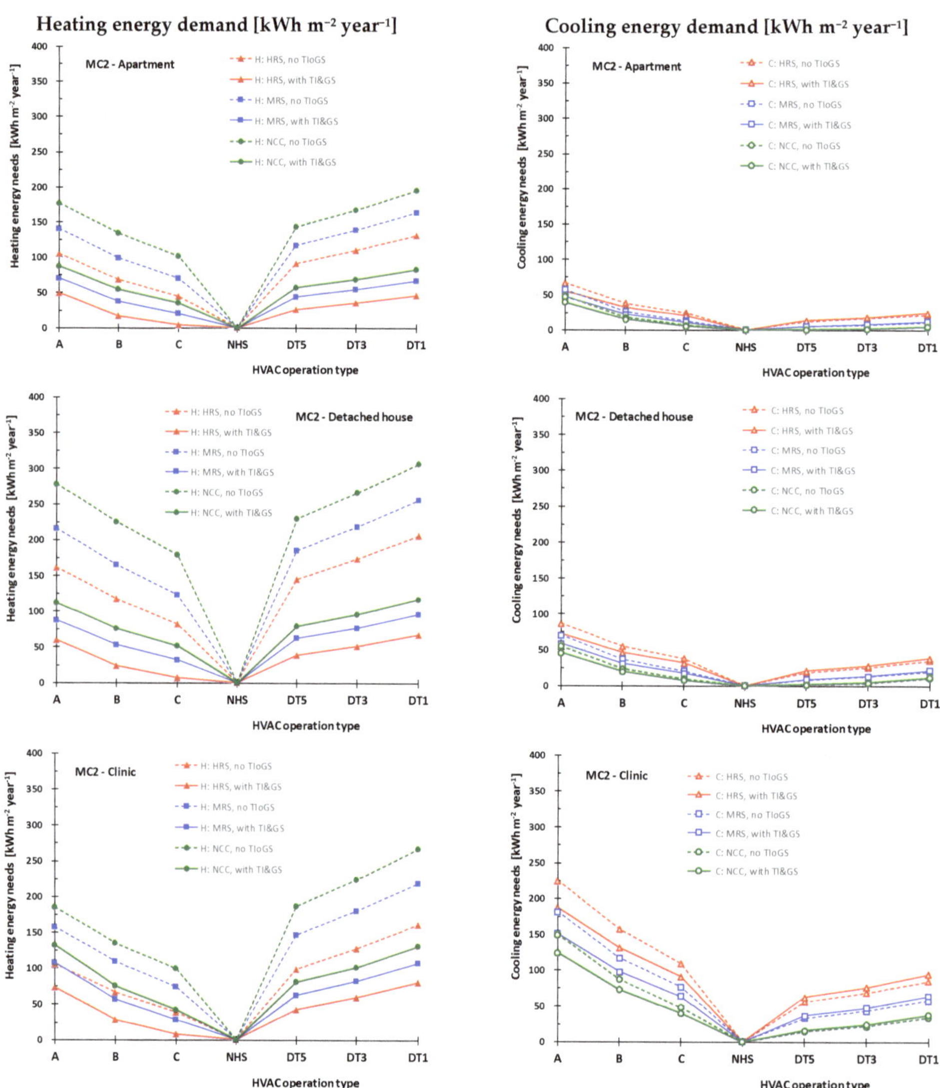

Figure 3. Annual energy demand for heating and for cooling of buildings with permanent use, in Mediterranean climate type 2 (MC2): NCC—no further climatic changes; MRS—mid-range scenario (RCP 4.5); HRS—high-range scenario (RCP 8.5); no TIoGS—without thermal insulation or glazing shading; with TI&GS—with energy-optimal thermal insulation and glazing shading.

In general, the results indicate that across all buildings and situations, the thermal energy requirements for heating are higher when the HVAC system control relies on indoor air temperature (T_{air}) setpoints compared to when it is based on predicted mean vote (PMV) setpoints. Conversely, for energy demands related to cooling, the situation is reversed; values are higher when the HVAC system control uses PMV setpoints rather than T_{air} setpoints. Therefore, from an energy demand perspective, the ideal HVAC control system operates based on PMV setpoints during the heating function and switches to T_{air} setpoints during cooling periods. However, in terms of thermal comfort, the energy-optimal HVAC control system consistently relies on PMV setpoints [7].

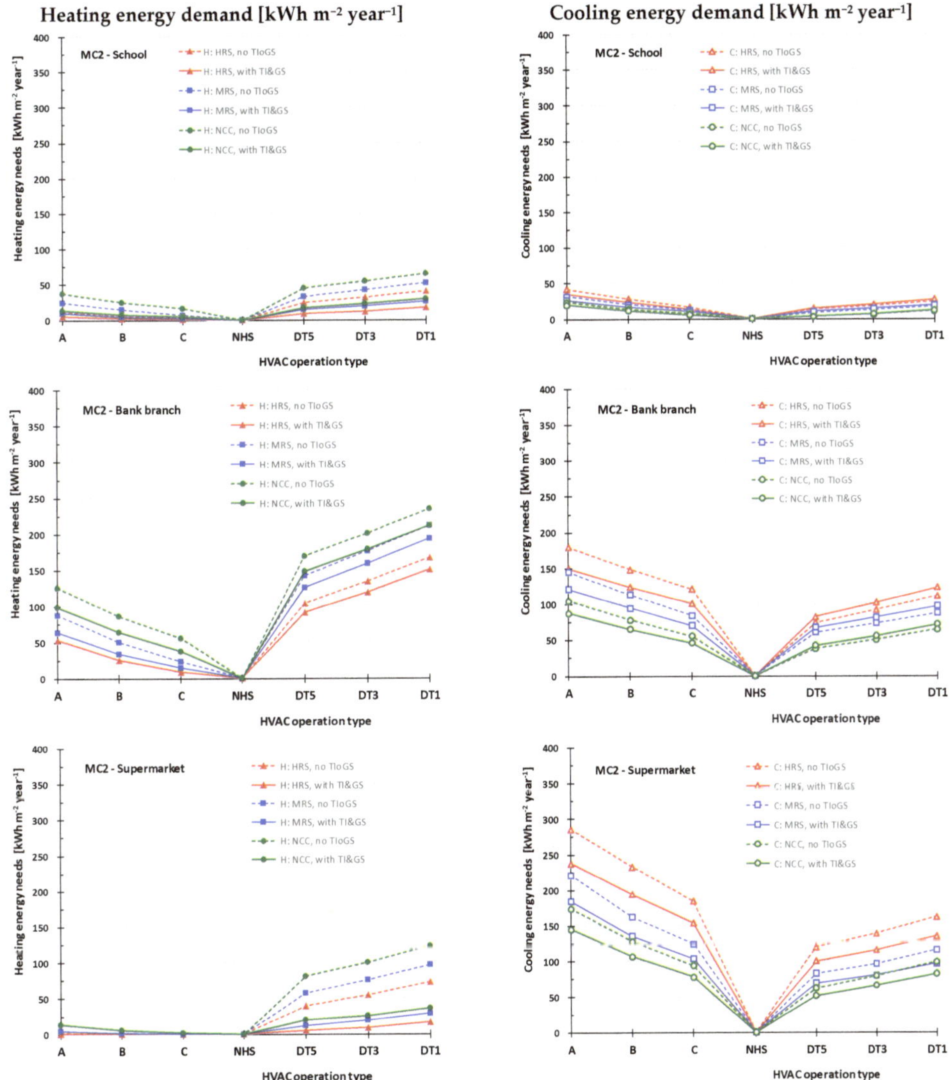

Figure 4. Annual energy demand for heating and for cooling of buildings with intermittent use, in Mediterranean climate type 2 (MC2): NCC—no further climatic changes; MRS—mid-range scenario (RCP 4.5); HRS—high-range scenario (RCP 8.5); no TIoGS—without thermal insulation or glazing shading; with TI&GS—with energy-optimal thermal insulation and glazing shading.

As stated in previous studies [7,10,20], increasing the thickness of thermal insulation applied to a building's opaque elements leads to improved thermal comfort indoors, to a substantial decrease in annual energy requirements for heating, and to a slight increase in those for cooling. Conversely, the installation of external glazing shades contributes to better thermal comfort (by eliminating excess indoor air temperature peaks) and reduces energy demands for cooling, albeit with an insignificant increase in heating requirements [34–37]. Thus, the energy impact of applying thermal insulation becomes evident through the graphs depicting heating demands (left column of Figures 3 and 4). Similarly, the significance of installing glazing shades is apparent from the graphs related to cooling needs (right column of Figures 3 and 4).

The results depicted in Figures 3 and 4 consistently demonstrate that, regardless of the climate change scenario, the application of thermal insulation to opaque elements of the building envelope consistently reduces energy demands for heating, which is particularly relevant for buildings with permanent use (apartment, detached house, and clinic) and has limited relevance for buildings that are intermittently occupied (school, bank branch, and supermarket). Conversely, when considering energy terms, the installation of exterior glazing shades is almost always beneficial, although it never becomes highly significant.

4.3. Energy Consumption by the HVAC System

The Heating, Ventilation, and Air Conditioning (HVAC) system in all buildings is based on a chiller/heat-pump of European class A+, with a seasonal coefficient of performance (*SCOP*) of 4.30 in heating mode, a seasonal energy efficiency ratio (*SEER*) of 5.85 in cooling mode, and on ventilation equipment with a performance coefficient of 70%.

The annual energy consumption by the HVAC system of the six buildings, when placed in the three Mediterranean climate types (mild (MC1), moderate (MC2), and intense (MC3)), exposed to the three climate change scenarios (no further climate change (NCC), mid-range (MRS), and high-range (HRS)), and considering energy-optimal values for thermal insulation thickness and glazing shading length, is depicted in Figure 5. The data are presented in the form of boxplot graphs, and each case including all setpoints (A, B, C, DT1, DT3, DT5).

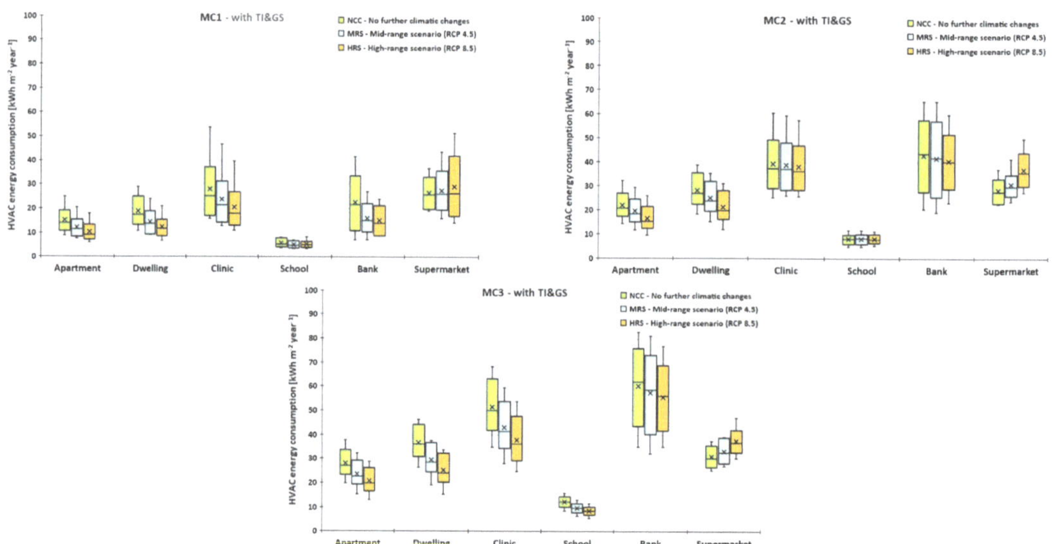

Figure 5. Annual energy consumption by the HVAC system (for heating, cooling, and ventilation) of buildings with the energy-optimal values of thermal insulation thickness and glazing shading length (with TI&GS), for Mediterranean climates mild (MC1), moderate (MC2), and intense (MC3). Each case shown includes all setpoints (A, B, C, DT1, DT3, DT5).

The results from Figure 5 underscore that, irrespective of climate type or climate change scenario, the school consistently exhibits significantly lower annual energy consumption for climatization compared to other buildings. In the mild climate (MC1), the clinic has the highest annual energy consumption for climatization under the present climate scenario (NCC). Meanwhile, the supermarket takes the lead in both the mid-range (MRS) and high-range (HRS) climate change scenarios. Regardless of the climate change scenario, the bank branch consistently demonstrates the highest annual energy consumption for climatization in temperate (MC2) and intense (MC3) climates.

Regardless of the climate change scenario, energy consumption for climatization in buildings generally increases with the intensity of the climate (MC1 → MC2 → MC3), except for the supermarket. Conversely, the supermarket's energy consumption decreases as the climate intensity rises. The reason behind this divergence lies in the supermarket's unique characteristics compared to other buildings. Specifically, the supermarket experiences higher internal heat gains (due to factors such as lighting, people, and devices), and occupants have higher clothing insulation and activity levels (as indicated in Table 4). Consequently, the supermarket's energy demand for cooling outweighs its heating requirements.

Irrespective of climate intensity, and except for the supermarket, the energy consumption for climatization in buildings decreases as the severity of climate changes increases (NCC → MRS → HRS). For the supermarket, this energy consumption rises with the increasing severity of climate change. The underlying reason is that, in this building type, the energy demand for cooling outweighs that for heating.

The results obtained with a Student's *t*-test for the level of statistical difference in the energy consumption by buildings' HVAC systems for scenarios mid-range (MRS) and high-range (HRS), and Mediterranean climates mild (MC1), moderate (MC2), and intense (MC3), are summarized in Table 9. These results reveal that, regardless of building and climate types, the difference in energy consumption for climatization relative to the NCC scenario is not statistically significant in the case of the mid-range scenario (MRS). In the case of the high-range scenario (HRS), whatever building type, the difference remains not statistically significant for buildings placed in the mild (MC1) and moderate (MC2) climates. Therefore, the difference becomes statistically significant only when both the intense climate (MC3) and the extreme climate change scenarios (HRS) coincide.

Table 9. Student's *t*-test statistical significance of the difference relative to the NCC scenario of energy consumption by the HVAC system, for scenarios MRS and HRS and climate types MC1, MC2, and MC3. Legend: — → no statistical difference ($p > 0.05$), * → significant difference with $p < 0.05$, ** → significant difference with $p < 0.01$.

Building	Climate Type	HVAC MRS	Consumption HRS
Apartment	MC1	—	—
	MC2	—	—
	MC3	—	*
Detached house	MC1	—	—
	MC2	—	—
	MC3	—	**
Clinic	MC1	—	—
	MC2	—	—
	MC3	—	*
School	MC1	—	—
	MC2	—	—
	MC3	—	*
Bank branch	MC1	—	—
	MC2	—	—
	MC3	—	*
Supermarket	MC1	—	—
	MC2	—	—
	MC3	—	*

Comparing the statistical differences related to energy consumption by the HVAC system (shown in Table 9) with those related to energy demands for heating and cooling (presented in Table 8), leads to the conclusion that the energy efficiency of the HVAC system plays a decisive role in determining the significance of the differences in energy consumption associated with various climate change scenarios. This fact highlights the critical importance of HVAC system energy performance in the energy consumption for climatization.

Figures 6 and 7 illustrate the annual energy consumption for heating, cooling, and ventilation in buildings with permanent use (Figure 6) and in buildings with intermittent use (Figure 7), when placed in the Mediterranean climates MC1 (mild; left column) and MC3 (intense; right column). As shown, the energy consumption by the HVAC system depends on the building type of use, the existence of thermal insulation of opaque elements of the building's envelope and of external glazing shading, the type of control of climatization system operation, the climate change scenario, and the Mediterranean climate type.

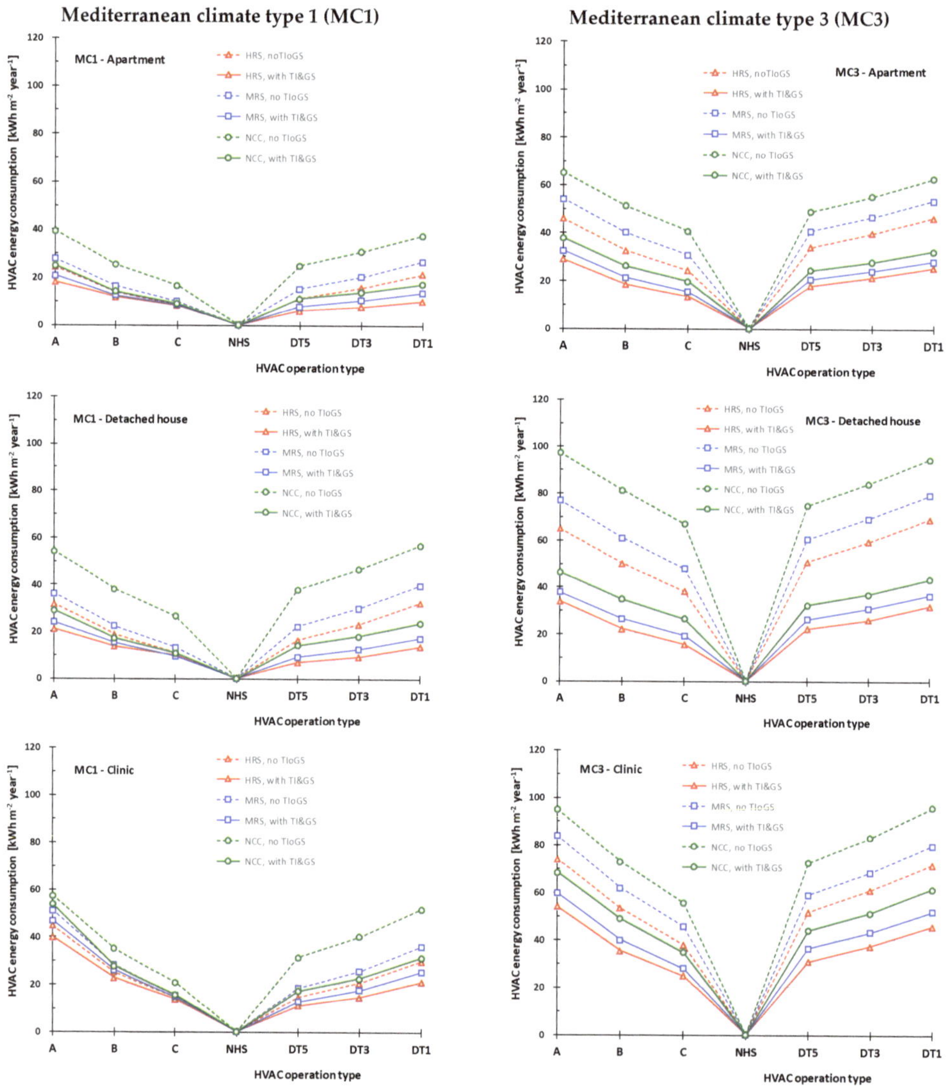

Figure 6. Annual energy consumption by the HVAC system (for heating, cooling, and ventilation), in Mediterranean climates type 1 and 3, of buildings with permanent use. Legend: NCC—no further climatic changes; MRS—mid-range scenario (RCP 4.5); HRS—high-range scenario (RCP 8.5); no TIoGS—without thermal insulation or glazing shading; with TI&GS—with energy-optimal thermal insulation and glazing shading.

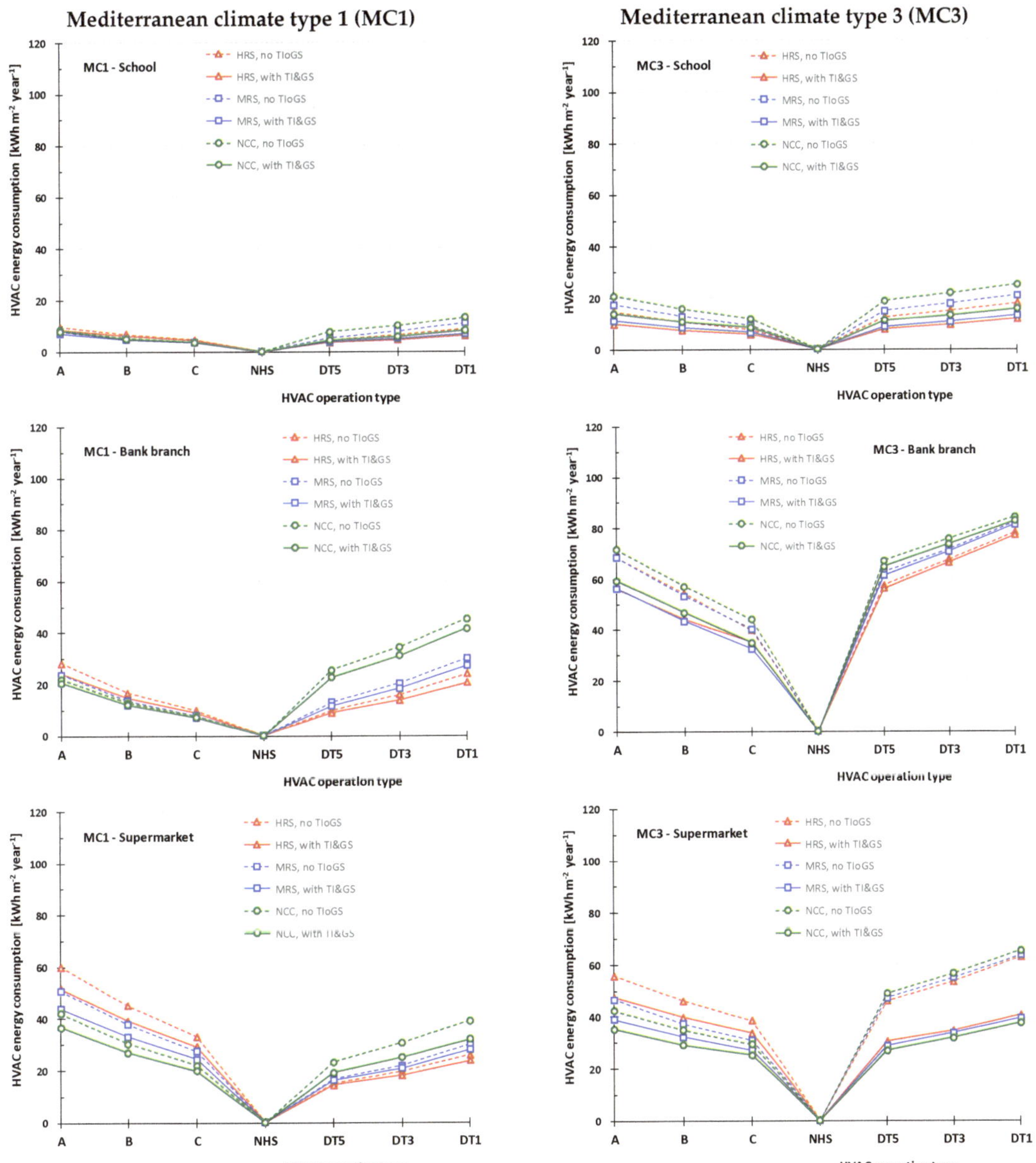

Figure 7. Annual energy consumption by the HVAC system (for heating, cooling, and ventilation), in Mediterranean climates type 1 and 3, of buildings with intermittent use. Legend: NCC—no further climatic changes; MRS—mid-range scenario (RCP 4.5); HRS—high-range scenario (RCP 8.5); no TIoGS—without thermal insulation or glazing shading; with TI&GS—with energy-optimal thermal insulation and glazing shading.

In general, the results highlight that the influence of climate type and the presence of thermal insulation in opaque elements and external glazing shades is much more pronounced for buildings with permanent use (apartment, detached house, and clinic) than for buildings with intermittent use (school, bank branch, and supermarket).

The benefits of having thermal insulation in opaque elements and external glazing shades become more pronounced as the climate intensity increases (MC1 → MC2 → MC3). Interestingly, these benefits remain consistent across all climate change scenarios (NCC, MRS, HRS). These statements emphasize the importance of energy-efficient building design regardless of the specific climate change context.

As mentioned earlier, the type of control of the climatization system operation is directly linked to the desired thermal comfort. The results depicted in Figures 6 and 7 reveal that, as expected, energy consumption by the air conditioning system increases as the quality of thermal comfort improves (from C → B → A; and from DT5 → DT3 → DT1). This increase is more pronounced in buildings with permanent use than in those with intermittent occupancy.

4.4. Influence of HVAC System Type of Control

Different types of HVAC system control led to different levels of thermal comfort indoors. The relation between the type of setpoint and the thermal comfort level is not addressed here, but it was analyzed in another work of the authors [7], where the same buildings and locations were considered, but for the present climate (NCC scenario).

In this study, six control types were considered (as detailed in Table 5), three of them by the traditional setpoints of indoor air temperature (T_{air})—referred as DT1, DT3, and DT5—and the remaining three by setpoints of the predicted mean vote (PMV)—designed as A, B, and C. In terms of thermal comfort quality, controls A and DT1 guarantee a high level, B and DT3 a good level, and C and DT5 a moderate level. These traits highlight the importance of selecting the appropriate control strategy to ensure the desired occupants' thermal comfort.

A comprehensive comparison was conducted to explore the relationship between the two control types of air conditioning system operation (PMV or T_{air} setpoints), the three Mediterranean climate types (MC1, MC2, MC3), and the three climate change scenarios (NCC, MRS, HRS), considering the buildings with the energy-optimal values of thermal insulation thickness and of glazing shading length. Table 10 summarizes the findings of this assessment, where the type of HVAC system control associated with lower annual energy consumption for air conditioning is identified.

As mentioned earlier, across all buildings and situations, the energy consumption by the air conditioning system increases as the quality of thermal comfort improves (from C → B → A; and from DT5 → DT3 → DT1), but this increase is more pronounced in buildings with permanent use than in those with intermittent occupancy. Additionally, the thermal energy requirements for heating are lower when the HVAC system control relies on PMV setpoints, whereas the energy demands for cooling are lower when the control is based on T_{air} setpoints. This inference emphasizes the trade-off between thermal comfort, type of control of HVAC system operation and energy consumption for climatization that must be considered in building design.

The results presented in Table 10 reveal that the energy-optimal type of HVAC system control depends on the building and on the Mediterranean climate types. Interestingly, it is independent of the climate change scenario. In the case of intense climate (MC3), regardless of the building type, controlling the HVAC system using PMV setpoints consistently leads to lower energy consumption for climatization. In the case of climate types MC1 (mild) and MC2 (moderate) and for buildings with permanent use (apartment, detached house, and clinics) and the supermarket, it is preferable to control the HVAC system using T_{air} setpoints. For the school and the bank branch, the better option is always to control the HVAC system using PMV setpoints.

Previous insights emphasize the importance of tailored HVAC control strategies based on specific building contexts and present climate conditions. Conversely, within temperate Mediterranean climates, the severity of climate change is unlikely to significantly affect the better control type of HVAC systems operation.

Table 10. HVAC system control type that leads to lower energy consumption for climatization, for the three Mediterranean climate types (MC1, MC2, MC3) and the three climate change scenarios (NCC, MRS, HRS), considering the buildings with the energy-optimal values of thermal insulation thickness and of glazing shading length.

Building	Climate Type	Climate Change NCC	Climate Change MRS	Scenario HRS
Apartment	MC1	T_{air}	T_{air}	T_{air}
	MC2	T_{air}	T_{air}	T_{air}
	MC3	PMV	PMV	PMV
Detached House	MC1	T_{air}	T_{air}	T_{air}
	MC2	T_{air}	T_{air}	T_{air}
	MC3	PMV	PMV	PMV
Clinic	MC1	T_{air}	T_{air}	T_{air}
	MC2	T_{air}	T_{air}	T_{air}
	MC3	PMV	PMV	PMV
School	MC1	PMV	PMV	PMV
	MC2	PMV	PMV	PMV
	MC3	PMV	PMV	PMV
Bank Branch	MC1	PMV	PMV	PMV
	MC2	PMV	PMV	PMV
	MC3	PMV	PMV	PMV
Supermarket	MC1	T_{air}	T_{air}	T_{air}
	MC2	T_{air}	T_{air}	T_{air}
	MC3	PMV	PMV	PMV

5. Conclusions

This study aims to systematically assess how climate changes, properties of construction elements, and the type of control used in HVAC systems impact the energy requirements of six buildings (apartment, detached house, clinic, school, bank branch, and supermarket) situated in a temperate Mediterranean climate.

The buildings were situated in three different climate zones: mild (MC1), moderate (MC2), and intense (MC3). The buildings' envelopes incorporate a traditional External Thermal Insulation Composite System (ETICS) based on expanded polystyrene (EPS). Insulation thicknesses ranging from 0 (without insulation) to 12 cm, as well as horizontal external fixed shades with lengths varying from 0 (absence) to 150 cm, were tested. Six different setpoint ranges for the HVAC system control were evaluated: three based on the predicted mean vote (PMV) and three based on the indoor air temperature (T_{air}). For the year 2070, three climatic change circumstances were assumed: (i) maintaining the current climate (NCC); (ii) resulting from medium-intensity climate changes (mid-range scenario, MRS); and (iii) subsequent from extreme climate changes (high-range scenario, HRS).

Climate hourly dataset files were prepared by applying "coefficients" predicted by "global climatic models" to present-day climate data using the "morphing procedure" methodology. A Student's t-test was performed on air temperature (T_{air}), relative humidity (RH), and horizontal global solar radiation ($HGSR$). In relation to present climate (scenario NCC), the other two (MRS and HRS) exhibit a statistically significant difference ($p < 0.001$) for the parameters T_{air}, RH, and $HGSR$ across the three climate types (MC1, MC2, and MC3). Additionally, there is a statistically significant difference ($p < 0.001$) between the MRS and HRS scenarios for T_{air} and RH, but not for $HGSR$ ($p > 0.05$).

The energy-optimal values for thermal insulation thickness and the length of the shade tend to align with either the respective minimum (0 cm, 0 cm) or maximum values (12 cm, 150 cm, respectively) that were tested. Generally, this optimal insulation thickness is greater when the HVAC system is controlled using T_{air} setpoints compared to PMV setpoints. This thickness tends to increase with higher climate intensity and decrease with more severe climate change. Except for the supermarket, the use of additional glazing shades is energetically advantageous when the HVAC system is controlled by PMV setpoints, but not when it is performed by T_{air} setpoints. For the supermarket, the use of additional glazing shades is advantageous regardless of the HVAC control type.

As expected, irrespective of building and climate types, an escalation in the severity of climate changes reduces the energy requirements for heating and amplifies the energy demands for cooling. The relative magnitude of these fluctuations depends on both the specific building and the climate type.

When comparing the two types of HVAC system control, the thermal energy requirements for heating are lower when the control of the HVAC system is performed by PMV setpoints, and the energy demands for cooling are lower when this control is performed by T_{air} setpoints. Therefore, from an energy demand perspective, the ideal HVAC control system operates based on PMV setpoints during heating periods and switches to T_{air} setpoints during cooling periods.

As anticipated, energy consumption by the air conditioning system increases with improved thermal comfort, more pronounced in buildings with continuous occupancy than in those with intermittent use. Regarding energy consumption for climatization, the optimal type of HVAC system control varies based on the specific building and climate conditions, but not of the climate change scenario.

In all building types and climates, relative to the current climate (NCC scenario), the difference in energy demands for heating and cooling is statistically significant only in the case of extreme climate change (HRS). On the other hand, the energy efficiency of the HVAC system is also a determining factor in its energy consumption. Therefore, the statistical significance of the difference between energy needs cannot be directly extrapolated to energy consumption for air conditioning. If buildings are equipped with an HVAC system based on a class A+ chiller/heat-pump, compared to the NCC scenario, the difference in energy consumption for climatization is only statistically significant when the HRS scenario and climate type MC3 are simultaneously present.

For buildings equipped with an HVAC system based on a class A+ or higher chiller/heat-pump, the impact on energy consumption for air conditioning due to factors such as thermal insulation, external glazing shading systems and HVAC system control type depends very little on the climate change scenario. Consequently, a building designed for good energy performance in the current climate will likely maintain that efficiency when exposed to the climate resulting from future climate change. As the energy efficiency of the HVAC system plays a crucial role, so this assertion may not hold if the energy efficiency of the air conditioning system is significantly lower than the one considered in this study.

Author Contributions: Methodology, A.M.R.; Software, A.M.R.; Validation, A.M.R. and A.V.M.O.; Formal analysis, A.M.R. and A.V.M.O.; Investigation, A.M.R.; Data curation, A.M.R. and A.V.M.O.; Writing—original draft, A.M.R.; Writing—review & editing, A.M.R. and A.V.M.O. All authors have read and agreed to the published version of the manuscript.

Funding: This research received no external funding.

Data Availability Statement: The original contributions presented in the study are included in the article, further inquiries can be directed to the corresponding author.

Conflicts of Interest: The authors declare no conflict of interest.

References

1. IPCC. *Climate Change 2022: Mitigation of Climate Change. Working Group III Contribution to the Sixth Assessment Report of the Intergovernmental Panel on Climate Change*; Cambridge University Press: Cambridge, UK, 2022. Available online: https://www.ipcc.ch/report/sixth-assessment-report-working-group-3/ (accessed on 15 September 2023).
2. UNEP—United Nations Environment Programme. *Emissions Gap Report 2022: The Closing Window—Climate Crisis Calls for Rapid Transformation of Societies*; United Nations Environment Programme: Nairobi, Kenya, 2022. Available online: https://www.unep.org/resources/emissions-gap-report-2022 (accessed on 20 September 2022).

3. IPCC. *Climate Change 1995: The Science of Climate Change. Contribution of Working Group I to the Second Assessment Report of the Intergovernmental Panel on Climate Change*; Cambridge University Press: Cambridge, UK, 1996; 572p. Available online: https://www.ipcc.ch/site/assets/uploads/2018/02/ipcc_sar_wg_I_full_report.pdf (accessed on 13 September 2023).

4. IPCC. *Emissions Scenarios. Special Report of Working Group III of the Intergovernmental Panel on Climate Change*; Cambridge University Press: Cambridge, UK, 2000; 599p. Available online: https://archive.ipcc.ch/pdf/special-reports/emissions_scenarios.pdf (accessed on 11 September 2023).

5. IPCC. *Climate Change 2014: Mitigation of Climate Change. Contribution of Working Group III to the Fifth Assessment Report of the Intergovernmental Panel on Climate Change*; Cambridge University Press: Cambridge, UK, 2014. Available online: https://www.ipcc.ch/report/ar5/wg3/ (accessed on 12 September 2023).

6. Salimi, S.; Hammad, A. Optimizing energy consumption and occupants comfort in open-plan offices using local control based on occupancy dynamic data. *Build. Environ.* **2020**, *176*, 106818. [CrossRef]

7. Raimundo, A.M.; Oliveira, A.V.M. Analyzing thermal comfort and related costs in buildings under Portuguese temperate climate. *Build. Environ.* **2022**, *219*, 109238. [CrossRef]

8. Frontczak, M.; Wargocki, P. Literature survey on how different factors influence human comfort in indoor environments. *Build. Environ.* **2011**, *46*, 922–937. [CrossRef]

9. Hoyt, T.; Arens, E.; Zhang, H. Extending air temperature setpoints: Simulated energy savings and design considerations for new and retrofit buildings. *Build. Environ.* **2015**, *88*, 89–96. [CrossRef]

10. Raimundo, A.M.; Sousa, A.M.; Oliveira, A.V.M. Assessment of energy, environmental and economic costs of buildings' thermal insulation—Influence of type of use and climate. *Buildings* **2023**, *13*, 279. [CrossRef]

11. Bamdad, K.; Cholette, M.E.; Omrani, S.; Bell, J. Future energy-optimised buildings—Addressing the impact of climate change on buildings. *Energy Build.* **2021**, *23*, 110610. [CrossRef]

12. Congedo, P.M.; Baglivo, C.; Seyhan, A.K.; Marchetti, R. Worldwide dynamic predictive analysis of building performance under long-term climate change conditions. *J. Build. Eng.* **2021**, *42*, 103057. [CrossRef]

13. Baglivo, C.; Congedo, P.M.; Murrone, G.; Lezzi, D. Long-term predictive energy analysis of a high-performance building in a mediterranean climate under climate change. *Energy* **2022**, *238*, 121641. [CrossRef]

14. Jalali, Z.; Shamseldin, A.Y.; Ghaffarianhoseini, A. Impact assessment of climate change on energy performance and thermal load of residential buildings in New Zealand. *Build. Environ.* **2023**, *243*, 110627. [CrossRef]

15. *ASHRAE 55:2010*; Thermal Environment Conditions for Human Occupancy. American Society of Heating, Refrigeration and Air Conditioning Engineers, Inc.: Peachtree Corners, GA, USA, 2010.

16. *ISO 7730:2005*; Ergonomics of the Thermal Environment—Analytical Determination and Interpretation of Thermal Comfort Using Calculation of the PMV and PPD Indices. International Organization for Standardization: Geneva, Switzerland, 2005.

17. Zampetti, L.; Arnesano, M.; Revel, G.M. Experimental testing of a system for the energy-efficient sub-zonal heating management in indoor environments based on PMV. *Energy Build.* **2018**, *166*, 229–238. [CrossRef]

18. Xu, X.; Liu, W.; Lian, Z. Dynamic indoor comfort temperature settings based on the variation in clothing insulation and its energy-saving potential for an air conditioning system. *Energy Build.* **2020**, *220*, 110086. [CrossRef]

19. Giorgi, F.; Lionello, P. Climate change projections for the Mediterranean region. *Glob. Planet. Chang.* **2008**, *63*, 90–104. [CrossRef]

20. Raimundo, A.M.; Saraiva, N.B.; Oliveira, A.V.M. Thermal insulation cost optimality of opaque constructive solutions of buildings under Portuguese temperate climate. *Build. Environ.* **2020**, *182*, 107107. [CrossRef]

21. Kolaitis, D.I.; Malliotakis, E.; Kontogeorgos, D.A.; Mandilaras, I.; Katsourinis, D.I.; Founti, M.A. Comparative assessment of internal and external thermal insulation systems for energy efficient retrofitting of residential buildings. *Energy Build.* **2013**, *64*, 123–131. [CrossRef]

22. Lam, J.C.; Wan, K.K.W.; Tsang, C.L.; Yang, L. Building energy efficiency in different climates. *Energy Convers. Manag.* **2008**, *49*, 2354–2366. [CrossRef]

23. McGraw-Hill Construction. *Energy Efficiency Trends in Residential and Commercial Buildings*; U.S. Department of Energy, Office of Energy Efficiency and Renewable Energy: Washington, DC, USA, 2010. Available online: https://www.energy.gov/sites/prod/files/2013/11/f5/building_trends_2010.pdf (accessed on 6 February 2019).

24. Belcher, S.E.; Hacker, J.N.; Powell, D.S. Constructing design weather data for future climates. *Build. Serv. Eng. Res. Technol.* **2005**, *26*, 49–61. [CrossRef]

25. Guan, L. Preparation of future weather data to study the impact of climate change on buildings. *Build. Environ.* **2009**, *44*, 793–800. [CrossRef]

26. Larsen, M.A.D.; Petrovic, S.; Radoszynki, A.M.; McKenna, R.; Balyk, O. Climate change impacts on trends and extremes in future heating and cooling over Europe. *Energy Build.* **2020**, *226*, 110397. [CrossRef]

27. Aguiar, R.; Oliveira, M.; Gonçalves, H. Climate change impacts on the thermal performance of Portuguese buildings. Results of the SIAM study. *Build. Serv. Eng. Res. Technol.* **2002**, *23*, 223–231. [CrossRef]

28. Michałowski, B.; Marcinek, M.; Tomaszewska, J.; Czernik, S.; Piasecki, M.; Geryło, R.; Michalak, J. Influence of rendering type on the environmental characteristics of expanded polystyrene-based external thermal insulation composite system. *Buildings* **2020**, *10*, 47. [CrossRef]

29. Grazieschi, G.; Asdrubali, F.; Thomas, G. Embodied energy and carbon of building insulating materials: A critical rewiew. *Clean. Environ. Syst.* **2021**, *2*, 100032. [CrossRef]

30. Ounis, S.; Aste, N.; Butera, F.M.; Pero, C.D.; Leonforte, F.; Adhikari, R.S. Optimal balance between heating, cooling and environmental impacts: A method for appropriate assessment of building envelope's U-value. *Energies* **2022**, *15*, 3570. [CrossRef]
31. Wang, H.; Huang, Y.; Yang, L. Integrated economic and environmental assessment-based optimization design method of building roof thermal insulation. *Buildings* **2022**, *12*, 916. [CrossRef]
32. Vasconcelos, A.B.; Pinheiro, M.D.; Manso, A.; Cabaço, A. EPBD cost-optimal methodology: Application to the thermal rehabilitation of the building envelope of a Portuguese residential reference building. *Energy Build.* **2016**, *111*, 12–25. [CrossRef]
33. Raimundo, A.M.; Saraiva, N.B.; Dias, L.P.; Rebelo, A.C. Market-oriented cost-effectiveness and energy analysis of windows in Portugal. *Energies* **2021**, *14*, 3720. [CrossRef]
34. Al-Masrani, S.M.; Al-Obaidi, K.M.; Zalin, N.A.; Isma, M.I.A. Design optimisation of solar shading systems for tropical office buildings: Challenges and future trends. *Sol. Energy* **2018**, *170*, 849–872. [CrossRef]
35. Settino, J.; Carpino, C.; Perrella, S.; Arcuri, N. Multi-objective analysis of a fixed solar shading system in different climatic areas. *Energies* **2020**, *13*, 3249. [CrossRef]
36. Alrasheed, M.; Mourshed, M. Domestic overheating risks and mitigation strategies: The state-of-the-art and directions for future research. *Indoor Built Environ.* **2023**, *32*, 1057–1077. [CrossRef]
37. Shah, I.; Soh, B.; Lim, C.; Lau, S.-K.; Ghahramani, A. Thermal transfer and temperature reductions from shading systems on opaque facades: Quantifying the impacts of influential factors. *Energy Build.* **2023**, *278*, 112604. [CrossRef]
38. European Commission. Consolidated text of Commission Delegated Regulation (EU) 626/2011 of 4 May 2011 supplementing Directive 2010/30/EU with regard to energy labelling of air conditioners. *Off. J. Eur. Union* **2011**, *L178*, 1–72. Available online: https://eur-lex.europa.eu/legal-content/EN/TXT/?uri=CELEX:02011R0626-20200809 (accessed on 12 July 2018).
39. *ISO 13790:2008*; Energy Performance of Buildings—Calculation of Energy Use for Space Heating and Cooling. International Organization for Standardization: Geneva, Switzerland, 2008.
40. *EN 16798-1:2019*; Energy Performance of Buildings—Part 1: Indoor Environment Input Parameters for Design and Assessment of Energy Performance of Buildings Addressing Indoor Air Quality, Thermal Environment, Lighting and Acoustics. European Committee for Standardization: Brussels, Belgium, 2019.
41. CCCMA. IPCC Experiments SRES B1 and SRES A2 with CGCM3.1/T47 for Years 2001–2100, Initialized from the End of the 20C3M Experiment. Canadian Centre for Climate Modelling and Analysis (2007). Available online: http://climate-modelling.canada.ca/data/cgcm3/cgcm3.shtml (accessed on 17 April 2020).
42. Dylewski, R.; Adamczyk, J. Optimum thickness of thermal insulation with both economic and ecological costs of heating and cooling. *Energies* **2021**, *14*, 3835. [CrossRef]

MDPI AG
Grosspeteranlage 5
4052 Basel
Switzerland
Tel.: +41 61 683 77 34

Energies Editorial Office
E-mail: energies@mdpi.com
www.mdpi.com/journal/energies